NOBLE ELEMENTS

	3A	4A	5A	6A	7A	HELIUM 4.0026
						0.179 **He** 2
						$1s^2$
						3.57 HEX 1.633
						~1.0 (26Atm) 26^{LT}
	BORON 10.81	CARBON 12.01	NITROGEN 14.007	OXYGEN 15.999	FLUORINE 18.998	NEON 20.18
	2.34 **B** 5	2.26 **C** 6	1.03 **N** 7	1.43 **O** 8	1.97(α) **F** 9	1.56 **Ne** 10
	$1s^22s^22p^1$	$1s^22s^22p^2$	$1s^22s^22p^3$	$1s^22s^22p^4$	$1s^22s^22p^5$	$1s^22s^22p^6$
	8.73 TET 0.576	3.57 DIA	4.039 HEX 1.651	6.83 CUB	MCL	4.43 FCC
	2600 1250	(4300) 1860	63.3 (β)79^{LT}	54.7 (γ)46^{LT}	53.5	24.5 63
	ALUMINUM 26.982	SILICON 28.086	PHOSPHORUS 30.974	SULFUR 32.064	CHLORINE 35.453	ARGON 39.948
	2.70 **Al** 13	2.33 **Si** 14	1.82(white) **P** 15	2.07 **S** 16	2.09 **Cl** 17	1.78 **Ar** 18
	[Ne]$3s^23p^1$	[Ne]$3s^23p^2$	[Ne]$3s^23p^3$	[Ne]$3s^23p^4$	[Ne]$3s^23p^5$	[Ne]$3s^23p^6$
	4.05 FCC	5.43 DIA	7.17 CUB	10.47 ORC 2.339 / 1.229	6.24 ORC 1.324 / 0.718	5.26 FCC
	933 394	1683 625	317.3	386 172.2	83.9 85
1B	2B					
NICKEL 58.71	COPPER 63.55	ZINC 65.38	GALLIUM 69.72	GERMANIUM 72.59	ARSENIC 74.922	SELENIUM 78.96	BROMINE 79.91	KRYPTON 83.80

NICKEL 58.71	COPPER 63.55	ZINC 65.38	GALLIUM 69.72	GERMANIUM 72.59	ARSENIC 74.922	SELENIUM 78.96	BROMINE 79.91	KRYPTON 83.80
8.9 **Ni** 28	8.96 **Cu** 29	7.14 **Zn** 30	5.91 **Ga** 31	5.32 **Ge** 32	5.72 **As** 33	4.79 **Se** 34	4.10 **Br** 35	3.07 **Kr** 36
[Ar]$3d^84s^2$	[Ar]$3d^{10}4s^1$	[Ar]$3d^{10}4s^2$	[Ar]$3d^{10}4s^24p^1$	[Ar]$3d^{10}4s^24p^2$	[Ar]$3d^{10}4s^24p^3$	[Ar]$3d^{10}4s^24p^4$	[Ar]$3d^{10}4s^24p^5$	[Ar]$3d^{10}4s^24p^6$
3.52 FCC	3.61 FCC	2.66 HEX 1.856	4.51 ORC 1.695 / 1.001	5.66 DIA	4.13 RHL 54°10'	4.36 HEX 1.136	6.67 ORC 1.307 / 0.672	5.72 FCC
1726 375	1356 315	693 234	303 240	1211 360	1090 285	490 150^{LT}	266 73^{LT}	116.5
PALLADIUM 106.40	SILVER 107.87	CADMIUM 112.40	INDIUM 114.82	TIN 118.69	ANTIMONY 121.75	TELLURIUM 127.60	IODINE 126.90	XENON 131.30
12.0 **Pd** 46	10.5 **Ag** 47	8.65 **Cd** 48	7.31 **In** 49	7.30 **Sn** 50	6.62 **Sb** 51	6.24 **Te** 52	4.94 **I** 53	3.77 **Xe** 54
[Kr]$4d^{10}5s^0$	[Kr]$4d^{10}5s^1$	[Kr]$4d^{10}5s^2$	[Kr]$4d^{10}5s^25p^1$	[Kr]$4d^{10}5s^25p^2$	[Kr]$4d^{10}5s^25p^3$	[Kr]$4d^{10}5s^25p^4$	[Kr]$4d^{10}5s^25p^5$	[Kr]$4d^{10}5s^25p^6$
3.86 FCC	4.09 FCC	2.98 HEX 1.886	4.59 TET 1.076	5.82 TET 0.546	4.51 RHL 57°6'	4.45 HEX 1.330	7.27 ORC 1.347 / 0.659	6.20 FCC
1825 275	1234 215	594 120	429.8 129	505 170	904 200	723 139^{LT}	387	161.3 55^{LT}
PLATINUM 195.09	GOLD 196.97	MERCURY 200.59	THALLIUM 204.37	LEAD 207.19	BISMUTH 208.98	POLONIUM 210	ASTATINE 210	RADON 222
21.4 **Pt** 78	19.3 **Au** 79	13.6 **Hg** 80	11.85 **Tl** 81	11.4 **Pb** 82	9.8 **Bi** 83	9.4 **Po** 84	**At** 85	(4.4) **Rn** 86
[Xe]$4f^{14}5d^96s^0$	[Xe]$4f^{14}5d^{10}6s^1$	[Xe]$4f^{14}5d^{10}6s^2$	[Xe]$4f^{14}5d^{10}6s^26p^1$	[Xe]$4f^{14}5d^{10}6s^26p^2$	[Xe]$4f^{14}5d^{10}6s^26p^3$	[Xe]$4f^{14}5d^{10}6s^26p^4$	[Xe]$4f^{14}5d^{10}6s^26p^5$	[Xe]$4f^{14}5d^{10}6s^26p^6$
3.92 FCC	4.08 FCC	2.99 RHL 70°45'	3.46 HEX 1.599	4.95 FCC	4.75 RHL 57°14'	3.35 Sc		(FCC)
2045 230	1337 170	234.3 10^0	577 96	601 88	544.5 12^0	527	(575)	(202)

EUROPIUM 151.96	GADOLINIUM 157.25	TERBIUM 158.92	DYSPROSIUM 162.50	HOLMIUM 164.93	ERBIUM 167.26	THULIUM 168.93	YTTERBIUM 173.04	LUTETIUM 174.97
7.90 **Eu** 63	8.23 **Gd** 64	8.54 **Tb** 65	8.78 **Dy** 66	9.05 **Ho** 67	9.37 **Er** 68	9.31 **Tm** 69	6.97 **Yb** 70	9.84 **Lu** 71
[Xe]$4f^76d^06s^2$	[Xe]$4f^75d^16s^2$	[Xe]$4f^95d^06s^2$	[Xe]$4f^{10}5d^06s^2$	[Xe]$4f^{11}5d^06s^2$	[Xe]$4f^{12}5d^06s^2$	[Xe]$4f^{13}5d^06s^2$	[Xe]$4f^{14}5d^06s^2$	[Xe]$4f^{14}5d^16s^2$
4.61 BCC	3.64 HEX 1.588	3.60 HEX 1.581	3.59 HEX 1.573	3.58 HEX 1.570	3.56 HEX 1.570	3.54 HEX 1.570	5.49 FCC	3.51 HEX 1.585
1095 107^{LT}	1585 176^{LT}	1633 188^{LT}	1680 186^{LT}	1743 191^{LT}	1795 195^{LT}	1818 200^{LT}	1097 118^{LT}	1929 207^{LT}
AMERICIUM 243	CURIUM 247	BERKELIUM 247	CALIFORNIUM 251	EINSTEINIUM 254	FERMIUM 257	MENDELEVIUM 256	NOBELIUM 254	LAWRENCIUM 257
11.8 **Am** 95	**Cm** 96	**Bk** 97	**Cf** 98	**Es** 99	**Fm** 100	**Md** 101	**No** 102	**Lw** 103
[Rn]$5f^76d^07s^2$	[Rn]$5f^76d^17s^2$	[Rn]$5f^76d^27s^2$	[Rn]$5f^96d^17s^2$					
1267	1600							

Reprinted with permission from Ashcroft, N. W., and Mermin, N. D. (1976). *Solid State Physics*. Thomson Learning: Brooks/Cole.

Solid State Physics

DAN WEI

Tsinghua University, Beijing

Australia • Brazil • Japan • Korea • Mexico • Singapore • Spain • United Kingdom • United States

Solid State Physics
Dan Wei

Publishing Director: Paul Tan

Development Editor: Yang Liping

Copy Editor: Deborah Berger-North

Product Director: Janet Lim

Product Manager: Kevin Joo

Cover Designer: Vincent Lim

Compositor: ARK Imaging Services

© 2008 Cengage Learning Asia Pte Ltd and Tsinghua University Press

ALL RIGHTS RESERVED. No part of this work covered by the copyright hereon may be reproduced or used in any form or by any means graphic, electronic, or mechanical, including photocopying, recording, taping, web distribution, information networks, or information storage and retrieval systems without the prior written permission of the publisher.

For permission to use material from this text or product, email to asia.publishing@cengage.com

ISBN-13: 978-981-4227-97-1

ISBN-10: 981-4227-97-8

Cengage Learning Asia Pte Ltd
5 Shenton Way
#01-01 UIC Building
Singapore 068808

Cengage Learning products are represented in Canada by Nelson Education, Ltd.

For product information, visit **cengageasia.com**

Printed in Singapore
1 2 3 4 5 12 11 10 09 08

Preface

This is a solid state physics textbook written for undergraduates majoring in physics, materials science, electronics, or applied physics. The author has been teaching Solid State Physics to undergraduates at Tsinghua University in the People's Republic of China since 1997. This book is based on the first Chinese edition of *Solid State Physics*, published in 2003 and reprinted in 2004.

I first thought about writing a new Solid State Physics book in the late 1980s when I was a graduate student at the University of California, San Diego. I found that the knowledge I had acquired in the undergraduate Solid State Physics course did not fully prepare me for the graduate course or for research. For example, "neutron diffraction" was a term frequently mentioned by the professors; however, I had absolutely no idea what it meant at that time. I realized that a Solid State Physics book that emphasizes both the theory and the corresponding experiments would be really helpful for students.

When I started to write my lecture notes in 1997, I covered both the basic scientific theory and the experimental proofs for the structural, electrical, magnetic, sonic, optical, and thermal properties of matter. In the fall of 2001, Mr. Cheng-bin Song, who worked at the Tsinghua University Press, read my lecture notes and invited me to write a new Solid State Physics textbook. As I was writing the book, I realized that the establishment of solid state physics was deeply rooted in the work of the great philosophers of science and the subsequent development of modern physics in the 20th century. It is also important to have a good grounding in other relevant fields in physical science.

In July 2005, I was informed by Tsinghua University Press that Cengage Learning wanted to publish the English version of my book. I have since worked closely with development editor Yang Liping and copyeditor Deborah Berger-North in finalizing this book.

I would like to thank my husband, Prof. Chuan Liu of Peking University—without his influence, I wouldn't have such a deep interest in basic science. I also would like to thank my advisor Prof. Daniel P. Arovas, and Prof. Robert C. Dynes, Prof. Lu J. Sham, Prof. Harry Suhl, and Prof. Jorge Hirsch from the University of California, San Diego; Prof. Qin Guo-guang from Peking University; and Prof. Zhu Jin and Prof. Liu Bai-xin from Tsinghua University. Their knowledge and support have been invaluable.

Dan WEI
Tsinghua University, 2007

Acknowledgements

The author and publisher would like to gratefully credit or acknowledge the following:

Figure 2.9   Reprinted with permission from Cornell University Press.

Figure 3.37   Reprinted with permission from Schlenz, H., Neuefeind, J., and Rings, S., Phys. Rev. B, Vol. 32, 4892–4898, 2003.

Figure 3.37   Reprinted with permission from Schlenz, H., Neuefeind, J., and Rings, S., Phys. Rev. B, Vol. 32, 4892–4898, 2003.

Figure 3.39   Reprinted with permission from Shechtman, D., Blech, I., Gratias, D. and Cahn, J. W., Phys. Rev. Lett., Vol.53, 1951–1953, 1984.

Figure 3.41   Reprinted with permission from Elser, V., Phys. Rev. B, Vol.32, 4892–4898, 1985.

Figure 3.50(a)   Reprinted with permission from Prof. Lorenzo Marrucci.

Figure 4.6   Reprinted with permission from Kellermann, E. W., Proceedings of the Royal Society of London. Series A, Mathematical and Physical Sciences, Vol. 178, No. 972, 17–24, 1941.

Figure 4.13(a)   Reprinted with permission from Vocadlo, L., and Alfe, D., Phys. Rev. B, Vol.65, 214105, 2002.

Figure 5.7   Reprinted with permission from Coles, B. R., Rev. Mod. Phys., Vol. 36, 139–145, 1964.

Figure 5.8   Reprinted with permission from Dushman, S., Rev. Mod. Phys., Vol. 2, 381–476, 1930.

Figure 5.16   Reprinted with permission from Burns, G., Solid State Physics, Academic Press, Orlando.

Figure 5.17(b), 5.18(c), 5.19(b)   Reprinted with permission from Alex Choy.

Figure 5.18(a)   Reprinted with permission from Jan, J. P., and Skriver, H. L., J. Phys. F: Met. Phys., Vol. 1, 805–820, 1981.

Figure 5.18(b) Reprinted with permission from Condon, J. H., and Marcus, J. A., Phys. Rev., Vol. 134, A446–452, 1964.

Figure 5.19(a) Reprinted with permission from Singhal, S. P., and Callaway, J., Phys. Rev. B, Vol.16, 1744–1745, 1977.

Figure 5.20(a)(c) Reprinted with permission from Eastman, D. E., Himpsel, F. J., and Knapp, J. A., Phys. Rev. Lett., Vol. 44, 95–98, 1980.

Figure 5.20(b) Reprinted with permission from Callaway J. and Wang C. S., Phys. Rev. B, Vol.16, 2095–2105, 1977.

Figure 5.21(a) Reprinted with permission from Brener, N. E., Phys. Rev. B, Vol. 11, 929–934, 1975.

Figure 5.21(a) Reprinted with permission from Kahn, A. H., and Leyendecker, A. J., Phys. Rev., Vol.135, A1321–A1325, 1964.

Figure 6.3(b) Reprinted with permission from Khoshenevisan M., Pratt Jr. W. P., Schroeder P. A., and Steenwyk S. D., Phys. Rev. B, Vol. 19, 3873–3878, 1979.

Figure 6.8(a–c) Reprinted with permission from Chelikowsky, R., and Cohen, M. L., Phys. Rev. B, Vol. 14, 556–582, 1976.

Figure 6.8(d) Reprinted with permission from Burns, G., Solid State Physics, Academic Press, Orlando.

Figure 6.18 Reprinted with permission from Pearson, G. L., and Bardeen, J., Phys. Rev., Vol. 75, 865–883, 1949.

Figure 6.22(a) Reprinted with permission from Sze, S. M., Physics of Semiconductor Devices, John Wiley and Sons, New York.

Figure 6.22(b) Reprinted with permission from del Alamo, J. A., and Swanson, R. M., Solid State Electron, Vol. 30, 1127–1136, 1987.

Figure 6.33(b) Reprinted with permission from Little, W. A., and Parks, R. D., Phys. Rev. Lett., Vol. 9, 9–12, 1962.

Figure 7.4(b) Reprinted with permission from Fowler, A. B., Fang, F. F., Howard, W. E., and Stiles, P. J., Phys. Rev. Lett., Vol. 16, 901–903, 1966.

Figure 7.4(c–d) Reprinted with permission from Knobel, R., Samarth, N., Harris, J. G. E., and Awschalom, D. D., Phys. Rev. B, Vol. 65, 235327, 2002.

Figure 7.5(b) Reprinted with permission from Henry, W. E., Rev. Mod. Phys., Vol. 25, 163–164, 1953.

Figure 7.7(b)   Reprinted with permission from Weiner, D., Phys. Rev., Vol. 125, 1226–1238, 1962.

Figure 7.7(a)   Reprinted with permission from Vuillemin, J. J., Phys. Rev., Vol. 144, 396–405, 1966.

Figure 7.9,10   Reprinted with permission from Kaya, S., Sci. Repts. Tohoku Imp. Univ., Vol. 17, 1157, 1928.

Figure 7.11   Reprinted with permission from Burns, G., Solid State Physics, Academic Press, Orlando.

Figure 7.12   Reprinted with permission from Connolly, J. W. D., Phys. Rev., Vol. 159, 415–426, 1967.

Figure 7.13(b)   Reprinted with permission from Moyer, C. A., Arajs, S., and Hedman, L., Phys. Rev. B, Vol. 14, 1233–1238, 1976.

Figure 7.14   Reprinted with permission from Squire, C. F., Phys. Rev., Vol. 56, 922–925, 1939.

Figure 7.17   Reprinted with permission from Shull, C. G., Strauser, W. A., and Wollan, E. O., Phys. Rev., Vol. 83, 333–345, 1951.

Figure 7.19   Reprinted with permission from Aldred, A. T., Phys. Rev. B, Vol. 11, 2597–2601, 1975.

Figure 7.20(a)   Reprinted with permission from Brockhouse, B. N., Phys. Rev., Vol. 106, 859–864, 1957.

Figure 7.20(b)   Reprinted with permission from Sinclair, R. N., and Brockhouse, B. N., Phys. Rev., Vol. 120, 1638–1640, 1960.

Figure 7.21(a)   Reprinted with permission from Edmonds, D. T., and Petersen, R. G., Phys. Rev. Lett., Vol. 2, 499–500, 1959.

Figure 7.21(b)   Reprinted with permission from Pollack, S. R., and Atkins, K. R., Phys. Rev., Vol. 125, 1248–1254, 1962.

Figure 8.8(b)   Reprinted with permission from Hass, M., Phys. Rev., Vol. 117, 1497–1499, 1960.

Figure 8.14(b)   Reprinted with permission from Holonyak, N. and Bevacqua, S. F., Appl. Phys. Lett., Vol. 1, 82–83, 1962.

Every attempt has been made to trace and acknowledge copyright holders. Where the attempt has been unsuccessful, the publisher welcomes information that would redress the situation.

About the Author

Professor Dan Wei received her bachelor's degree in physics from Peking University in 1988 and her Ph.D. in Physics from the University of California at San Diego (UCSD) in 1993 under the supervision of Professor Daniel P. Arovas. From 1993 to 1996, she worked as a postdoctoral researcher with Professor H. Neal Bertram in the Center for Magnetic Recording Research (CMRR) at UCSD, doing research in the field of micromagnetic theory for magnetic tape recording systems. From May to October 1996, she worked as a visiting researcher at the Data Storage Institute (DSI) of the National University of Singapore (NUS), where she built a model to simulate the read/write process in longitudinal recording hard disk drives. She joined the Department of Materials Science and Engineering, Tsinghua University in November 1996 as an associate professor. A full professor since 2000, she is a senior IEEE member, a senior member of Chinese Institute of Electronics, and an oversea member of IEICE.

Contents

1 Exordium 1

1.1 Atomism 1
1.2 History of Solid State Physics 4
1.3 Solids in Nature and Solid State Physics 8
Summary 12
References 12

2 Chemical Bonds and Crystal Formation 15

2.1 Quantum Model of Atoms 16
2.2 Ionic Bonds and Ionic Crystals 20
2.3 Covalent Bonds and Covalent Crystals 25
2.4 Metallic Bonds and Typical Metals 30
2.5 Atomic and Molecular Solids 33
Summary 38
References 39
Exercises 40

3 Structure of Solids 41

3.1 Geometrical Description of Crystals 41
3.2 Symmetry and Classification of Crystal Structures 46
    3.2.1 *Symmetry and Classification of 2D Lattices* 48
    3.2.2 *Point Groups and Classification of 3D Bravais Lattices* 50
3.3 Natural Structures of Crystals 54
    3.3.1 *Structures of Element Crystals* 54
    3.3.2 *Structures of Compounds: Pauling's Rules* 58
3.4 Reciprocal Lattices and Brillouin Zones 63
    3.4.1 *Reciprocal Lattices* 63
    3.4.2 *Brillouin Zones* 66
3.5 Measurement of Crystal Structure by Diffraction 69
    3.5.1 *X-ray, Electron, and Neutron Diffraction* 71
    3.5.2 *Diffraction Theory* 79

- 3.6 Disordered Solid Structure 87
  - 3.6.1 *Amorphous Materials* 89
  - 3.6.2 *Quasicrystals* 92
  - 3.6.3 *Liquid Crystals* 95
- Summary 104
- References 105
- Exercises 106

4 Thermal Properties of Solids and Lattice Dynamics 111

- 4.1 Albert Einstein and his Phonon Model 113
- 4.2 Debye Model of Specific Heat 117
- 4.3 Lattice Dynamics and Neutron Diffraction 122
  - 4.3.1 *Lattice Harmonic Theory* 123
  - 4.3.2 *Optical Branches and Acoustic Branches* 126
  - 4.3.3 *Measurement of Phonon Spectra by Neutron Diffraction* 131
- Summary 135
- References 135
- Exercises 136

5 Solid State Electronics Theory 139

- 5.1 Drude Model: Free Electron Gas 141
- 5.2 Sommerfeld Model: Free Electron Fermi Gas 147
  - 5.2.1 *Specific Heat of Electrons* 152
  - 5.2.2 *Electrical and Thermal Conductivity* 155
  - 5.2.3 *Thermionic Emission of Electrons from Metal Surfaces* 157
  - 5.2.4 *Hall Effect* 159
- 5.3 Band Theory 162
  - 5.3.1 *Bloch's Theorem* 164
  - 5.3.2 *Tight-Binding Model* 167
  - 5.3.3 *Weak Potential Approximation* 172
  - 5.3.4 *Introduction to DFT and Computational Methods* 176
  - 5.3.5 *Real Bands and Fermi Surfaces* 178
  - 5.3.6 *Semiclassical Model and Effective Mass* 184
- Summary 188
- References 188
- Exercises 190

6 Electrical Transport Properties of Solids 193

- 6.1 Conductors 194

- **6.2** Semiconductors 199
  - 6.2.1 *Characteristics of Semiconductors* 201
  - 6.2.2 *Carrier Density and Mobility* 209
  - 6.2.3 *Basic Concepts of Semiconductor Devices* 223
- **6.3** Superconductors 235
  - 6.3.1 *Characteristics of Superconductors* 237
  - 6.3.2 *Phenomenological Theories* 241
  - 6.3.3 *Microscopic BCS Theory* 247

Summary 251
References 252
Exercises 254

7 Magnetic Properties of Solids 257

- **7.1** Quantum Mechanical Origin of Magnetism 260
  - 7.1.1 *Single Atom Approximation: Atomic Moments* 262
  - 7.1.2 *Free Electron Approximation: Landau Levels* 266
- **7.2** Categories of Magnetism 269
  - 7.2.1 *Diamagnetism* 270
  - 7.2.2 *Paramagnetism* 272
  - 7.2.3 *Ferromagnetism* 278
  - 7.2.4 *Antiferromagnetism and Ferrimagnetism* 285
- **7.3** Spins Interacting with Fundamental Particles 289
  - 7.3.1 *Magnetic Neutron Diffraction and Magnetic Structure* 289
  - 7.3.2 *Spin Wave and Neutron Inelastic Scattering* 291
  - 7.3.3 *Electron Spin Resonance and Neutron Magnetic Resonance* 296

Summary 300
References 302
Exercises 303

8 Optical and Dielectric Properties of Solids 307

- **8.1** Unification of Optical, Electrical, and Magnetic Properties 309
- **8.2** Lorentz Optical Model and Polarization Process 312
- **8.3** The Laser: Einstein's Stimulated Radiation Theory 320
  - 8.3.1 *Quantum Mechanical Theory of Radiation* 321
  - 8.3.2 *Masers and Lasers* 323

Summary 327
References 328
Exercises 329

Index 331

Exordium

ABSTRACT

- History of Atomism
- History of Solid State Physics

The foundation of solid state physics is the atomic and electronic structure of crystals. The structure of matter has long been a subject of study by natural philosophers and scientists. The first major breakthrough came with the development of atomism in Ancient Greece. It is believed that Thales, who lived around 600 B.C. in the city of Miletus (then the most important Greek city in Asia Minor), was the first person to ask the key philosophical question "What is matter composed of?"

Figure 1.1 Thales of Miletus (624 B.C.–547 B.C.)

1.1 Atomism

Since Thales first posed the question "What is matter composed of?" around 2,600 years ago, many of the world's greatest philosophers and scientists have suggested answers. The first description, given by Thales

Figure 1.2 Ancient Greece: Miletus, Ephesus, Samos, and Athens

himself, was quite simple. Thales lived in a coastal city where he could see the clouds rising up from the sea everyday. Therefore Thales decided that matter is condensed or dispersed from water. Water is vaporized into air, and air is condensed into water or solids, so the essence of matter is simply water.

After Thales, many philosophers from Ancient Greece provided various answers to the same question. For example, Anaximander of Miletus (ca. 610 B.C.–546 B.C.) thought that the universe was composed of infinite primordial matter. Anaximenes of Miletus (ca. 570 B.C.– ca. 500 B.C.) deemed that the essence of matter was air. As the Ancient Greek culture entered its golden period around 510 B.C., Heraclitus (540 B.C.– 480 B.C.), who lived in Ephesus, another important Greek city in Asia Minor, declared the essence of matter to be fire: "A fire, when mixed with other spices, is named after the spice since that is the smell we perceive. The fire remains unchanged; it, in fact, remains constant" (Russell, 1992). Empedocles of Akragas in Sicily (ca. 492 B.C.–ca. 432 B.C.), who was a philosopher, a poet, a material physicist, and a subscriber to the Pythagorean school, summarized his predecessors' thoughts about nature by formulating the "four elements of matter" theory: "Matter is composed of four elements: water, air, fire, and earth" (Russell, 1992).

It is interesting to note that around 600 years before Thales' time, in ancient China, the book *Hong Fan*, (literally meaning "Vast Pattern"), presented the following "five elements" theory: "Matter is composed of metal, wood, water, fire, and earth" (Hu, 1936). However Chinese

Figure 1.3 The archaeological locale of Plato's Academy

philosophy also offered many other theories about the universe, such as the Yin-Yang (Feminine-Masculine) and Ba-Gua (Eight Diagrams) theories, and the five elements theory did not become the mainstream school of thought among the ancient Chinese scholars.

One of the most important schools in Ancient Greece was the Pythagoreans. The founder of the school, Pythagoras (ca. 580 B.C.–500 B.C.), was born in Samos in Asia Minor. He met Thales when he was young, and eventually settled in Croton on the eastern coast of Italy. The Pythagorean School was a religious group worshipping Apollo and the Muses. They also studied mathematics, music, and physics, and, significantly, introduced the concept of the vacuum. They refined the four elements of matter theory by suggesting that all four elements were composed of cubic elementary particles.

Leucippus of Miletus (480 B.C.– 420 B.C.) is generally credited as the founder of atomism. He and his student, Democritus of Abdera (460 B.C.–370 B.C.), theorized that matter was composed of innumerable indivisible particles which they called atoms ($\alpha\tau o\mu$). They thought that atoms could be of different sizes and shapes and could move around in space. This notion of atomism, in which nature consists only of two things, namely atoms and the void that surrounds them, challenged the ideas of Parminedes of Elea (who in essence said that everything was one and a void could not exist because something could not be nothing) and Anaxagoras of Athens (who stated that matter could be divided infinitely), and proved astonishingly far-sighted.

The Greek philosopher and mathematician, Plato (ca. 428 B.C.–ca. 347 B.C.), founded his Academy in Athens around 387 B.C. There atomism was taught for many years, and Plato himself devised a geometrical construction for the elements in which the cube, tetrahedron, octahedron, and icosahedron were given as the shapes of the atoms of earth, fire, air, and water respectively. Plato's theory of atoms was wrong; however his idea paved the way for numeric descriptions of atoms which eventually led to the modern periodic table of elements.

The development of atomism was one of the great achievements of the Ancient Greeks, and it has become one of the philosophical fundamentals of science. As R. P. Feynman said in his book, *The Feynman Lectures on Physics*: "If, in some cataclysm, all scientific knowledge were to be destroyed, and only one sentence passed on to the next generations of creatures, what statement would contain the most information in the fewest words? The sentence is: 'All things are made of atoms...'" (Feynman, 1963).

1.2 History of Solid State Physics

About two thousand years passed following the discovery of atomism before further great achievements in science led to the birth of modern physics. Aristotle (384 B.C.–322 B.C.) did not make any major contributions to physics; however, in his book, *Physics*, which was kept in monasteries during the Middle Ages, the discussions about space, time, and motion formed a basis for the future development of physics. In the 17th century, Isaac Newton established the foundation of mechanics, astronomy, and, in part, optics, basing his work on the revolutionary breakthroughs made by Galileo Galilei, Johannes Kepler, and Christian Huygens.

In the 19th century, following on from the experiments of Cavendish, Coulomb, Oersted, Ampere, and especially Faraday, James Clerk Maxwell formulated the classic electrodynamic theory, which eventually became widely accepted after 30 years of controversy. Maxwell's theory and Hertz's electromagnetic wave experiment seemed to indicate the "triumph" of the wave theory of light following 200 years of post-Newtonian wave-versus-particle debate. Maxwell's equations, together with Lorentz's electron theory, perfectly explained the complex macroscopic electromagnetic phenomena and unified the fields of electronics, magnetics, and optics. In the late 19th century, Ludwig Boltzmann developed the field of statistical mechanics and clearly explained Maxwell's molecular speed distribution law. By 1900, physicists realized that Newtonian mechanics, Maxwell's Electrodynamics, and Boltzmann's statistical physics were independent from each other; most of the physical phenomena at that time could be explained, and the great laws of classical physics were hence completed.

Figure 1.4 Thomson discovered electrons

During the later stages of the development of classical physics, the ancient concept of atomism was revived by chemists. It could be said that both physics and chemistry participated in creating the modern concepts of atoms.

In 1803, John Dalton explained his atomism at the University of Manchester. He made two main points: (1) that the weights of the same kind of atoms were identical, therefore the concept of atomic weight could be defined; (2) that different atoms were chemically combined with an integer ratio, therefore the chemical formulae could be written. In 1811, Amedeo Avogadro proposed the atom-molecule theory. He said that if temperature and pressure were fixed, the total number of molecules per unit volume would be the same for all ideal gases. The branch of physics in which modern atomism first led to a breakthrough was thermodynamics. In the 1850s and 1860s, based on Newtonian mechanics and atomism, the kinetic theory of gases was gradually developed by Rudolf Clausius, Maxwell, and Boltzmann.

In 1865, Joseph Loschmidt found the exact order of the molecular radius (10^{-8} cm) and molecular number in one mole (10^{23}), based on the mean free path of gas molecules and the liquid volume respectively. The Loschmidt number gradually evolved into the Avogadro number through various experimental and theoretical proofs. Atomism also influenced electronic theory. In 1834, Michael Faraday formulated the law of electrolysis, based on the idea that each ion has an electric charge one or several times the "fundamental charge." Joseph Larmor and Lorentz addressed the idea that the carrier in matter should have a fundamental charge in their theory of electrons. In 1897, Wilhelm Wien, Joseph John Thomson, and their colleagues finally found the negative carrier—the electron—by measuring the charge-to-mass ratio in a mass spectrograph.

Another significant breakthrough in physics was the development of quantum theory in the early 20th century. In 1900, Max Planck proposed the energy quantum hν of an electromagnetic wave to explain his own heat radiation formula. In 1905, Albert Einstein developed Planck's energy quanta for electromagnetic waves into the concept of

Figure 1.5 Quartz crystals: the crystal planes

the "photon," and explained the photoelectric phenomena in terms of the wave-particle duality of light, which finally concluded the 240-year-old debate about the essence of light. In 1913, Niels Bohr formulated the quantum theory of atoms, based on the Rutherford model of the atom and the formula of the optical spectrum. The wave-particle duality of all fundamental particles, proposed by L. V. de Broglie in 1924, was a very important contribution to basic physics theory. Quantum mechanics and quantum statistics were established in 1926 by Schrodinger, Heisenberg, Pauli, Fermi, and Dirac. Wave-particle duality could be viewed as a fusion of the particle concept in classical physics and the wave concept in quantum physics. The phonon, band electron, cooper pair, and magneton are examples of quasi-particles with wave-particle duality.

Solid state physics is a recently developed branch of physics. It is hard to determine the starting point, but the X-ray diffraction of crystals in 1912 might be a suitable choice. Historically, crystallography had little to do with mainstream physics, because physicists had few opportunities to see well-formed crystals, while mineralogists were not really interested in physics. The concepts of symmetry and crystal structure were first mentioned by Johannes Kepler in his book *Hexagonal Snow*. Kepler was perceptive enough to predict the symmetry of solid structures simply from observing the hexagonal shape of a snowflake, and he even imagined that snow was formed from closely-packed spheres. In 1669, Niels Stenson found that, for different quartz crystals, the angles between different crystal planes were always the same. After that, the word "kristall" gradually came to be used to describe solids with natural fixed shapes. In the 19th century, geometrical crystallography was gradually developed by Christian Samuel Weiss, Franz Ernst Neumann, and William Hallows Miller. Miller defined the indices of crystal planes using Haüy's law of rational intercepts. In 1830, a mineralogy professor, J. F. C. Hessel, studied the symmetry in crystals and derived the 32 crystal classes, also using Haüy's law of rational intercepts. In 1850, a physicist, A. Bravais, proved that there are only 14 classes of lattice for three-dimensional crystals. In 1890 and 1891 respectively, two mathematicians, E. C. Feodorov and A. M. Schoenflies, employed different methods to prove that there are 230 space groups, which led to the establishment of mathematic tools for defining crystal structure.

In the early 20th century, physicists had conflicting ideas about the distribution of atoms in solids. Some believed that atoms stayed in a lattice; others thought that atoms were distributed randomly. At the University of Munich, Ludwig August Seeber's book, *Atomism in Crystals*, which said that a crystal lattice is formed from chemical atoms, was the accepted teaching. In 1912, in line with Max von Laue's prediction, W. Friedrich and P. Knipping from A. Sommerfeld's group discovered the beautiful X-ray diffraction patterns of crystals, which confirmed both the wave nature of X-rays and the lattice structure of crystals. In 1913, W. H. Bragg, (using his X-ray spectroscopy) and his son, W. L. Bragg (using his famous formula), accurately determined the lattice constant and the X-ray wavelength. In 1927 and 1928, C. J. Davisson and G. P. Thomson observed the reflecting and tunneling electron diffraction patterns of crystals respectively, which clearly proved the de Broglie wave-particle duality of electrons. After World War II, neutron diffraction was developed as the third diffraction method, which was especially important for the study of quasi-particles in solids.

There are many branches of solid state physics, including the basic study of the electrical, magnetic, acoustic, optical, and thermal properties of solids. Early studies of the electrical properties of solids can be traced back to George Simon Ohm's law of metal resistance in 1826. In 1926, the establishment of quantum mechanics and quantum statistics opened the door to modern solid state electronics theory. In 1928, A. Sommerfeld used the Fermi statistics to describe the free electron Fermi gas in metals. In the same year, F. Bloch formulated the energy band theory for electrons in solids. The density functional theory (DFT), proposed by W. Kohn in 1963, became the foundation of band calculation, quantum chemistry, and computational materials science. Solid state electronics is an important branch of modern physics. Many Nobel prizes have been awarded to those who have made advances relating to semiconductors, superconductors, and DFT. It has also enabled the development of solid state electronic devices, such as transistors in integrated circuits.

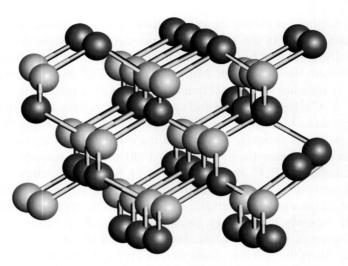

Figure 1.6 Lattice structure: periodic arrangements of atoms

Regarding the magnetic properties of solids, the magnetism of matter was classified by M. Faraday in the 1840s; phase transition in magnetism was analyzed by P. Curie in 1895; and ferromagnetism was phenomenologically explained by P. E. Weiss in 1907 using the molecular field theory. With respect to the thermal properties of solids, the Dulong-Petit law in 1820 and its deviation in diamonds were explained by A. Einstein in 1907 when he identified the energy quanta—phonons—of atomic vibration in crystals. In 1911, P. Debye modified Einstein's phonon model using Planck's and his own radiation theories, and predicted specific heats at low temperatures. The lattice vibration theory was formulated by M. Born and T. von Karman in 1913 as a result of their investigations into the sonic properties of solids, and the phonon spectrum was later proved by means of neutron diffraction. An understanding of the interactions of the optical, electrical, and magnetic properties of solids has been fundamental to the development of nuclear magnetic resonance, laser and semiconductor optical devices. More recently, the scope of research in solid state physics has been extended to include condensed matter physics and materials science.

1.3 Solids in Nature and Solid State Physics

The structure of matter in nature is complex. Different disciplines have distinct classifications of matter, that is to say, each discipline emphasizes one particular representation of atomism. In biology, the universe is classified into living things and others. In the living things, several kinds

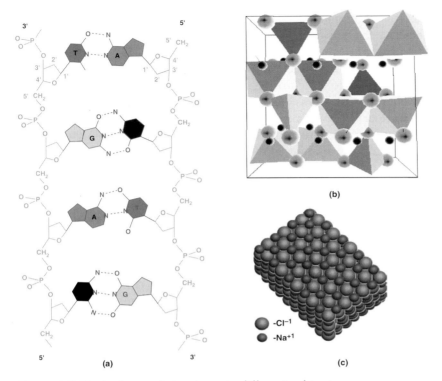

Figure 1.7 Typical crystal structures in different subjects:
(a) biology—DNA structure; (b) chemistry—Topaz crystal structure;
(c) physics—NaCl structure

of atoms—carbon, hydrogen, oxygen, nitrogen, and phosphorus—are organized in complex ways to form the essential factors of metabolism and procreation: genes and proteins. In chemistry, substances are classified into minerals and organic compounds. Organic compounds consist of living things and the relics of life-forms, such as oil and coal. In physics, matter is classified into four states: solids, liquids, gases, and plasmas.

In fact, the words "solid" and "liquid" have existed since ancient times, and the word "gas" was first introduced by a pioneer chemist, J. P. van Helmont, in the early 17th century. The word "plasma" comes from the ancient Greek word $\pi\lambda\alpha\sigma\mu\alpha$ ("jelly"), and was first used in 1927 by the Nobel prize-winning chemist I. Langmuir to describe ionized gas. Physicists are more interested in the essence and general properties of the four states, and they prefer solids with simple structures, as it is easier to study their physical properties.

The concepts "solid," "liquid," and "gas" were originally defined under normal everyday conditions of room temperature and air pressure. When temperature or pressure changes, the solid, liquid, and gas states can transform into one another. Hence any natural matter or man-made material can become solid under certain conditions. The scope of solid state physics is therefore quite broad.

Since 1912, diffraction methods have been widely used to explore the microscopic structure of solids. Every solid has its own type of arrangement of atoms, and it usually has multiform microstructures on the macroscopic, mesoscopic, and microscopic scales. Well-grown large crystals are rare, but polycrystals containing randomly orientated crystal grains are quite common. In a crystal or crystal grain, atoms have a startlingly perfect distribution. From very early times it was thought that metals and jewels were crystals. However, after diffraction methods were developed, organic compounds, such as wood, cloth, muscles, and nerves, were also shown to have micro-crystalline structures. Therefore, the crystal state is the "normal" state of solids; only a few amorphous solids, such as glass, have a microscopically random atomic structure.

As a branch of physics, the main objective of solid state physics is to explain the macroscopic properties of matter in terms of its microstructure and the fundamental laws of the microscopic world, especially quantum physics. The first solid state physics book, *The Modern Theory of Solids*, was written in 1940 by F. Seitz, who was the first Ph.D. student of the 1963 Nobel Prize winner in Physics, E. P. Wigner. In his book, Seitz discussed the important properties of metals, ionic crystals, valence crystals, semiconductors, and molecular crystals. Seitz paid great attention to the special properties of crystals of different elements and compounds. C. Kittel's book, *An Introduction to Solid State Physics*, was first published in 1958, and there have been seven editions to date. His book is a comprehensive text on solid state physics, and it is one of the classics in this field. Also in 1958, Prof. Huang Kun started a solid state physics course at Peking University, and his classes had a profound impact on several generations of young students in China. Ashcroft & Mermin's book, *Solid State Physics*, had a new structure and focused on solid state electronics theory. Since 1980 this book has been widely used as a graduate textbook on advanced solid state physics in the United States.

The original aim of this book was to focus on applied physics, especially materials science and electronics. However, during the writing process, the author found it necessary to state many of the fundamental laws of physics. This is because solid state physics is derived from a wide range of disciplines, namely classical physics, quantum physics, and crystallography. In Chapter Two of this book, based on the Bohr atomic model and the Hartree-Fock method, the physical essence of five kinds of chemical bonds, especially the physical source of the related binding energy, are discussed. In Chapter Three, the structure of crystals is analyzed, based on the symmetry of lattices; also the origins of X-ray, electron, and neutron diffraction are discussed to introduce the experimental methods of solid structure analysis; finally amorphous solids, quasi-crystals, and liquid crystal structures are analyzed using diffraction theory.

In Chapters Four to Eight, the essence of the physical properties of crystals is discussed. Only in strictly ordered crystals is it possible to analyze the physical properties by the first principles. In Chapter Four, the thermal and sonic properties of solids are explained using Einstein's phonon model and the lattice vibration theory of atoms. In Chapter Five, the electronic theories of solids, including the Drude model, the Sommerfeld model, and the energy band theory, are introduced to illustrate electronic structure in solids. In Chapter Six, the transport properties of metals, semiconductors, and superconductors are analyzed based on solid state electronics theory and some advanced theories, such as the BCS theory of superconductors. In Chapter Seven, the essence of the magnetic properties of solids, including paramagnetism, diamagnetism, ferromagnetism, and antiferromagnetism, are discussed based on the quantum theory of atomic moments. In Chapter Eight, the dielectric and optical properties are analyzed based on the Lorentz model, and the basics of laser are also introduced. Some optical properties are also discussed in Chapters Four, Five, Six, and Seven.

Solid state physics forms a bridge between the microscopic and the macroscopic world, therefore it is important to both physical science and the modern information industry. Some branches of solid state physics have grown into independent research fields, such as semiconductor physics, magnetic physics, and diffraction physics. The ideas

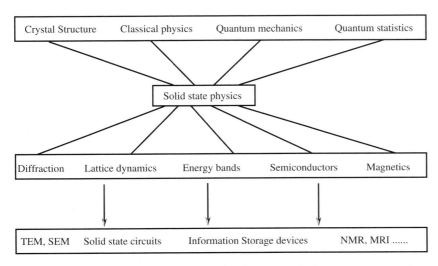

Figure 1.8 Solid State Physics: a bridge between the microscopic and macroscopic world

in solid state physics have also influenced cross-field research areas such as physical chemistry, structural chemistry, and biological physics. Spectroscopy and nuclear magnetic resonance are examples of the benefits of this multi-disciplinary approach in physics, chemistry, biology, materials science, and medical science. Magnetic physics is important in the research and development of computer hard drives. Semiconductor physics has underpinned the development of solid state electronics, which is one of the greatest 20th century examples of the way in which science can improve people's standard of living.

This book is of particular use to senior undergraduates majoring in the physical sciences, as well as certain non-physics major graduate students and engineers who are interested in this field, both as a text book and as a reference.

Summary

This chapter reviews the history of solid state physics. Solid state physics is a relatively recent branch of physics, built upon a foundation of scientific theory and experiment developed over hundreds of years. The key developments are covered as shown below.

1. Atomism: from Ancient Greek atomism to modern atomism.
2. Crystallography and diffraction: from Kepler's idea of lattice structure to the development of X-ray and electron diffraction based on quantum mechanical wave-particle duality.
3. More branches in solid state physics: exploring the acoustic, electrical, magnetic, optical, and thermal properties of solids.
4. The relationship between solid state physics and other branches of science and engineering: how solid state physics has played a central role in science and engineering since the mid-20th century.

References

Ashcroft, N. W., and Mermin, N. D. (1976). *Solid State Physics*, Holt, Rinehart and Winston, New York.
Encyclopedia of China: Physics. (1987). Encyclopedia of China Publishing House, Beijing.
Feynman, R. P., Leighton, R. B., and Sands, M. (1963). *The Feynman Lectures on Physics*, Addison-Wesley, New York.
Gu, Z. (1982). *Greek City-State System* (in Chinese), China Social Sciences Press, Beijing.
Hu, Y. (1936). *Hong Fan Kou Yi* (in Chinese), Commercial Press, Shanghai.

Huang, K., and Han, R. (1988). *Solid State Physics* (in Chinese), Higher Education Press, Beijing.

Kittel, C. (1986). *Introduction to Solid State Physics*, Wiley, New York.

von Laue, M. (1978), translated by Fan, D., and Dai, N. *History of Physics* (in Chinese), Commercial Press, Beijing.

Russell, B. (1992), translated by Ma, J., and He, L. *Wisdom of the West: A Historical Survey of Western Philosophy in its Social and Political Setting* (in Chinese), World Affairs Press, Beijing.

Seitz, F. (1940). *The Modern Theory of Solids*, McGraw-Hill, New York.

Xu, G. (1966). *An Introduction to the Structure of Matter* (in Chinese), Higher Education Press, Beijing.

Chemical Bonds and Crystal Formation

2

ABSTRACT

- History of atomic physics and Hartree-Fock theory
- Essence of chemical bonds

The French chemist, A. L. Lavoisier, published the first table of elements in his book on elementary chemistry in 1789. Lavoisier's table was a combination of some true atomic elements and some other "elements," such as the calorie. In 1816, a London doctor, W. Prout, proposed the following first principle of chemistry: the mass of any atom is an integer times the mass of a hydrogen atom, which became the rule by which the periodic table is arranged. In 1869, D. I. Mendeleev in Russia and J. L. Meyer in Germany independently published the periodic table of elements. This became the basis of the development of chemistry and physics in the 20th century.

With the guidance of the periodic table, chemists were able to devise methods of measuring the binding energy of solids. Binding energy is defined as the energy required to disassemble a whole into free atoms, and it can be measured via various chemical reactions. The binding energies of elements in crystals are listed in Table 2.1. Positive binding energy indicates that a crystal has lower energy than the total energy of free atoms. Higher binding energy indicates a more stable solid. Tungsten has the highest binding energy per atom. Helium has the lowest binding energy, which is too low to be listed in Table 2.1.

In this chapter, the physical essence of chemical bonds is discussed, based on atomic physics and quantum mechanics. The concept of the chemical bond, especially the covalent bond between a shared pair of electrons, was proposed by the physical chemist, G. N. Lewis, in 1916. In simple structural models, the chemical bonds are represented by small sticks between neighboring atoms. In fact, chemical bonds are not necessarily attractive interactions between nearest neighboring atoms, but can have various sources. If r is the distance

Table 2.1 Binding energy of crystalline elements at 1 atm and near 0 K (based on Encyclopedia of China: Physics, 1987)

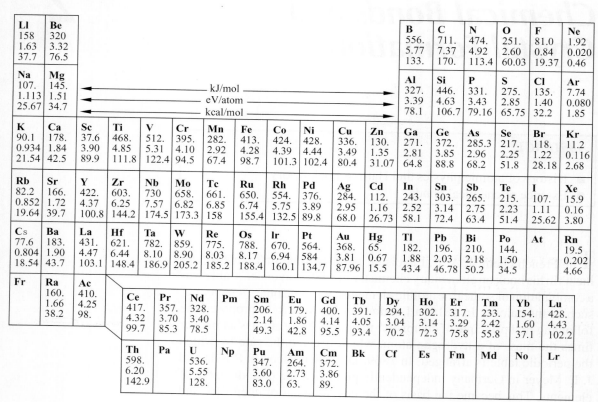

between nearest neighboring atoms and $W(r)$ is the binding energy, then $U(r) = -W(r)$ (sometimes $E_b = U(r) - U(\infty)$) is defined as the cohesive energy of a crystal; and the attractive force with respect to the chemical bond is $F(r) = -dU/dr = dW/dr$, which is contributed by the electromagnetic interactions and quantum effects.

2.1 Quantum Model of Atoms

A Dane, Niels Henrik David Bohr, was the third pioneer in quantum physics after Planck and Einstein. In 1911, Bohr received his Ph.D. in electron theory of metals from Copenhagen University. In the fall of 1911 he spent some time at Cambridge University, where he benefited from experimental work on electrons at the Cavendish Laboratory under the guidance of Sir J. J. Thomson. From March to July in 1912, Bohr worked in Prof. Ernest Rutherford's laboratory

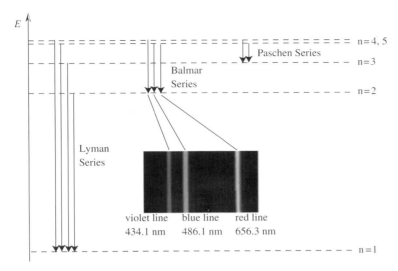

Figure 2.1 The atomic spectrum of hydrogen and the Bohr model

at Manchester University. He was deeply impressed by Rutherford's discovery of the atomic nucleus, and developed a life-long friendship with him.

In the fall of 1912, Bohr went back to Copenhagen University as an assistant professor. By chance, he learned from a colleague that several laws had been summarized from the complicated spectra of gases. The spectrum of hydrogen gas, as shown in Figure 2.1, was discovered in the late 19th century. In 1885, J. J. Balmer summarized the wavelength of the visible lines as the "Balmar series": $\lambda = Bn^2/(n^2 - 2^2)$ where $n = 3, 4, 5 \ldots$. In 1890, J. R. Rydberg simplified Balmer's formula to the form $\lambda^{-1} = R_H(2^{-2} - n^{-2})$ where the Rydberg constant $R_H = 4/B = 1.096775854 \times 10^7 \text{m}^{-1}$. Bohr was inspired by these formulae.

In 1913, Bohr published the paper "On the Constitution of Atoms and Molecules" in the *Philosophical Magazine*. J. J. Thomson believed that electrons in the Rutherford model moved in concentric orbits around the nucleus. However, according to classical electrodynamics, the acceleration of a circulating electron would cause radiation, and the atom would not be stable. To solve this puzzle, Bohr realized that an alternative theory was required to that of Max Planck, who had tried to derive quantum properties from classical physics. He raised two significant new concepts: "static state" and "transition." An electron staying at the n'th static state at energy level E_n would not radiate. A spectrum line corresponds to a transition from the initial n'th state to the final m'th state and the energy of light is $h\nu = E_n - E_m$. The energy levels in a hydrogen atom could be derived directly from Rydberg's formula:

$$\frac{1}{\lambda} = \frac{E_1}{hc}\left(\frac{1}{2^2} - \frac{1}{n^2}\right); \quad E_n = -\frac{Ry}{n^2} = -\frac{R_H hc}{n^2} = -\frac{e^2}{2a_B}\frac{1}{n^2} \quad (2.1)$$

where the Rydberg unit $Ry = 13.6 eV$ is the binding energy, and $a_B = 0.53 Å$ is the Bohr radius. Bohr related the Rydberg unit and Bohr radius directly to the Planck constant: $Ry = R_H hc = 2\pi^2 me^4/h^2$ and $a_B = \hbar^2/me^2$. The radius of the n'th orbit is $a_n = n^2 a_B$; therefore, the energy in Eq. (2.1) perfectly corresponds to the oscillating energy of the classical circular motion. Moreover, the quantum orbit of a static state is intrinsically different from a classical orbit. The Bohr model constituted one of the pivotal steps from classical physics to quantum physics, and proved the correctness of Rutherford's atomic model.

The next breakthrough in quantum theory came in 1924, when L. V. de Broglie proposed the idea of wave-particle duality for fundamental particles. In 1926, Erwin Schrodinger, who was working at the University of Zurich, was inspired by his colleague, Peter Debye, to formulate an equation which the "de Broglie wave" of electrons in atoms obeys. He thus established quantum wave mechanics. Schrodinger found that Bohr's static states in a hydrogen atom could be derived from the wavefunction $\psi_{nlm}(\vec{r}) = R_n(r) Y_{lm}(\theta, \phi)$ and that they were the common eigenstates of the operators hamiltonian $\mathcal{H}$ (quantum number n), angular momentum $\vec{L}^2$ (quantum number l), and its component L_z (quantum number m). In 1926, Max Born gave a statistical explanation of the wavefunction: the probability density of an electron at $\vec{r}$ was $\rho(\vec{r}) = |\psi(\vec{r})|^2$; therefore the Bohr radius a_B was not the radius of the electron orbit, but the one where $\rho(a_B) = e^{-1}\rho(0)$ in a hydrogen atom. Bohr, Schrodinger, and Born won the Nobel Prize in Physics in 1922, 1933, and 1954 respectively for their contributions to the quantum model of atoms.

In the hydrogen atom there is only one electron, and the eigenstates can be derived accurately. In other atoms with multiple electrons, the electron-electron interactions have to be included and the electron many-body effects occur; therefore, the Bohr model cannot not be directly applied to find the quantum states of electrons.

Issac Newton said that if all gravitational interactions between the planets in the solar system had to be included, the accurate calculation of orbits would be almost impossible. Similarly, in an atom with multiple electrons, it is also impossible to solve the Schrodinger equation analytically, and the variational principle has to be used to find the quantum states. The energy functional of a differentiable trial wavefunction Ψ should be minimized:

$$E[\Psi] = \frac{\int d\tau \Psi^* \mathcal{H} \Psi}{\int d\tau \Psi^* \Psi} \geq E_0; \quad \mathcal{H} = \sum_\alpha \left(\frac{\vec{p}_\alpha^2}{2m} - \frac{Ze^2}{r_\alpha} + \frac{1}{2} \sum_{\beta \neq \alpha} \frac{e^2}{r_{\alpha\beta}} \right)$$

(2.2)

where $d\tau$ is the infinitesimal volume of electrons in the atom. The electrons are fermions whose quantum states obey the Pauli's exclusion principle; therefore, one of the best choices of the trial wavefunction is the anti-symmetric determinant of Z electrons in Z orbits proposed by J. C. Slater:

$$\Psi = \frac{1}{\sqrt{Z!}} \begin{vmatrix} \psi_1(\zeta_1) & \psi_1(\zeta_2) & \cdots & \psi_1(\zeta_Z) \\ \psi_2(\zeta_1) & \psi_2(\zeta_2) & \cdots & \psi_2(\zeta_Z) \\ \vdots & \vdots & & \vdots \\ \psi_Z(\zeta_1) & \psi_Z(\zeta_2) & \cdots & \psi_Z(\zeta_Z) \end{vmatrix} \qquad (2.3)$$

where $\psi_\alpha(\zeta_\beta)$ is a single-electron trial wavefunction. The initial quantum states $|\alpha\rangle$ could choose the Bohr atomic orbits $|nlm\sigma\rangle$ (σ for spin) of a single electron. ζ_β represents the coordinates x, y, z, s ($s = \pm\frac{1}{2}$ for spin) of the β'th electron. Therefore Z identical electrons in an atom will fill in Z quantum orbits.

The Hartree-Fock single-electron effective Hamiltonian can be derived from a set of energy minimum conditions $\delta E[\Psi]/\delta \psi_\alpha = 0$ ($\alpha = 1, 2, \ldots, Z$) following the functional principle in Eq. (2.2), where the multi-electron wavefuntion Ψ is chosen as the Slater determinant in Eq. (2.3):

$$\begin{aligned}\mathcal{F}_\alpha(\vec{r}, \vec{p}) = {}& \frac{\vec{p}^2}{2m} - \frac{Ze^2}{r} \\ & + \frac{1}{|\psi_\alpha^2(\vec{r})|} \sum_{\beta \neq \alpha} \iiint d^3\vec{r}' \Big[\psi_\alpha^*(\vec{r}) \psi_\beta^*(\vec{r}') \\ & \times \frac{e^2}{|\vec{r} - \vec{r}'|} \psi_\alpha(\vec{r}) \psi_\beta(\vec{r}') - \delta_{s_\alpha s_\beta} \psi_\alpha^*(\vec{r}) \psi_\beta^*(\vec{r}') \\ & \times \frac{e^2}{|\vec{r} - \vec{r}'|} \psi_\beta(\vec{r}) \psi_\alpha(\vec{r}') \Big]; \quad \delta_{s_\alpha s_\beta} = 0 \text{ or } 1. \end{aligned} \qquad (2.4)$$

The four terms in Eq. (2.4) are the kinetic energy, the electric potential of a nucleus, the electric repulsion from other electrons, and the quantum exchange interaction.

In an atom, the quantum state $|\psi_\alpha\rangle$ and its eigen-energy level ε_α can be found by solving a set of self-consistent Hartree-Fock equations numerically:

$$\mathcal{F}_\alpha(\vec{r}, \vec{p}) \psi_\alpha(\vec{r}) = \varepsilon_\alpha \psi_\alpha(\vec{r}); \quad \alpha = 1, 2, \ldots, Z \qquad (2.5)$$

The Hartree-Fock equations, formulated by D. R. Hartree, B. A. Fock, and J. C. Slater in 1930, represent a fairly accurate *ab initio* method in physics and chemistry. In chemical molecules and soft condensed matter, the Hartree-Fock method can also be used to solve problems, such as the bending in macromolecules and protein molecules, by choosing a proper set of orthogonal eigenstates. In solids, it is almost impossible to solve all electron eigenstates directly using the Hartree-Fock method because the total number of atoms and electrons is huge. However, in the later sections of this chapter, the Hartree-Fock Hamiltonian will be used to analyze the physical essence of chemical bonds in solids.

At one time, J. C. Slater worked with Niels Bohr as a post-doctoral researcher. D. R. Hartree was a theoretical physicist, and as an undergraduate he was much influenced by lectures given by Niels Bohr in Cambridge in 1921. Hartree was also among the first to use computers in physics. Therefore, the Hartree-Fock method is actually a computational physics method for multi-electron atomic wavefunction which significantly influenced the fields of quantum physics, quantum chemistry, materials science, and particularly computational science. The accurate energy bands have to be calculated using the density functional theory (DFT). One of the founders of DFT, Walter Kohn (winner of the 1998 Nobel Prize in Chemistry), pointed out that DFT is simply an accurate form of the Hartree-Fock method.

2.2 Ionic Bonds and Ionic Crystals

Here, the sodium chloride (NaCl) ionic crystal is used to illustrate the formation of ionic bonds.

The closed-shell structure has the most stable electronic state in an atom or ion. A sodium (Na) atom transfers an electron to a chlorine (Cl) atom; both Na^+ and Cl^- ions become the closed-shell structure. Na^+ and Cl^- ions will form the sodium chloride ionic crystal structure, as shown in Figure 2.2. The binding energy W of ionic bonds is mainly contributed by two terms:

1. Coulomb Interactions (Positive contribution to W): Coulomb interactions between N sodium ions and N chlorine ions, including Na^+–Na^+ repulsions, Cl^-–Cl^- repulsions, and Na^+–Cl^- attractions. The major contributors are the Coulomb attractions.
2. Exchange Potential (Negative contribution to W): the exchange potential $-\langle \alpha\beta | \frac{e^2}{r} | \beta\alpha \rangle$ due to the overlapping of the $2p-3p$ orbits of neighboring Na^+ and Cl^- ions. It is always positive and increases rapidly when the ion-ion distance becomes shorter. It is the balancing force of the Coulomb attractions.

The Coulomb interactions between the Na^+ ions and the Cl^- ions are the main sources of the "attractive force" forming the ionic bond, which is known as the classical Madelung energy. The Madelung energy

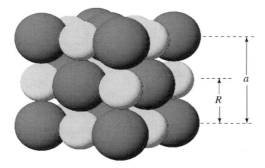

Figure 2.2 Schematics of sodium chloride crystal structure

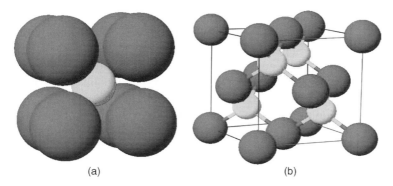

Figure 2.3 Schematics of (a) cesium chloride (CsCl) structure (b) zinc blende structure

of the sodium chloride crystal with $2N$ ions is:

$$E_{\text{Madelung}} = \frac{1}{2} \sum_{i=1}^{2N} \sum_{j=1}^{2N} \frac{q_i q_j}{|\vec{r}_i - \vec{r}_j|} = N \left(-\frac{\alpha q^2}{R} \right);$$

$$\alpha = -\sum_{n_1 n_2 n_3} \frac{(-1)^{n_1+n_2+n_3}}{(n_1^2 + n_2^2 + n_3^2)^{1/2}} \qquad (2.6)$$

where $q = e$ is the ionic charge in sodium chloride, α is the Madelung constant, and the sum over the three integers $n_1 n_2 n_3$ should omit the origin. It should be emphasized that the Madelung constant varies with different ionic crystal structures. The Madelung constant, which was calculated analytically by Madelung in 1910 and can now be calculated numerically, equals 1.74757 for the sodium chloride structure, 1.76268 for the cesium chloride structure, and 1.63806 for the zinc blende structure.

If the Madelung energy was the only form of cohesive energy, the sodium chloride crystal would collapse due to the Coulomb attractions.

The stability of the ionic crystal indicates that there is a short-range strong repulsive force between nearest neighboring Na^+ and Cl^- ions. The short-range repulsive force can be explained by quantum mechanics. Both Na^+ and Cl^- ions have closed-shell electronic structures. When the nearest neighboring ions approach one another, the "electron clouds" start to overlap and, according to Pauli's exclusion principle, the electrons of one ion push the electrons from the nearest neighboring ions into orbits with higher energies. High cohesive energy at a short distance R results in the repulsive force: $F = -dU/dR > 0$, which is the "hard sphere repulsion." This hard sphere repulsion can be explained by exchange interaction.

The binding energy of the sodium chloride crystal was calculated by R. Landshoff in 1936 based on the Hartree-Fock theory. Landshoff found that there were two primary contributors: the Madelung energy and the exchange interaction. In quantum mechanics, the positions of electrons in an ion are not fixed and the electron clouds of nearest neighbor ions will overlap; however, after taking into account the probability density of electrons, the total electrostatic energy differs little from the classical Madelung energy given in Eq. (2.6). The exchange potential $-\langle\alpha\beta|\frac{e^2}{|\vec{r}-\vec{r}'|}|\beta\alpha\rangle$ is mainly contributed by the exchange of quantum orbits $|\alpha\rangle$ and $|\beta\rangle$ in nearest neighboring ions. When $R = a/2 = 2.82$Å, the Madelung energy is -204.1 kcal/mol and the exchange interaction is 25.2 kcal/mol. The difference between the calculated binding energy 178.9 kcal/mol and the measured one 182.6 kcal/mol of sodium chloride is caused by the van der Waals force, which is a weak attractive force between atoms and ions which will be discussed later in this chapter.

The cohesive energy of the sodium chloride crystal calculated by R. Landshoff could fit a formula with the following three terms: Madelung energy $g(R)$, exchange interaction $f(R)$, and the van der Waals term, as shown in Figure 2.4:

$$U(R) = -205.4 \times \frac{2.804\text{Å}}{R} + 26.4 \times \left(\frac{2.804\text{Å}}{R}\right)^{8.4}$$
$$- 3.0 \times \left(\frac{2.804\text{Å}}{R}\right)^6 \text{ (kcal/mol)} \quad (2.7)$$

Generally, the cohesive energy of ionic crystals with N molecules can be described quite well using the Born-Mayer potential, a classical formula published in 1924:

$$U(R) = N\left(-\frac{\alpha q_+ q_-}{R} + \frac{b}{R^n}\right) \quad \text{or} \quad U(R) = N\left(-\frac{a}{R} + z\lambda e^{-R/\rho}\right)$$

(2.8)

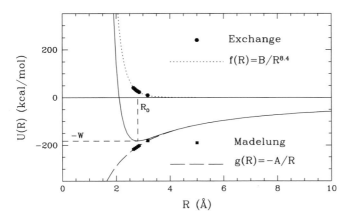

Figure 2.4 Madelung energy (dashed line), exchange interaction (dotted line), and the cohesive energy (solid line) versus distance in an NaCl crystal. The hexagonal and square dots stand for the analytical calculation of R. Landshoff, 1936.

where α is the Madelung constant, q_+ and q_- are the charges of the positive and negative ions respectively, and n is the repulsive index; z is the number of nearest neighboring ions and λ is the repulsive potential constant; the constant term $z\lambda$ has the dimension of energy and the repulsive range ρ has the dimension of length. The parameters n, $z\lambda$, and ρ for ionic crystals with the sodium chloride structure are listed in Table 2.2. The repulsive index for a sodium chloride crystal is $n = 8$, which is a little smaller than the power 8.4 of the exchange interaction term in Eq. (2.7), because the van der Waals term was included in the repulsive term in the Born-Mayer potential.

The lattice constant a of a crystal with the sodium chloride structure is twice the equilibrium ion-ion distance, R_0, as shown in Figure 2.2. At equilibrium, the cohesive energy reaches the minimum:

$$\frac{dU}{dR}\bigg|_{R_0} = N\left(\frac{\alpha q_+ q_-}{R_0^2} - \frac{nb}{R_0^{n+1}}\right) = 0; \quad b = \frac{R_0^n}{n}\frac{\alpha q_+ q_-}{R_0} \quad (2.9)$$

$$U(R_0) = -W = -N\frac{\alpha q_+ q_-}{R}\left(1 - \frac{1}{n}\right) \quad (2.10)$$

The equilibrium distance $R_0 = 2.82$Å and the binding energy $W = 181.8$ kcal/mol of the NaCl crystal are indicated in Figure 2.4.

The potential curve $U(R)$ is very useful for predicting the mechanical properties of a solid. The bulk modulus, K_V, listed in Table 2.2, can be calculated by the curvature of the cohesive energy around the

Table 2.2 Binding energy and parameters in ionic crystals with NaCl structure (based on the Encyclopedia of China: Physics, 1987, and Seitz, 1940)

Salt	n	$R_0/\text{Å}$	$K_V/(10^{11}\text{dyn/cm}^2)$	$z\lambda/(10^{-8}\text{erg})$	$\rho/\text{Å}$	Binding Energy W/(kcal/mol)	
						Experiment	Calculation
LiF	6.0	2.014	6.71	0.296	0.291	242.3[246.8]	242.2
LiCl	7.0	2.570	2.98	0.490	0.330	198.9[201.8]	192.9
LiBr	7.5	2.751	2.38	0.591	0.340	189.8	181.0
LiI	8.5	3.000	(1.71)	0.599	0.366	177.7	166.1
NaF	7.0	2.317	4.65	0.641	0.290	214.4[217.9]	215.2
NaCl	8.0	2.820	2.40	1.05	0.321	182.6[185.3]	178.6
NaBr	8.5	2.989	1.99	1.33	0.328	173.6[174.3]	169.2
NaI	9.5	3.237	1.51	1.58	0.345	163.2[162.3]	156.6
KF	8.0	2.674	3.05	1.31	0.298	189.8[194.5]	189.1
KCl	9.0	3.147	1.74	2.05	0.326	165.8[169.5]	161.6
KBr	9.5	3.298	1.48	2.30	0.336	158.5[159.3]	154.5
KI	10.5	3.533	1.17	2.85	0.348	149.9[151.1]	144.5
RbF	8.5	2.815	2.62	1.78	0.301	181.4	180.4
RbCl	9.5	3.291	1.56	3.19	0.323	159.3	155.4
RbBr	10.0	3.445	1.30	3.03	0.338	152.6	148.3
RbI	11.0	3.671	1.06	3.99	0.348	144.9	139.6

Note: all experimental data were measured at room temperature and 1 atm except the data in [ ], which were measured at 0 K and 0 atm.

equilibrium point:

$$K_V = V\left.\frac{\partial^2 U}{\partial V^2}\right|_{V_0} = \frac{R^2}{9V}\left.\frac{d^2 U}{dR^2}\right|_{R_0} = \frac{N\alpha q_+ q_-}{V}\frac{n-1}{R_0}\frac{1}{9} = \frac{n}{9}\frac{W}{V} \quad (2.11)$$

Therefore the rigidity of a solid is proportional to the binding energy. The bulk modulus unit 1 dyn/cm^2 equals 0.1 Pa = 0.1 N/m^2 in MKS unit. The bulk modulus K_V of solid sodium chloride can be calculated by Eq. (2.11) and Eq. (2.8):

$$K_V = \frac{n}{9}\frac{W}{V}$$
$$= \frac{8}{9}\frac{178.9 \times 4.18 \times 10^3}{6.022 \times 10^{23} \times 44.85 \times 10^{-30}} = 2.46 \times 10^{10}\,\text{Pa} \quad (2.12)$$

which basically agrees with the experimental data in Table 2.2.

The thermal expansion of solids is a statistical phenomenon. When the temperature rises, not every nearest neighbor distance R_{ij} increases, but statistically, the distance increases. Actually R_{ij} fluctuates around the equilibrium distance R_0. The potential $U(R)$ is asymmetric around R_0; therefore, at certain temperatures, R_{ij} is likely to be larger, but not smaller, than R_0, which explains why most solids have thermal expansion phenomena.

The binding energy of a bond can be defined as $w = W/N = -U(R_0)/N$, which is about 7.9 eV in a sodium chloride crystal. The binding energy of most of ionic crystals is larger than 5 eV, which corresponds to a very high temperature $T_i = w/k_B = 57971$ K. However, the actual melting temperature $T_m \sim 1000$ K of the ionic crystal is much lower than T_i because the number of point defects will increase exponentially with temperature and eventually the ionic crystal will be destroyed by the electrostatic forces around the defects. At high temperatures, ionic crystals become conductive due to migrating defects; while at very low temperatures, ionic crystals are perfect insulators because electrons cannot travel through closed-shell ions.

An ionic crystal must be a compound of typical metallic and nonmetallic elements. The IA-VIIA, IIA-VIA, and IIB-VIA compounds are typical ionic crystals with the structures of sodium chloride (NaCl), cesium chloride (CsCl) and zinc blende (ZnS). Many natural minerals, silicates, and ceramics are ionic crystals.

2.3 Covalent Bonds and Covalent Crystals

Silicon (Si) has become the most significant material used in electronics since monolithic integrated circuits were first produced in the late 1950s. Here it is chosen as the example crystal illustrating the formation of covalent crystals.

In 1874, the tetrahedral chemical bonds of carbon atoms in organic compounds were discovered by two chemists, J. H. van't Hoff and J. A. Le Bel. The tetrahedral bonds formed a crystal with a diamond structure as shown in Figure 2.5. The silicon and germanium (Ge) single crystals in the 4A family also have a diamond structure. In a crystal with a diamond structure, identical atoms obviously cannot become ions either by losing or gaining electrons; instead they share electrons to form covalent bonds. The binding energy W of covalent bonds is mainly contributed by the following three terms:

1. Formation of tetragonal bonds (negative contribution to W): it costs some energy to form the tetragonal sp^3 hybrid orbits from the $s - p$ quantum orbits in a free atom.

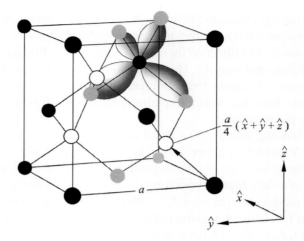

Figure 2.5 Four covalent bonds around an atom in crystals with the diamond structure

2. Coulomb Interactions (negative contribution to W): the Coulomb repulsions between the ions or between the inner shell electrons (such as the $1s^2 2s^2 2p^6$ shells in a silicon atom).
3. Exchange Potential (positive contribution to W): the exchange potential (non-Hartree-Fock form) $\langle \alpha_1 \beta_2 | V | \beta_1 \alpha_2 \rangle < 0$, where V includes electron-electron repulsion in the covalent bond of two electrons with opposite spins and the valence electron-ion attraction.

The precise quantum mechanical calculation of the binding energy in covalent crystals, such as diamond, silicon, or carbon nano-materials, is an important current research topic. However, this is quite difficult as it is a many-body quantum mechanical problem. The first step in analyzing the binding energy of a silicon crystal is the hybridization of valence electrons in an atom. It costs several eV to form the tetrahedral sp^3 hybrid orbits from the 3s,3p single-electron eigenstates in a silicon atom:

$$\begin{aligned}
\Psi_{[111]} &= (\psi_{3s} + \psi_{3p_x} + \psi_{3p_y} + \psi_{3p_z}); \\
\Psi_{[1\bar{1}\bar{1}]} &= (\psi_{3s} + \psi_{3p_x} - \psi_{3p_y} - \psi_{3p_z}); \\
\Psi_{[\bar{1}1\bar{1}]} &= (\psi_{3s} - \psi_{3p_x} + \psi_{3p_y} - \psi_{3p_z}); \\
\Psi_{[\bar{1}\bar{1}1]} &= (\psi_{3s} - \psi_{3p_x} - \psi_{3p_y} + \psi_{3p_z})
\end{aligned} \quad (2.13)$$

The hybrid orbits in Eq. (2.13) are for the silicon atoms at sites "A" (on the surface of the cubic cell); while the hybrid orbits for atoms at sites "B" (inside the cubic cell) should be $\Psi_{[\bar{1}\bar{1}\bar{1}]}$, $\Psi_{[\bar{1}11]}$, $\Psi_{[1\bar{1}1]}$, and $\Psi_{[11\bar{1}]}$ instead.

2.3 Covalent Bonds and Covalent Crystals

In 1927, the Heitler-London theory of hydrogen molecules was proposed by W. Heitler and F. London, which explained the bound (singlet) and unbound (triplet) states based on quantum mechanics. In a covalent bond, there are two electrons (labeled 1 and 2 respectively). The two electrons stay in the hybrid orbit $\Psi^i_{[111]}$ of the i'th atom (labeled $|\alpha\rangle$) and the hybrid orbit $\Psi^j_{[\bar{1}\bar{1}\bar{1}]}$ of the nearest neighboring j'th atom (labeled $|\beta\rangle$) respectively before bonding. The eigenstate of the covalent bond can be written as:

$$|\Psi\rangle = |\alpha_1 \beta_2\rangle + \eta |\beta_1 \alpha_2\rangle \rightleftharpoons$$
$$\Psi = \Psi^i_{[111]}(\vec{r}_1)\Psi^j_{[\bar{1}\bar{1}\bar{1}]}(\vec{r}_2) + \eta \Psi^j_{[\bar{1}\bar{1}\bar{1}]}(\vec{r}_1)\Psi^i_{[111]}(\vec{r}_2) \quad (2.14)$$

where the single electron hamiltonian is $h_l = \vec{p}_l^2/2m - Ze^2/|\vec{r}_l - \vec{R}|$ ($l=1,2$; $\vec{R}$ the position of nucleus), the eigen-energy can be found by $h_l|\alpha\rangle = \varepsilon_\alpha|\alpha\rangle$ or $h_l|\beta\rangle = \varepsilon_\beta|\beta\rangle$. The bi-electron hamiltonian is $\mathcal{H} = \mathcal{H}_1 + \mathcal{H}_2 + \frac{e^2}{r_{12}} = h_1 + h_2 + V$, where the "new" single electron hamiltonian becomes $\mathcal{H}_l = \vec{p}_l^2/2m - Ze^2/|\vec{r}_l - \vec{R}_\alpha| - Ze^2/|\vec{r}_l - \vec{R}_\beta|$. The energy of the two electrons in the bond can be found by $\mathcal{H}\Psi = \varepsilon\Psi$:

$$(\varepsilon_\alpha + \varepsilon_\beta + \langle\alpha_1\beta_2|V|\alpha_1\beta_2\rangle) + \eta\langle\alpha_1\beta_2|V|\beta_1\alpha_2\rangle$$
$$= (\varepsilon_\alpha + \varepsilon_\beta + \bar{V}) + \eta V_{ex} \simeq \varepsilon$$
$$\langle\beta_1\alpha_2|V|\alpha_1\beta_2\rangle + \eta(\varepsilon_\alpha + \varepsilon_\beta + \langle\beta_1\alpha_2|V|\beta_1\alpha_2\rangle)$$
$$= V^*_{ex} + \eta(\varepsilon_\alpha + \varepsilon_\beta + \bar{V}) \simeq \varepsilon\eta$$
$$\varepsilon_\pm = \varepsilon_\alpha + \varepsilon_\beta + \bar{V} \mp |V_{ex}|; \quad \eta_\pm = \mp|V_{ex}|/V_{ex} \quad (2.15)$$

Here the Heitler-London approximation $\langle\alpha_1\beta_2|\beta_1\alpha_2\rangle \simeq 0$ has been used, which is more accurate for larger ion-ion distance. The "extra" Coulomb potential is $V = -Ze^2/|\vec{r}_1 - \vec{R}_\beta| - Ze^2/|\vec{r}_2 - \vec{R}_\alpha| + e^2/|\vec{r}_1 - \vec{r}_2|$. The exchange term $V_{ex} = \langle\alpha_1\beta_2|V|\beta_1\alpha_2\rangle$ is always negative, therefore in the bond state η_+ equals 1, and $|\Psi_+\rangle = |\alpha_1\beta_2\rangle + |\beta_1\alpha_2\rangle$ is symmetric. The bonding and antibonding energy equal to $\varepsilon_+ = \bar{V} - |V_{ex}|$ and $\varepsilon_- = \bar{V} + |V_{ex}|$ respectively, is shown in Figure 2.6.

There are a huge number of covalent bonds in a covalent crystal and the bond-bond interactions are very strong, therefore the accurate calculation of the binding energy in a covalent crystal is very complex. Two methods are used: the first principle *ab initio* method, and the phenomenological molecular dynamics method. Both methods are quite complex and cannot be covered in detail here. To summarize, the potential energy between atoms or ions can be substituted directly in the

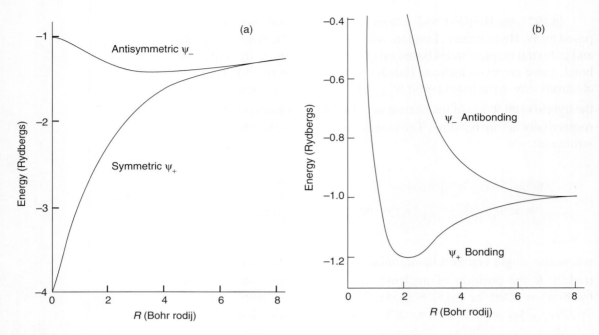

Figure 2.6 Bonding and antibonding energy of covalent bonds: (a) energy without Coulomb and Pauli repulsion among ions; (b) total energy versus distance (based on Slater, 1963)

molecular dynamics method, and has to be calculated in the *ab initio* method.

When free atoms with hybrid orbits form a silicon crystal, the total Coulomb repulsion between electrons increases, but the total Coulomb attraction between electrons and ions decreases more, as indicated by the binding energy in Eq. (2.15). The equilibrium energy of electrons is about $-100\,\text{eV}$ per silicon atom in the crystal, which is 4.63 eV lower than that of electrons in a free silicon atom. It is the joint effect of the hybridization energy, the electrostatic energy between electrons, the Coulomb repulsion, and the exchange. There are many forms of cohesive energy or potential in silicon crystals (based on Balamane, Halicioglu, Tiller, 1992), and the most commonly used one is the Stillinger-Weber potential (based on Stillinger, Weber, 1985) with the classical two-body potential and the three-body potential:

$$E/E_b = \frac{1}{2!}\sum_{ij} V_2(r_{ij}) + \frac{1}{3!}\sum_{ijk} V_3(\vec{r}_{ij}, \vec{r}_{ik}, \vec{r}_{jk}) \qquad (2.16)$$

$$V_2(r) = \frac{1}{2} f_c(r)\left[\frac{AB}{r^p} - \frac{A}{r^q}\right]; \quad f_c(r) = \exp(-1/(a-r))\ (r < a)$$

(2.17)

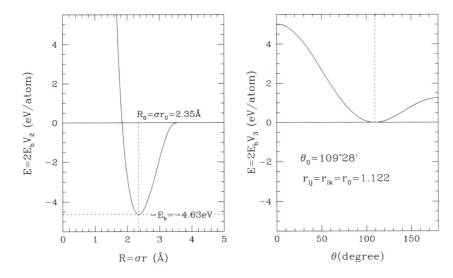

Figure 2.7 Stillinger-Weber potential of silicon crystal: (a) cohesive energy versus distance with ideal bond angle; (b) cohesive energy versus bond angle at $r = R_0/\sigma$

$$V_3(\vec{r}_{ij}, \vec{r}_{ik}, \vec{r}_{jk}) = \frac{\lambda}{2} f_c^\gamma(r_{ij}) f_c^\gamma(r_{ik}) \left(\cos\theta_{ijk} + \frac{1}{3}\right)^2 ; \quad \theta_{ijk} = \theta(\vec{r}_{ij}, \vec{r}_{ik})$$

(2.18)

where $E_b = U(R) - U(\infty) = 4.63\,eV$ is the binding energy per atom in the silicon crystal; the nearest neighbor distance R is normalized as $r = R/\sigma$; the cut-off function $f_c(r)$ is zero when $r > a$; and $-\frac{1}{3} = \cos(\theta_{ijk})$, where $\theta_{ijk} = 109°28'$ is the ideal bond angle in the diamond structure. The eight parameters of the Stillinger-Weber potential for the silicon crystal are $\sigma = 2.0951\,\text{Å}$, $A = 7.4096$, $B = 0.6222$, $p = 4$, $q = 0$, $a = 1.8$, $\lambda = 21$, and $\gamma = 1.2$ (based on Stillinger, Weber, 1985). The minimum two-body potential, V_2, which is located at $R_0 = 1.122\sigma = 2.35\,\text{Å}$, is simply $-\frac{1}{2}$. The three-body potential, V_3, is related to the bond angle, and it is important for evaluating the mechanical properties of silicon.

The *ab initio* potential $E_{ab}(R)$ is similar to the Stillinger-Weber potential $E_{SW}(R)$ in Figure 2.7 when $R < R_0$; however, $E_{ab}(R)$ has a slower increase when $R > R_0$ and turns to zero at $R = 5.2\,\text{Å}$, which is larger than the Stillinger-Weber cut-off radius $R = a\sigma = 3.77\,\text{Å}$. The Stillinger-Weber potential does not consider the interaction of next nearest neighbors, but it does explain the diamond structure and general properties of covalent crystals, as given in Table 2.3, fairly well.

Table 2.3 General properties of covalent crystals: R_0 is the equilibrium distance; E_b the binding energy; K the bulk modulus; R_h the mineral hardness; T_m the melting temperature; α_T the thermal expansion; q^* the charge transfer

Covalent Crystal	R_0 (at 300 K)	E_b (eV)	K (GPa)	R_h (degree)	T_m (K)	α_T (at 300 K)	q^* (e)
Diamond	1.5445Å	7.37(atom)	443	10	3773	$1.05 \times 10^{-6} \mathrm{K}^{-1}$	0
Si	2.3517Å	4.63(atom)	102	6.5	1687	$2.6 \times 10^{-6} \mathrm{K}^{-1}$	0
Ge	2.4500Å	3.85(atom)	77	6	1211	$5.8 \times 10^{-6} \mathrm{K}^{-1}$	0
GaAs	2.448Å	6.5(pair)	70	6	1513	$6.4 \times 10^{-6} \mathrm{K}^{-1}$	0.46

Covalent bonds are strong chemical bonds. The binding energy per atom in a covalent crystal is lower than that in an ionic crystal with a similar atomic number; however the melting temperature of the covalent crystal is much higher due to the much lower density of defects. Covalent bonds are deemed to be coordination bonds, and the filling factor of atoms is low (only 0.34 in diamond); as a result, covalent crystals are hard, relatively brittle, and cannot be bent.

In the periodic table, group IVa elements are typical covalent crystals with a diamond structure, except tin, which will change from the diamond structure to a complex structure at 18°C, and lead, which is not a covalent crystal at all. The group IIIa and Va elements can form III-V covalent compounds, and among them, gallium arsenide (GaAs) is the most important. Their basic properties are also listed in Table 2.3. In gallium arsenide, a 0.46 e electric charge is transferred from the arsenic (As) atom to the gallium (Ga) atom; therefore, besides the dominant covalent bonds, there are also some ionic bonds in gallium arsenide, which makes its melting temperature higher than that of pure germanium. The group II and group VI elements can form II-VI covalent compounds, which are polarized covalent crystals with both covalent and ionic bonds.

2.4 Metallic Bonds and Typical Metals

Most elements are metals. Here the alkali metals are chosen to illustrate the essence of metal bonds.

In alkali metals there is only one valence electron in an atom, therefore atoms cannot lose or gain electrons to form ionic bonds, and nearest neighboring atoms cannot share valence electrons to form covalent bonds. In fact, the valence electrons will escape from their "own" atoms and become almost free electrons in the crystal, known as "free electron gas." The binding energy W of metallic bonds is mainly contributed by three terms:

1. Kinetic Energy (positive contribution to W): the kinetic energy K of valence electrons in free electron gas is less than that in free atoms.
2. Exchange Interactions (positive contribution to W): the exchange interaction between valence electrons $U_{\text{ex}} = -\frac{1}{2}\sum_\alpha \sum_\beta \langle \alpha\beta|\frac{e^2}{r}|\beta\alpha\rangle$ is negative in free electron gas.
3. Coulomb Interactions (negative contribution to W): the Coulomb interactions among ions, valence electrons and inner shell electrons (such as the $1s^2 2s^2 2p^6$ shells in a sodium atom), including ion-ion and valence electron-electron repulsion U_r, and the valence electron-ion attraction U_a. Coulomb repulsions and attractions both increase in metals.

Both the momentum and displacement of electrons follow the Heisenberg uncertainty principle: $\Delta p \Delta x > \hbar$; thus the kinetic energy is in the order of $K \sim \frac{\hbar^2}{2m\Delta x^2}$. The displacement Δx of the valance electron in metals is obviously larger than that in a free atom, therefore, the kinetic energy K of valence electrons is much lower. Meanwhile, the Coulomb attraction U_a is higher for larger electron-ion distance Δx.

Potential energy can be expressed in different forms. The semi-experiential Bardeen potential is useful for the alkali metals since it is compact and has clear physical meaning (based on Seitz, 1940). In the early 1930s, E. P. Wigner and his graduate student, F. Seitz, tried to calculate the binding energy of metallic sodium quantum mechanically. In 1938, another graduate student of Eugene Wigner, John Bardeen, was a post-doctoral researcher at MIT under the supervision of J. C. Slater. He calculated the cohesive energy E_b^0 of the valence electron in a free alkali atom, which is between -0.39Ry and -0.29Ry from lithium to cesium respectively. The total energy per atom E_{tot}, which was later named the Bardeen potential, equals the sum of E_b and the cohesive energy:

$$E_{\text{tot}}(v) = E_b(v) + U(v) = A\left(\frac{v_0}{v}\right) + B\left(\frac{v_0}{v}\right)^{2/3} - C\left(\frac{v_0}{v}\right)^{1/3}$$

(2.19)

where v is the volume of an atom. The kinetic energy K is proportional to $bv^{-2/3} + b'v^{-1}$; the Coloumb attraction U_a has the form of $-\frac{1}{\Delta x} \sim -v^{-1/3}$; and the repulsion U_r is expressed by the term $\frac{1}{\Delta x^3} \sim v^{-1}$. The exchange interaction U_{ex} is quite complex and usually has the form $-c_{\text{ex}} v^{-1/3} - b_{\text{ex}} v^{-2/3}$.

The parameters A, B, C in the Bardeen potential can be determined by the equilibrium volume $v_e = v_0$, the energy in an free atom E_b^0, the binding energy $W = -(E_0 - E_b^0)$, and the compressibility

Table 2.4 Empirical parameters in the Bardeen potential of alkali metals (based on Seitz, 1940, and Ashcroft and Langreth, 1967)

Alkali Metal	E_b^0 (eV)	W (eV)	v_0 (Å³)	$1/\beta$ (10 GPa)	A (eV/atom)	B (eV/atom)	C (eV/atom)
Li	−5.32	1.63	21.3	1.293	0.79	5.37	13.11
Na	−5.15	1.11	37.8	0.850	2.78	0.71	9.74
K	−4.37	0.93	71.5	0.413	3.01	−0.73	7.58
Rb	−4.13	0.85	87.3	0.305	2.54	−0.09	7.43
Cs	−3.89	0.80	110.7	0.225	2.31	0.08	7.07

coefficient $1/\beta$:

$$\begin{cases} -v\,\dfrac{dE_{tot}}{dv}\Big|_{v_0} = 0 = A + \tfrac{2}{3}B - \tfrac{1}{3}C \\ E_{tot}|_{v_0} = E_0 = -W + E_b^0 = A + B - C \\ v^2\,\dfrac{d^2 E_{tot}}{dv^2}\Big|_{v_0} = v_0/\beta = 2A + \tfrac{10}{9}B - \tfrac{4}{9}C \end{cases}$$

$$\rightarrow \begin{cases} A = \tfrac{9}{2}v_0/\beta - |E_b^0 - W| \\ B = 3|E_b^0 - W| - 9v_0/\beta \\ C = 3|E_b^0 - W| - \tfrac{9}{2}v_0/\beta \end{cases} \qquad (2.20)$$

The empirical parameters A, B, C in the Bardeen potential of alkali metals are listed in Table 2.4, and the respective Bardeen potentials are plotted in Figure 2.8. The parameter B is negative in potassium and rubidium crystals, because the parameter b_{ex} in the $-b_{ex}v^{-2/3}$ term of the exchange interaction is larger than that in the kinetic energy term. It should be emphasized that the Bardeen potential also applies to an isolated atom. When v approaches v_{max}=140, 150, 260, 330Å³ in lithium, sodium, potassium, and rubidium, respectively, the total energy $E_{tot}(v)$ per atom tends to the free atom's energy E_b^0. When $v > v_{max}$, the Bardeen potential is meaningless, and $E = E_b^0$ should be used instead.

The metallic bond is not a strong chemical bond, as has been illustrated by the Bardeen potential of alkali metals in Figure 2.8. However, the metallic bond should not be neglected as it still has a significant influence on the properties of metals. The high electrical and thermal conductivity of metals stems from the motion of valence electrons in the whole space of the crystal, which is the key to the metallic bond. The Coulomb attraction term $-C(v_0/v)^{1/3}$ decreases for a smaller volume v; therefore, the filling factor of atoms and the density of metals are usually very high. The ductility and the machinability of metals is due to the fact that the metallic bond is isotropic in space (which is why R is replaced

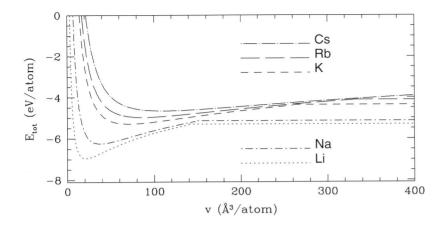

Figure 2.8 Bardeen potential of alkali metals

by v in the potential). Metallic alloys can be formed because the valence electrons from different metals can share metallic bonds and the metallic bond does not care where the electrons come from.

Alkali metals are crystals which only have metallic bonds. The transition metals have both metallic and covalent bonds, where d-electrons are mostly localized and s-electrons will form the metallic bonds. These are the basic physical reasons why the transition metals are so useful: good conductivity from the metallic bonds and good mechanical properties due to the covalent bonds. The use of metals dates back to the ancient civilizations and it is just as important in today's modern industries.

2.5 Atomic and Molecular Solids

Most organisms are molecular solids, and hydrogen bonds play an important role in their structure. The lifeblood of living things—water—is also formed from hydrogen bonds. Another essential ingredient for life—air—would become molecular solids of nitrogen and oxygen at low temperatures, and at very low temperatures, the inert gases would become atomic solids. The intermolecular force, which is the source of the binding energy of molecular liquids or solids, was first hypothesized by a French mathematician, Claude Clairault (1713–1765), as he tried to explain the capillary phenomenon, where the liquid near the wall of a vessel rises up against gravity.

In atomic and molecular solids, the wavefunction of valence electrons would almost be the same after bonding, which is different from ionic, covalent, and metal bonds. In crystalline ice there are covalent

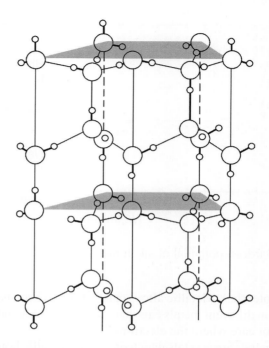

Figure 2.9 One kind of structure of crystalline ice (based on Pauling, 1960)

bonds and hydrogen bonds. In a water molecule, one oxygen atom and two hydrogen atoms are bonded together by two covalent bonds, where oxygen and hydrogen have negative and positive charge respectively. The positive charge center has a displacement to the negative charge center, therefore there is a dipole in the water molecule. When water molecules condense into crystalline ice, the dipole-dipole attractive interactions play the primary role, where the dipole $\vec{p}$ is related to the charge distribution $\rho(\vec{r})$ in a molecule:

$$U_{d-d}(\vec{r}_1, \vec{r}_2) = \frac{\vec{p}_1 \cdot \vec{p}_2 - 3(\vec{p}_1 \cdot \hat{r}_{12})(\vec{p}_2 \cdot \hat{r}_{12})}{|\vec{r}_1 - \vec{r}_2|^3}; \quad \vec{p} = \iiint d^3\vec{r}' \rho(\vec{r}')\vec{r}'$$

(2.21)

where $\vec{r}_1$ and $\vec{r}_2$ are coordinates of the charge centers ($\hat{r}_{12} = (\vec{r}_1 - \vec{r}_2)/|\vec{r}_1 - \vec{r}_2|$), $\vec{p}_1$ and $\vec{p}_2$ are the electric dipoles in molecule 1 and molecule 2, respectively. The dipole-dipole interactions between molecules are the basis of the hydrogen bond.

In organic compounds or living things, molecules or macromolecules are joined by hydrogen bonds. In DNA, the A,T,C,G base pairs in one chain are linked to the A,T,C,G base pairs in another chain by hydrogen bonds. DNA duplication and other basic processes

Table 2.5 Dipoles of atoms and molecules in debyes (1 debye = 10^{-18} esu · cm)

Molecule	Dipole	Molecule	Dipole
Ar	0	HCl	1.12
CO	0.10	SO_2	1.7
CO_2	0	CH_4	0
H_2	0	CH_2CH_2	0
O_2	0	CH_3CH_3	0
O_3	0.53	C_2H_5OH	1.68
N_2	0	CH_3CHO	2.68
NO	0.10	$(CH_3)_2CO$	2.85
NO_2	0.30	$C_3H_6O_3$	2.08
H_2O	1.94	$(C_2H_5)_2CO$	2.75
NH_3	1.468	C_6H_6	0

in organisms can be viewed as the binding or unbinding of hydrogen bonds. An organism could form molecular crystals at low temperatures or under high pressure, however the binding energy of molecular crystals is in the range of 0.1–0.5 eV, which corresponds to a temperature of 1000–5000 K. Therefore an organism is very stable at room temperature. Enzymes are required to enable hydrogen bonds to unbind and metabolism to take place at room temperature.

In Table 2.5, the dipoles of atoms and molecules occurring in nature are listed, where the unit of dipole 1debye equals $\frac{1}{3} \times 10^{-29}$ C · m. The inert elements and molecules with symmetrical structures have no dipoles because their positive and negative charge distributions share the same center. It is interesting to note that most molecules in air have no dipoles, therefore these molecules cannot form liquids or solids by hydrogen bonds. It was J. D. van der Waals who assumed there should be a force between the molecules in air, which was subsequently named "van der Waals force." The hydrogen bond is a special example of van der Waals force.

There are three basic types of van der Waals force: (1) the dipole-dipole interaction or Keesom force, where two interacting dipoles have $p_1 p_2 \neq 0$, which accords with the definition of the hydrogen bond; (2) the dipole-induced dipole interaction or Debye force, where in the static state two dipoles have $p_1 p_2 = 0$, and the effect of the induced dipole on the original dipole should be calculated self-consistently; and (3) the dispersion interaction or London force, which has a purely quantum mechanical origin and needs QED to be fully understood. All three interactions are attractive, and follow the -6 power law: $-A/R^6$.

Normally the London interaction dominates, but in water, the Keesom interaction dominates.

In 1930, Fritz London used quantum mechanics to explain the hydrogen bond and van der Waals force. According to the second order perturbation theory with respect to the dipole-dipole or equivalent interaction $U(R) = C/R^3$, the energy of the hydrogen bond or van der Waals force should be a function of distance $R = |\vec{r}_1 - \vec{r}_2|$:

$$E_{\text{vdW}}(R) = E_0 + \sum_\alpha \frac{|\langle 0|U(R)|\alpha\rangle|^2}{E_0 - E_\alpha} = E_0 - \frac{A_1}{R^6} - \frac{A_2}{R^8} - \frac{A_3}{R^{10}} \quad (2.22)$$

where $|0\rangle$ is the ground state and $|\alpha\rangle$ is the excited state, therefore $E_0 - E_\alpha$ is less than zero. The R^{-6} term is van der Waals force, and A_1 term is in the order of 0.1 eV; the dipole-quadrapole R^{-8} term and quadrapole-quadrapole R^{-10} term can be neglected. A_1 can be non-zero even if both molecules (atoms) have zero average dipoles (corresponding to the Debye force), because there are instantaneous dipoles due to the charge fluctuation with respect to the electron motions in an atom. Van der Waals force also exists in ionic crystals, covalent crystals, and metals, but it only plays a significant role in atomic solids, molecular solids, mesoscopic colloidal particles, and photonic materials.

In 1924, the Lennard-Jones potential was formulated by an Englishman, John Edward Lennard-Jones, the father of modern computational chemistry. It includes the attractive van der Waals force and the repulsive exchange interaction term. The Lennard-Jones potential is a two-body potential:

$$U(R) = \varepsilon \left[\left(\frac{\sigma}{R}\right)^{12} - \left(\frac{\sigma}{R}\right)^6 \right] \quad (2.23)$$

The cohesive energy of atomic solids with the FCC structure can be calculated by the sum of all other atoms (12 nearest neighbors, six next nearest neighbors, etc.):

$$\begin{aligned}E &= \frac{1}{2} N \varepsilon \left[\left(\frac{\sigma}{R}\right)^{12} \sum_j \left(\frac{R}{R_j}\right)^{12} - \left(\frac{\sigma}{R}\right)^6 \sum_j \left(\frac{R}{R_j}\right)^6 \right] \\ &= \frac{1}{2} N \varepsilon \left[12.132 \left(\frac{\sigma}{R}\right)^{12} - 14.454 \left(\frac{\sigma}{R}\right)^6 \right]\end{aligned} \quad (2.24)$$

In an atomic solid, the equilibrium energy per atom is $E_0 = -2.15\varepsilon$ and the equilibrium distance is $R_0 = 1.09\sigma$.

2.5 Atomic and Molecular Solids

Table 2.6 Empirical parameters in the Lennard-Jones potential of inert elements

Inert Element	W (eV/atom)	σ (Å)	R_0^{exp}/σ	ε (eV/atom)	$-E_0$ (eV/atom)	T_m (K)
He	–	2.56	–	0.0035	0.008	~1.0
Ne	0.020	2.74	1.14	0.0124	0.027	24.5
Ar	0.080	3.40	1.11	0.0418	0.090	83.9
Kr	0.116	3.65	1.10	0.0562	0.121	116.5
Xe	0.160	3.98	1.09	0.0800	0.172	161.3

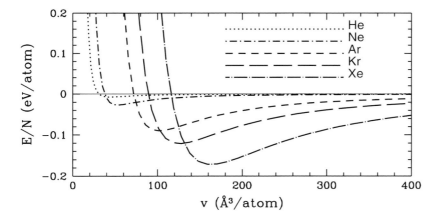

Figure 2.10 Binding energy per atom versus atomic volume in inert solids

In atomic solids of inert elements, the empirical parameters of the Lennard-Jones potential are given in Table 2.6. The corresponding binding energy per atom E/N in Eq. (2.24) versus the average volume per atom is illustrated in Figure 2.10.

The difference between the measured binding energy W and the calculated equilibrium potential $-E_0$ comes from the correction of kinetic energy. Van der Waals force is a weak chemical bond, and its binding energy is about 1–2% of the metallic bond. It is interesting to note that, for elements in a family with a larger value of Z, metallic bonds are weaker but van der Waals forces are stronger, because van der Waals force comes from the fluctuated dipole and will increase with Z.

In this chapter, the essence of five basic chemical bonds has been discussed. In Figure 2.11, it can be seen that the charge distribution is quite different in crystals formed by different chemical bonds. In an atomic solid (argon), the total charge of the atom is zero. In an ionic

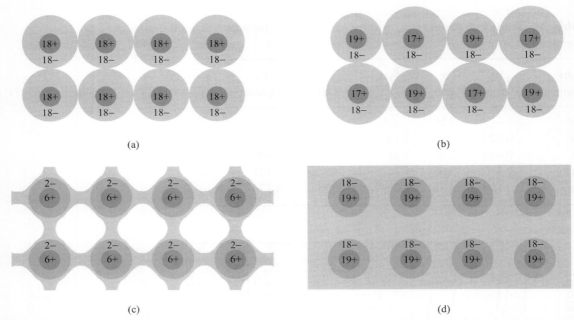

Figure 2.11 Charge distribution in crystals: (a) Ar; (b) KCl; (c) Diamond; (d) K (based on Ashcroft and Mermin, 1976)

crystal (potassium chloride), there are electron transfers between neighbors. In a metal (potassium), valence electrons become free electron gas and are dispersed in space. In a covalent crystal (diamond), valence electrons form a network connecting all the ions, which is similar to the original picture of chemical bonds. Therefore, the crystalline structure is determined by the electronic structure.

The electronic structures of crystals are so complicated that, in chemistry and materials science, phenomenological atomic potentials are often used to solve practical problems. Accurate calculations of binding energy require the use of the density functional theory of energy bands, which is introduced in Chapter Five. However the calculation methods are too detailed to cover fully in this book.

Summary

This chapter has explained the chemical bonds in the language of quantum mechanics and quantum chemistry so that the formation of different classes of solids can be understood:

1. Bohr model and Hartree-Fock theory: showing how the single-electron quantum orbits in the Bohr model are combined with the

Slater determinant into the multi-electron orbits in a variety of matter other than hydrogen atoms. The effective Hartree-Fock Hamiltonian is thus given for a many-body quantum mechanical electronic system.
2. Ionic bonds: formed by the balance between the long-range Coulomb interactions and the short-range Hartree-Fock exchange potential. The Born-Mayer potential and the Landshoff theory give the first and second order description of binding energy.
3. Covalent bonds: formed by the counterbalance among the hybridization of valence electron orbits, Coulomb repulsion of ions, and Heitler-London-type exchange potential of two electrons in a covalent bond. The Heitler-London theory of a single covalent bond can help us understand the covalent bonds of huge amount of electrons in a crystal. The more complicated binding energy of covalent crystals can be given by the sum of the two-body potential and the three-body potential, such as the Stillinger-Weber potential for silicon.
4. Metallic bonds: formed by the balance among the kinetic energy of electrons, the Coulomb interactions between valence electrons and ions, and the Hartree-Fock exchange potential of valence electrons. The Bardeen potential gives quite a good empirical description of the alkali metals.
5. Van der Waals force: showing how the dipole-dipole interactions between molecules are key to understanding the formation of atomic and molecular solids. The London theory gives the quantum mechanical version of van der Waals force. The Lennard-Jones potential gives a useful two-body potential between the molecules or inert atoms.

References

Ashcroft, N. W., and Langreth, D. C. (1967). "Compressibility and Binding Energy of the Simple Metals," *Phys. Rev.*, Vol. 155, 682.
Ashcroft, N. W., and Mermin, N. D. (1976). *Solid State Physics*, Holt, Rinehart and Winston, New York.
Balamane, H., Halicioglu, T., and Tiller, W. A. (1992). "Comparative Study of Silicon Empirical Interatomic Potentials", *Phys. Rev. B*, Vol. 46, 2250.
Encyclopedia of China: Physics (1987). Encyclopedia of China Publishing House, Beijing.
Huang, K., and Han, R. (1988). *Solid State Physics* (in Chinese), Higher Education Press, Beijing.
Partington, J. R. (2003), translated by Z. Hu. *A Short History of Chemistry* (in Chinese), Guangxi Normal University Press, Guilin.
Pauling, L. (1960). *The Nature of The Chemical Bond*, Cornell University Press, Ithaca, New York.
Seitz, F. (1940). *The Modern Theory of Solids*, McGraw-Hill, New York.

Slater, J. C. (1963). *Quantum Theory of Molecules and Solids*, McGraw-Hill, New York.
Stillinger, F. H., and Weber, T. A. (1985). "Computer Simulation of Local Order in Condensed Phases of Silicon", *Phys. Rev. B*, Vol. 31, 5262.
Xu, G., and Li, L. (2001). *Quantum Chemistry* (in Chinese), Science Press, Beijing.

Exercises

2.1 Find the exchange potential between two hydrogen atoms by the Heitler-London approximation. The detailed expression of an exchange potential is

$$\langle \alpha\beta | V | \beta\alpha \rangle = \iiint d^3\vec{r} \iiint d^3\vec{r}' \psi_\alpha^*(\vec{r}) \psi_\beta^*(\vec{r}') V(\vec{r},\vec{r}') \psi_\beta(\vec{r}) \psi_\alpha(\vec{r}')$$

(2.25)

Calculate the following exchange potential versus the atom-atom distance $R = |\vec{R}_\alpha - \vec{R}_\beta|$ numerically, where $|\alpha\rangle$ and $|\beta\rangle$ are both $1s$ hydrogen electron states: (a) the electron-electron repulsion $\langle \alpha\beta | \frac{e^2}{|\vec{r}-\vec{r}'|} | \beta\alpha \rangle$; (b) the electron-ion attraction $-\langle \alpha\beta | \frac{e^2}{|\vec{r}-\vec{R}_\alpha|} | \beta\alpha \rangle$ and $-\langle \alpha\beta | \frac{e^2}{|\vec{r}-\vec{R}_\beta|} | \beta\alpha \rangle$; (c) which term is dominant, the exchange of the Coulomb repulsion or Coulomb attraction?

2.2 Prove the Madelung constant in a one-dimensional lattice is $\alpha = 2\ln 2$.

2.3 Calculate the Madelung constant in the cesium chloride (CsCl) crystal. The numerical integral can be utilized.

2.4 If the space between atoms in sodium chloride (NaCl) is filled with a liquid with the dielectric constant ϵ,

   a. use the Born-Mayer potential to calculate the binding energy and the equilibrium distance;
   b. given ϵ of water is 81, use this data to explain why the NaCl is easily dissolved in water.

2.5 Use the Stillinger-Weber potential to calculate the bulk modulus K_V of silicon, and compare with experimental data.

2.6 Explain why the compressibility $1/\beta$ of alkali metals used in the Bardeen potential calculation is a little larger than the bulk modulus of alkali metals.

2.7 Numerically calculate the two data 12.132 and 14.454 in Eq. (2.24) approximately by the sum of nearest neighbors and next nearest neighbors in a FCC lattice (FCC structure: nearest neighbor displacements $\pm\frac{a}{2}(\hat{e}_x \pm \hat{e}_y)$, $\pm\frac{a}{2}(\hat{e}_y \pm \hat{e}_z)$, $\pm\frac{a}{2}(\hat{e}_z \pm \hat{e}_x)$, next nearest neighbor vectors $\pm a\hat{e}_x$, $\pm a\hat{e}_y$, $\pm a\hat{e}_z$).

3

Structure of Solids

> **ABSTRACT**
> - Crystal structure and reciprocal lattices
> - X-ray, electron, and neutron diffraction
> - Amorphous solids, quasi-crystals, and liquid crystals

Analysis of solid structure is the basis of solid state physics. In traditional solid state physics, the term solid structure mainly refers to crystal structure. This chapter takes a broader view. In the first part, crystal structure is discussed based on the symmetry of lattices, and the hard-sphere model of crystal formation is also mentioned. In the second part, the reciprocal lattice and the Brillouin zone are described, based on the universal view of waves in crystals. In the third part, experimental measurements of solid structures are introduced by comparing X-ray, electron, and neutron diffraction methods. In the fourth part, disordered solid structures, including amorphous materials, quasi-crystals, and liquid crystals, as well as the basics of the related diffraction theory, are analyzed briefly.

3.1 Geometrical Description of Crystals

A crystal is a solid with a periodic atomic structure in space. Generally speaking, the word "crystal" can mean either a single crystal or a poly-crystal. An infinitely large single crystal can be seen as a perfect crystal. Strictly speaking, due to the existence of surface, atomic vibration, and other defects, there is no perfect crystal in the real world. In solid state physics, the geometrical description of a crystal is based on the assumption that many types of imperfection have little influence on the focusing property. In fact, in some cases, these imperfections are important research topics, such as the resistance caused by atomic vibrations

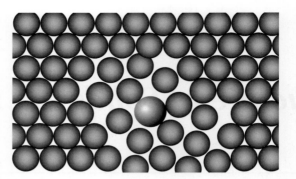

Figure 3.1 Imperfectness in crystal structure: a point defect

and the effects of impurities in semiconductors; then more complicated geometrical descriptions of solids are required for the analysis.

In 1824, a German physicist, Ludwig August Seeber, proposed the idea of equivalence of lattice sites and chemical atoms. In a geometrical description of a crystal, the word "crystal lattice" or simply "lattice" is used to represent a perfect crystal structure. In a perfect crystal, every atom is replaced by a geometrical point located at the equilibrium position of the atom, and by doing so, a lattice with the same geometrical characters but no physical properties can be obtained. These geometrical points in the lattice are called lattice sites. It should be emphasized that the geometrical points should be different if the original atoms are different chemical elements, so, for example, there are two kinds of lattice sites in the sodium chloride (NaCl) lattice.

In 1850, the French mathematician, A. Bravais, classified the three-dimensional lattice by symmetry. He proposed an important concept which was later named the "Bravais lattice." The Bravais lattice is an infinite lattice with no edges, and every lattice site is identical. Although it is a simple geometrical concept, it is one of the most important concepts in crystal geometry.

Real crystal structures can be more complicated than the Bravais lattice. The concept of the non-Bravais lattice, or the complex lattice, was introduced to describe the structure of compound crystals and some element crystals. The non-Bravais lattice is defined as an infinite lattice with no edges, and some lattice sites are inequivalent to others due to different elements or local geometries. The non-Bravais lattice could be regarded as a Bravais lattice where each site is replaced by a basis, as illustrated in Figure 3.2. The choice of the basis is not unique in space. In a compound crystal such as sodium chloride, the basis is an assembly of a set of atoms, and the assembly corresponds to the chemical formula NaCl. In a pure element crystal with a complex lattice, the basis is an assembly of atoms with different local orientations of chemical bonds, such as the "A" and "B" sites in the diamond structure.

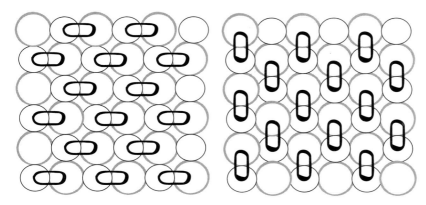

Figure 3.2 Different ways of choosing the basis in a non-Bravais lattice

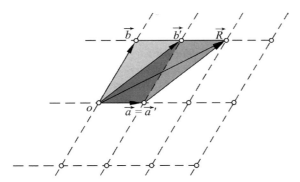

Figure 3.3 Two sets of primitive vectors in a 2D Bravais lattice, where $\vec{R} = 2\vec{a} + \vec{b} = \vec{a}' + \vec{b}'$

In the Bravais lattice, the position vector $\vec{R}$ between any two sites can be expressed by an integer linear combination of a set of primitive vectors:

$$\vec{R} = n\vec{a} + m\vec{b} \quad (2D); \quad \vec{R} = n_1\vec{a}_1 + n_2\vec{a}_2 + n_3\vec{a}_3 \quad (3D) \quad (3.1)$$

where $\vec{a}$ and $\vec{b}$ are primitive vectors in the two-dimensional (2D) Bravais lattice; $\vec{a}_1$, $\vec{a}_2$, and $\vec{a}_3$ are primitive vectors in the three-dimensional (3D) Bravais lattice. The choice of the primitive vector is not unique, as shown in Figure 3.3, and the only requirement is that the volume (or area) of the parallelepiped (or parallelogram) formed by the primitive vectors is the same.

The parallelepiped (parallelogram) formed by the 3D (2D) primitive vectors is called the primitive unit cell (PUC). The PUC is the minimum unit that can fill the whole lattice without overlapping. The choice of the PUC is not unique, but any PUC in a certain lattice has the same volume. The relationship between the PUC and the primitive

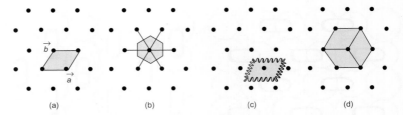

Figure 3.4 Primitive unit cell and conventional unit cell of a 2D triangular lattice: (a) PUC formed by primitive vectors; (b) Wigner-Seitz cell; (c) PUC with arbitrary shape; (d) conventional unit cell

vector is:

$$\Omega = |\vec{a} \times \vec{b}| \quad (2D); \quad \Omega = |\vec{a}_1 \cdot (\vec{a}_2 \times \vec{a}_3)| \quad (3D) \qquad (3.2)$$

The two PUCs in Figure 3.3 obviously have the same volume. Generally speaking, a PUC can have an arbitrary shape, as long as its volume is the same as Ω in Eq. (3.2). In a Bravais lattice, there is only one site in a PUC. In a non-Bravais lattice, there are several sites in a PUC, and these sites form the basis.

A special kind of PUC, called a Wigner-Seitz cell, is very useful in solid state theory, because the Wigner-Seitz cell has a central symmetry about a lattice site. The wavevector $\vec{k}$ space with respect to the energy spectrum of quasi-particles $E(\vec{k})$ is simply the Wigner-Seitz cell of a crystal's reciprocal space, which is discussed later in this chapter. In a 2D lattice, the Wigner-Seitz cell is surrounded by the planes perpendicularly bisecting the vectors $\vec{R}$ between nearest neighbors, as shown in Figure 3.4. In a 3D lattice, the Wigner-Seitz cell can be formed by the planes perpendicularly bisecting the vectors $\vec{R}$ between nearest neighbors or next nearest neighbors.

To illustrate the symmetry of a lattice more clearly, a conventional unit cell (CUC) is often used. The lattice constant "a" is usually defined as the length of the side edge of the CUC. The volume of the CUC is one or several times that of the PUC Ω. For example, in Figure 3.4, the area of the CUC is three times that of the PUC in the 2D triangular lattice. Note that the concept of the CUC should not be confused with the non-Bravais lattice. In the CUC of a Bravais lattice, there might be more than one site, and all these sites are equivalent to each other. In the PUC of a non-Bravais lattice, there must be more than one site and these sites are non-equivalent.

The "crystal direction" and "crystal plane" are useful concepts for the geometrical description of crystals. In a lattice, if all sites are connected by a set of parallel lines, the direction of the parallel lines can be defined as the crystal direction; if all sites are located in a set of parallel planes, the set of planes is called the crystal planes. It should be

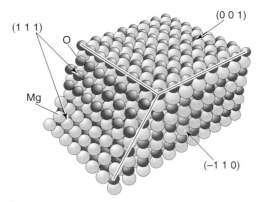

Figure 3.5 The crystal planes in MgO (rock salt)

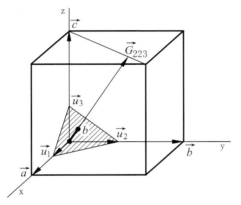

Figure 3.6 Definition of Miller indices of crystal planes: three intercept vectors $\vec{u}_1 = \vec{a}/2$, $\vec{u}_2 = \vec{b}/2$, and $\vec{u}_3 = \vec{c}/3$; vector $\vec{G}_{223}$ perpendicular to crystal plane (223)

emphasized that, for a better view of the crystal symmetry, the notation of crystal orientation or crystal plane is usually defined with respect to the CUC but not the PUC.

In a lattice with a cubic CUC, the crystal direction is labeled [nml], where $\vec{R} = \frac{1}{M}(n\hat{e}_x + m\hat{e}_y + l\hat{e}_z)a$ (M is an integer; $\hat{e}_x, \hat{e}_y, \hat{e}_z$ is the unit vector in three directions; a is the lattice constant) is the vector between nearest neighbor sites along the crystal direction. Equivalent crystal directions [nml], [mnl]... are labeled $\langle nml \rangle$. The most important crystal directions in a cubic CUC are $\langle 100 \rangle$ (six equivalents), $\langle 110 \rangle$ (12 equivalents), and $\langle 111 \rangle$ (eight equivalents).

In a 3D lattice with a CUC formed by vectors $\vec{a}, \vec{b}, \vec{c}$, the set of crystal planes are represented by the Miller indices (nml), where $\vec{u}_1 = \vec{a}/m$, $\vec{u}_2 = \vec{b}/n$, and $\vec{u}_3 = \vec{c}/l$ are three intercept vectors of a crystal plane in the CUC, as shown in Figure 3.6. In a lattice with a

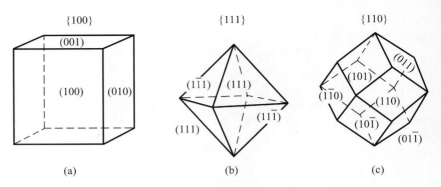

Figure 3.7 The crystal planes in a lattice with a cubic conventional unit cell

cubic or cuboid CUC, the distance between lattice planes equals $d_{nml} = (n^2/a^2 + m^2/b^2 + l^2/c^2)^{-1/2}$ for (nml) crystal planes. Its relationship with the perpendicular-to-planes reciprocal vector $\vec{G}_{nml}$ is discussed later in this chapter.

In a lattice with a cubic CUC, equivalent crystal planes, (mnl) (nml)... are labeled {nml}. The six equivalent {100} planes, 12 equivalent {110} planes, and eight equivalent {111} planes form a cube, a regular dodecahedron, and a regular octahedron respectively, as shown in Figure 3.7. Furthermore, the crystal plane and crystal direction have one-to-one correspondence: the equivalent crystal directions ⟨100⟩, ⟨110⟩, and ⟨111⟩ are perpendicular to the regular polyhedrons formed by crystal planes {100}, {110}, and {111} respectively.

3.2 Symmetry and Classification of Crystal Structures

Symmetry is an important fundamental concept in physics. In a system with complicated physical phenomena, a higher symmetry means a simpler description of the system. The essence of symmetry is the equivalence of some factors in the system, therefore a system with higher symmetry needs fewer independent factors for clarification.

In this section, the symmetry of lattices is discussed to determine the correct method of lattice classification. The symmetry of a lattice is determined by the assembly of all possible rearrangements after which all the sites in the lattice are unaltered, and these rearrangements are called symmetry transformations. A symmetry transformation in a lattice can be represented as a combination of one or more of the three fundamental types of transformation, namely the translation, rotation, and reflection of a lattice.

The basic characteristic of a lattice structure is the finite translational symmetry. An "infinite" translational symmetry operation refers to translation of the lattice by an arbitrary vector $\vec{r}$; a finite translational symmetry operation means that after moving a lattice by the position vector $\vec{R} = n_1 \vec{a}_1 + n_2 \vec{a}_2 + n_3 \vec{a}_3$, the lattice is unaltered. There are an infinite number of translations obtainable from different combinations of arbitrary integers $n_1 n_2 n_3$. In other words, the translation simply corresponds to the definition of a crystal lattice.

The rotational symmetry of a lattice is also a finite symmetry. A rotational symmetry operation rotates a lattice by an angle ϕ about a rotational symmetry axis, after which all sites in the lattice are unaltered. It will be proved later in this section that ϕ can only be one of eight special angles due to the limits of the translational symmetry of a lattice. In a 2D lattice, the rotational symmetry axis must pass one site and be perpendicular to the lattice plane; in a 3D lattice, the rotational axis should pass one site and be parallel to one crystal orientation.

The symmetry of a mirror image means a lattice is unaltered after a reflection operation with respect to a symmetry plane. In a 2D lattice, the symmetry plane must pass one site and be parallel to one crystal orientation; in a 3D lattice, the reflecting plane can be any crystal plane.

The symmetry of a lattice can be described accurately by point groups and space groups. In 1830, J. F. C. Hessel first derived the 32 three-dimensional point groups, or the 32 crystal classes, from Haüy's law of rational intercepts. Hessel also proposed the category of space groups, as well as four-dimensional groups. Hessel's research pioneered the systematic analysis of lattice symmetry.

The concept of a point group is quite simple. Firstly, a point group is a group. A group has many elements $\{e_i\}$, and a "product" operation between the elements $e_i \times e_j = e_k$ still belongs to the group. In a point group, the following rules apply:

1. An element in a point group is a series of symmetry transformations of a lattice. The identical transformation e_0 of a lattice is also an element.
2. The name "point group" comes from the restriction on the elements in the group: at least one lattice site should be fixed during the symmetry transformation. Therefore only the results of the rotational, reflectional symmetry operation or their linear combination can be elements in the point group.
3. The "product" in a point group is defined as two successive symmetry transformations of a lattice. The "product" of any two elements in a point group must be equivalent to an element in it.

In 1850, based on the 32 point groups, A. Bravais classified 3D Bravais lattices as seven crystal systems and 14 different lattices. In a Bravais lattice, all lattice sites are equivalent; therefore the point group is good enough to describe the lattice. Bravais' proof is too complex to be described in this book, however, later on in this section, his idea is demonstrated in a derivation of the classification of a 2D Bravais lattice from the 2D point groups.

The space group is an extension of the point group, and is a full symmetry group for the Bravais lattice and the non-Bravais lattice. A space group is a group whose elements are the translation, rotation, reflection symmetry transformations of lattice, or their linear combination. In 1890 and 1891, E. C. Feodorov in Russia, A. M. Schoenflies in Germany, and W. Barlow in England classified the two-dimensional space groups into 17 wallpaper groups, and the three-dimensional space groups into 230 crystallography groups. Schoenflies also proposed a set of Shoenflies notations for point groups, such as C,D,O,T groups. The Shoenflies notations (often used in solid state theory or quantum chemistry) and the international symbols (used in crystal structure measurement) for the 32 point groups will be introduced later.

3.2.1 Symmetry and Classification of 2D Lattices

The simplest point groups are the C_n groups, in which the elements are a set of rotational operations about an n-fold rotational symmetry axis. The discussion of C_n groups for 2D lattices can be extended to the rotational symmetry of 3D lattices, because the crystal planes perpendicular to the rotational axis in 3D lattices are analogous to 2D lattices. A C_n group contains n rotational symmetry transformations, and usually has the notation:

$$C_n = \left\{ \phi_m = \frac{2\pi m}{n} \middle| m = 0, 1, \ldots, n-1 \right\} \quad (3.3)$$

where the angle ϕ_m represents the m'th rotational symmetry transformation about the n-fold rotational symmetry axis.

It was J. F. C. Hessel who proved that only two-, three-, four-, and six-fold axes of rotational symmetry are possible in minerals, due to the restriction of the translational symmetry of the lattice. This can be proved by following Figure 3.8(a), where the distance a between nearest neighboring lattice sites 1 and 2 is the minimum length of primitive vectors. If a rotational transformation is made about site 1, site 2 will be rotated to site 1' by a counter-clockwise rotation with angle ϕ; then if a rotational transformation is made about site 2, site 1 will be rotated to site 2' by a clockwise rotation with angle ϕ. Following Eq. (3.1), the

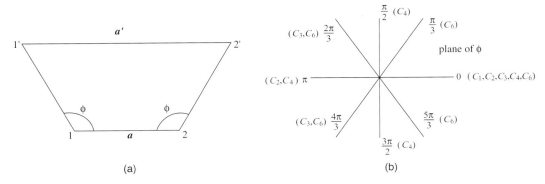

Figure 3.8 (a) restriction on symmetry angle ϕ of a lattice by parallel displacement transformation (based on Feng and Jin, 2003) (b) allowed ϕ angle and related C_n groups

distance a' must be an integer times a:

$$a' = a + 2a \sin\left(\phi - \frac{\pi}{2}\right) = ha \rightarrow \cos\phi = (1-h)/2 \qquad (3.4)$$

Note that $\cos\phi \in [-1, 1]$; therefore, when $h = -1, 0, 1, 2, 3$, the rotational angle ϕ equals $0, \pm\frac{\pi}{3}, \pm\frac{\pi}{2}, \pm\frac{2\pi}{3}$, and π, respectively. Accordingly the eight possible angles of rotational symmetry transformation are $0, \frac{\pi}{3}, \frac{\pi}{2}, \frac{2\pi}{3}, \pi, \frac{4\pi}{3}, \frac{3\pi}{2}$, and $\frac{5\pi}{3}$ and as a result, the only possible C_n groups are C_1, C_2, C_3, C_4, and C_6, as shown in Figure 3.8(b).

The 2D Bravais lattices can be classified based on the rotational point group C_n and the reflectional symmetry transformation σ. The clinic, square, or triangular lattice is just an arbitrary 2D Bravais lattice with a two-fold, four-fold, or six-fold rotational symmetry axis respectively.

The square and triangular lattices naturally satisfy the reflection symmetry along the $\langle 10 \rangle$ or $\langle 11 \rangle$ direction. When an extra reflectional symmetry transformation is applied to a clinic lattice, more restrictions will be added at the two primitive vectors $\vec{a}$ and $\vec{b}$. If $\vec{a} = a\hat{e}_x$, $\vec{b} = b_x\hat{e}_x + b_y\hat{e}_y$, and the lattice has a reflection symmetry along $\langle 10 \rangle$ direction, $\vec{b}' = \sigma\vec{b}$ is still a position vector after reflection:

$$\vec{b}' = b_x\hat{e}_x - b_y\hat{e}_y = n\vec{a} + m\vec{b} = (na + mb_x)\hat{e}_x + mb_y\hat{e}_y \qquad (3.5)$$

There are two possible solutions for Eq. (3.5). One solution is $n = 0$, $m = -1$, and $b_x = 0$, which corresponds to a rectangular lattice with $\vec{a} = a\hat{e}_x$ and $\vec{b} = b_y\hat{e}_y$. The other solution is $n = 1$, $m = -1$, and $b_x = a/2$, which represents a centered-rectangular lattice with $\vec{a} = a\hat{e}_x$ and $\vec{b} = \frac{a}{2}\hat{e}_x + b_y\hat{e}_y$.

In summary, there are five kinds of 2D Bravais lattices (crystal surfaces in reality): clinic, square, triangular, rectangular, and centered-rectangular lattices. There are only four different kinds of 2D CUCs,

Table 3.1 Symmetry groups and classification of 2D Bravais lattices

Lattice	Point group	Lattice	Point group
Clinic	C_1, C_2	Rectangular	C_1, C_2, σ
Square	C_1, C_2, C_4, σ	Centered-rectangular	C_1, C_2, σ
Triangular	$C_1, C_2, C_3, C_6, \sigma$		

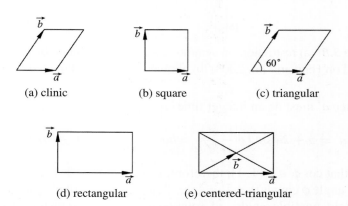

Figure 3.9 Five two-dimensional Bravais lattices and their primitive vectors

because the CUCs of the rectangular and centered-rectangular lattice are the same. Since the CUC and the crystal system have a one-to-one corresponding relationship, 2D lattices can be classified as clinic, square, triangular, and rectangular crystal systems.

3.2.2 Point Groups and Classification of 3D Bravais Lattices

In 1830, when the 32 point groups were first proposed, J. F. C. Hessel was a physician and a professor of mineralogy at Marberg in Germany. From 1849 to 1850, A. Bravais was a professor of physics at the École Polytechnique in Paris, where he formulated the classification of 3D lattices.

There are two notation systems for the point group: the Shoenflies notations and the international symbols (also called the Hermann-Mauguin symbols). The Shoenflies notations and the international symbols for the 32 point groups for 3D lattices are listed in Table 3.2.

Based on the 32 point groups, A. Bravais proved that there are seven crystal systems: triclinic, monoclinic, orthorhombic, tetragonal, cubic, trigonal, and hexagonal. The 14 Bravais lattices belong to the

Table 3.2 Thirty-two point groups for 3D lattices (based on Landau and Lifshitz, 1987)

Shoenflies notation	International symbol	Characteristics
C_1, C_2, C_3, C_4, C_6	1,2,3,4,6	a single n-fold axis
C_i, C_{2i}, C_{3i}, S_4	$\bar{1}, \bar{2}, \bar{3}, \bar{4}$	an n-fold rotation-inversion axis
(S_2, C_{2i}, S_6, S_4)	$(\bar{1}, m, \bar{3}, \bar{4})$	(inversion means $\vec{R} \to -\vec{R}$)
$C_{2h}, C_{3h}, C_{4h}, C_{6h}$	$\frac{2}{m}, \frac{3}{m}, \frac{4}{m}, \frac{6}{m}$ $(\frac{2}{m}, \bar{6}, \frac{4}{m}, \frac{6}{m})$	an n-fold rotation-reflection axis the reflection plane $\perp$ the axis
$C_{2v}, C_{3v}, C_{4v}, C_{6v}$	2mm, 3m, 4mm, 6mm	an n-fold rotation-reflection axis n planes passing through the axis
D_2, D_3, D_4, D_6	222, 32, 422, 622	an n-fold axis plus n 2-fold axes n 2-fold axes $\perp$ the n-fold axis
$D_{2h}, D_{3h}, D_{4h}, D_{6h}$	$\frac{2}{m}\frac{2}{m}\frac{2}{m}, \frac{3}{m}2m, \frac{4}{m}\frac{2}{m}\frac{2}{m}, \frac{6}{m}\frac{2}{m}\frac{2}{m}$ (mmm, ., 4/mmm, 6/mmm)	D_n symmetry plus a reflection a plane passing through 2-fold axes
D_{2d}, D_{3d}	$\bar{4}2m, \bar{3}\frac{2}{m}$	D_n symmetry plus n reflections n planes through n-fold axes midway between two 2-fold axes
T	23	tetrahedral ($\langle 111 \rangle$) 3-fold axes plus three ($\langle 100 \rangle$) 2-fold axes connecting centers of opposite edges
T_d	$\bar{4}3m$	T symmetry plus 6 reflections reflection symmetry planes {110}
T_h	$\frac{2}{m}\bar{3}$	T symmetry plus 3 reflections reflection symmetry planes {100}
O	432	three 4-fold ($\langle 100 \rangle$), four 3-fold ($\langle 111 \rangle$) and six 2-fold ($\langle 110 \rangle$) axes
O_h	$\frac{4}{m}\bar{3}\frac{2}{m}$	O symmetry plus an inversion

seven crystal systems and the Bravais lattice notation (BLN for short) is used to specify different Bravais lattices.

The simple Bravais lattices in a crystal system can be described geometrically using the three primitive vectors $\vec{a}, \vec{b}, \vec{c}$: the lengths are a, b, c, and the angles between a pair of primitive vectors are $\alpha = \langle \vec{b}, \vec{c} \rangle$, $\beta = \langle \vec{c}, \vec{a} \rangle$, and $\gamma = \langle \vec{a}, \vec{b} \rangle$.

In each crystal system, some extra sites can be added to the simple Bravais lattice to form a different Bravais lattice. The BLN is labeled (uvw) with respect to the positions of sites $\vec{r} = u\vec{a} + v\vec{b} + w\vec{c}$ in the PUC of the simple Bravais lattice in a crystal system.

Table 3.3 BLN labeled by sites in the simple Bravais lattice's primitive unit cell

BLN	Common name	Sites in PUC of simple Bravais lattice
P	simple	(000)
A	base-centered	(000), $\left(0\frac{1}{2}\frac{1}{2}\right)$
B	base-centered	(000), $\left(\frac{1}{2}0\frac{1}{2}\right)$
C	base-centered	(000), $\left(\frac{1}{2}\frac{1}{2}0\right)$
I	body-centered	(000), $\left(\frac{1}{2}\frac{1}{2}\frac{1}{2}\right)$
F	face-centered	(000), $\left(0\frac{1}{2}\frac{1}{2}\right)$, $\left(\frac{1}{2}0\frac{1}{2}\right)$, $\left(\frac{1}{2}\frac{1}{2}0\right)$

Table 3.4 Seven crystals systems, 14 3D Bravais lattices, and 32 point groups

3D Crystal systems	Geometrical Description of PUC	BLN & abbreviations	Point groups (two notations)
Triclinic	$a \neq b \neq c$ $\alpha \neq \beta \neq \gamma$	P	C_1, C_i $1, \bar{1}$
Monoclinic	$a \neq b \neq c$ $\alpha = \gamma = 90° \neq \beta$	P,A(C) MCL	C_2, C_{2i}, C_{2h} $2, \bar{2}, \frac{2}{m}$
Orthorhombic	$a \neq b \neq c$ $\alpha = \beta = \gamma = 90°$	P,A(BC),I,F ORC	D_2, D_{2h}, C_{2v} $222, \frac{2}{m}\frac{2}{m}\frac{2}{m}, 2mm$
Tetragonal	$a = b \neq c$ $\alpha = \beta = \gamma = 90°$	P,I TET	$D_4, D_{4h}, D_{2d}, C_4, C_{4v}, C_{4h}, S_4$ $422, \frac{4}{m}\frac{2}{m}\frac{2}{m}, \bar{4}2m, 4, 4mm, \frac{4}{m}, \bar{4}$
Cubic	$a = b = c$ $\alpha = \beta = \gamma = 90°$	P,I,F SC,BCC,FCC	O, O_h, T, T_h, T_d $432, \frac{4}{m}\bar{3}\frac{2}{m}, 23, \frac{2}{m}3, \bar{4}3m$
Trigonal	$a = b = c$ $\alpha = \beta = \gamma < \frac{2\pi}{3} \neq \frac{\pi}{2}$	P RHL	$D_3, D_{3d}, C_3, C_{3v}, C_{3i}$ $32, \bar{3}\frac{2}{m}, 3, 3m, \bar{3}$
Hexagonal	$a = b = c$ $\alpha = \beta = \frac{\pi}{2}, \gamma = \frac{2\pi}{3}$	P HEX	$D_6, D_{6h}, D_{3h}, C_6, C_{6v}, C_{6h}, C_{3h}$ $622, \frac{6}{m}\frac{2}{m}\frac{2}{m}, \frac{3}{m}2m, 6, 6mm, \frac{6}{m}, \bar{6}$

The relationships between point groups, Bravais lattices, and crystal systems for 3D lattices are listed in Table 3.4. The abbreviation RHL stands for the rhombohedral Bravais lattice, which is another name for the trigonal lattice. S, BC, and FC stand for the simple, body-centered, and face-centered BLN respectively.

A 3D Bravais lattice can be formed by piling up a series of 2D Bravais lattices. For example, if a 2D clinic lattice is piled up ($\alpha = \gamma = 90°$) by an arbitrary d, a monoclinic lattice is formed; if a 2D triangular lattice is piled up ($\alpha = \beta = 90°$) by an arbitrary d, a hexagonal lattice is formed.

In Figure 3.10, fourteen 3D Bravais lattices and their primitive vectors are shown. It should be emphasized that each unit shown in Figure 3.10 is a PUC of the simple Bravais lattice, and in most cases, it is also the CUC of all Bravais lattices in the same crystal system except that the CUC of the hexagonal lattice is a larger hexagonal prism.

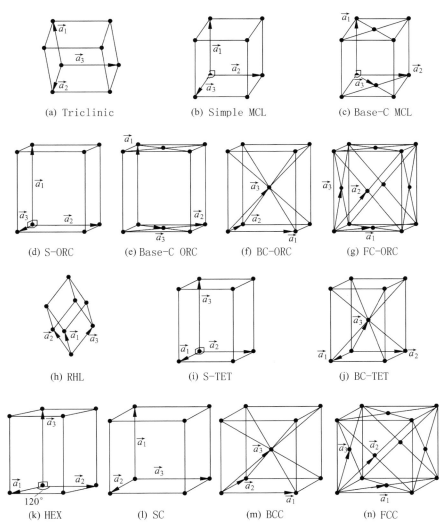

Figure 3.10 Fourteen three-dimensional Bravais lattices and their primitive vectors (based on Huang and Han, 1988)

The specific rotational axes with respect to a point group can be found in Figure 3.10. In the tetragonal lattice, a D_4 group has one [001] four-fold axis and four [100], [010], [110], [1$\bar{1}$0] two-fold axes. In the trigonal lattice, a D_3 group has one [111] three-fold axis and three [1$\bar{1}$0], [10$\bar{1}$], [01$\bar{1}$] two-fold axes. In the hexagonal lattice, a D_6 group has one [001] six-fold axis and six [100], [010], [110], [1$\bar{1}$0], [120], [210] two-fold axes.

In summary, the classification of 3D Bravais lattices can be performed by analyzing the symmetry of the lattices, i.e., the point groups. The crystal systems are distinguished by rotation symmetry, and the Bravais lattices in a crystal system correspond to different reflection symmetries.

3.3 Natural Structures of Crystals

In the last section, the possible 2D and 3D Bravais lattices were discussed based on symmetry of lattices. In reality, natural and man-made crystals usually have high symmetries. In this section, the typical structures of ordered solids are introduced, and the structures of minerals and ceramics are analyzed based on Pauling's rules.

3.3.1 Structures of Element Crystals

The most common structures of element crystals are BCC, FCC, HCP (hexagonal closed-packed structure), and DIA (diamond structure). BCC and FCC are cubic Bravais lattices, HCP is a non-Bravais hexagonal lattice, and DIA is a non-Bravais cubic lattice. All four structures have high symmetries.

The concept of the Wigner-Seitz cell (W-S cell) was formulated in the early 1930s, when E. P. Wigner and F. Seitz were studying the quantum ground state of metallic sodium. They found that the symmetry of a lattice could be illustrated by the W-S cell. The W-S cell of the BCC lattice is a truncated octahedron with 24 equivalent vertices W ($\vec{r}_W - \vec{r}_X$ along $\langle 100 \rangle$ direction, X is the face center) in Figure 3.11(a). The W-S cell of the FCC lattice is a regular dodecahedron, whose 14 vertices can be classified into six X vertices and eight P vertices ($\vec{r}_P - \vec{r}_\Gamma$ along $\langle 111 \rangle$ direction), as shown in Figure 3.11(b).

FCC and HCP lattices are close-packed structures with the highest atomic filling factors. The BCC lattice has a slightly lower atomic filling factor, but can still be viewed as a close-packed structure. The closed-packed structure as seen in Figure 3.12 is mainly due to the high density requirement of free electron gas in metallic bonds.

3.3 Natural Structures of Crystals

Table 3.5 Most common structures of element crystals

	BCC	FCC	HCP	DIA
Family	1A: Li,Na,K,Rb,Cs	2A: Ca,Sr	2A: Be,Mg	4A: C,Si,Ge
	2A: Ba	8B: Ni,Pd,Pt	3B: Sc,Y,La	Sn(<18°C)
	5B: V,Nb,Ta	1B: Cu,Ag,Au	4B: Ti,Zr,Hf	
	6B: Cr,Mo,W	3A: Al	7B-8B: Tc,Ru	
	8B: Fe	8A: Ne-Xe	2B: Zn,Cd	
Primitive vectors	$\vec{a}_1 = \frac{a}{2}(-\hat{e}_x + \hat{e}_y + \hat{e}_z)$ $\vec{a}_2 = \frac{a}{2}(+\hat{e}_x - \hat{e}_y + \hat{e}_z)$ $\vec{a}_3 = \frac{a}{2}(+\hat{e}_x + \hat{e}_y - \hat{e}_z)$	$\vec{a}_1 = \frac{a}{2}(\hat{e}_y + \hat{e}_z)$ $\vec{a}_2 = \frac{a}{2}(\hat{e}_z + \hat{e}_x)$ $\vec{a}_3 = \frac{a}{2}(\hat{e}_x + \hat{e}_y)$	$\vec{a}_1 = \frac{\sqrt{3}a}{2}\hat{e}_x - \frac{a}{2}\hat{e}_y$ $\vec{a}_2 = a\hat{e}_y$ $\vec{a}_3 = c\hat{e}_z$	Same as FCC
V_{PUC}	$\Omega = \frac{1}{2}a^3$	$\Omega = \frac{1}{4}a^3$	$\Omega = \frac{\sqrt{3}}{2}a^2c$	$\Omega = \frac{1}{4}a^3$
Basis			$0, \frac{2\vec{a}_1}{3} + \frac{\vec{a}_2}{3} + \frac{\vec{a}_3}{2}$	$0, \frac{\vec{a}_1}{4} + \frac{\vec{a}_2}{4} + \frac{\vec{a}_3}{4}$
W-S cell	Truncated octahedron	Dodecahedron	Hexagonal prism	Same as FCC

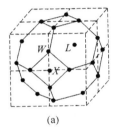 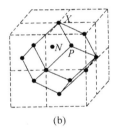

(a)　(b)

Figure 3.11 Wigner-Seitz cell and its drawing method: (a) W-S cell of BCC lattice: a truncated octahedron with 14 planes, where X stands for face centers and $|\vec{r}_W - \vec{r}_X| = a/4$; (b) W-S cell of FCC lattice: a regular dodecahedron, where X stands for face centers, Γ is the cubic center and $|\vec{r}_P - \vec{r}_\Gamma| = \sqrt{3}a/4$

In HCP lattices, a four-digit index $(uvwz)$, which describes the hexagonal symmetry more clearly, is introduced to label a crystal plane instead of the Miller indices (hkl), where $u = h$, $v = k$, $z = l$, and $u + v + w = 0$ are chosen to describe a plane perpendicular to the position vector $\vec{R}_{hkl} = u\vec{a}_1 + v\vec{a}_2 + z\vec{a}_3$, as seen in Figure 3.13.

The complicated non-Bravais non-close-packed structures, such as RHL, ORC, TET, MCL, and CUB, mostly occur in elements of the IIIA-VA families, and in non-closed-packed metals such as manganese and mercury. The position vectors in a non-Bravais lattice (HCP, DIA, or non-closed-packed structures) can be expressed as:

$$\vec{R} = (n_1\vec{a}_1 + n_2\vec{a}_2 + n_3\vec{a}_3) + \vec{\delta}_j; \quad (j = 1, 2, \ldots, n_a) \qquad (3.6)$$

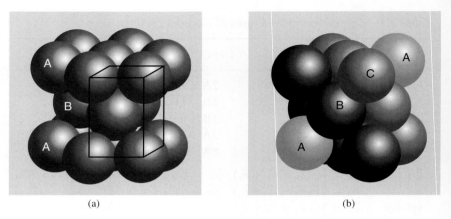

Figure 3.12 Close-packed structures: (a) HCP's CUC, with the ABABAB... type stacking of 2D atomic layers with a triangular lattice; (b) FCC's CUC, with the ABCABC... type stacking of 2D atomic layers with a triangular lattice

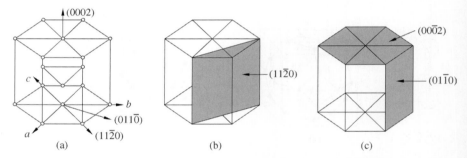

Figure 3.13 HCP structure: (a) conventional unit cell; (b)-(c) typical crystal planes

where n_a is the number of atoms in a basis, i.e., the total number of non-equivalent lattice sites in the PUC; and $\vec{\delta}_j = u_1\vec{a}_1 + u_2\vec{a}_2 + u_3\vec{a}_3$ represents the displacement of an atom in the basis, where the linear parameters u_1, u_2, and u_3 are fractional.

The CUC of α-gallium with a complicated ORC structure is a cuboid with lattice constants $a = 4.520\text{Å}$, $b = 7.663\text{Å}$, and $c = 4.526\text{Å}$. The PUC has a volume $\Omega = \frac{1}{2}abc$. The primitive vectors of a gallium crystal are

$$\vec{a}_1 = \frac{1}{2}(a\hat{e}_x - b\hat{e}_y); \quad \vec{a}_2 = \frac{1}{2}(a\hat{e}_x + b\hat{e}_y); \quad \vec{a}_3 = c\hat{e}_z. \quad (3.7)$$

$$\vec{\delta}_{1,2} = \mp[u(\vec{a}_1 - \vec{a}_2) - v\vec{a}_3];$$
$$\vec{\delta}_{3,4} = -\frac{1}{2}[(\vec{a}_1 - \vec{a}_2) - \vec{a}_3] \mp [u(\vec{a}_1 - \vec{a}_2) + v\vec{a}_3] \quad (3.8)$$

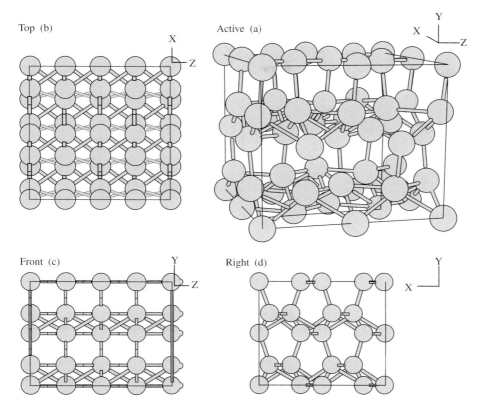

Figure 3.14 Eight conventional unit cells (3D view) of α-Ga with ORC structure

where $\vec{\delta}_1$ to $\vec{\delta}_4$ with $u = 0.1549$ and $v = 0.081$ are the four basis vectors of the four sites in a PUC.

Arsenic, antimony, bismuth, and rhombohedral graphite have similar trigonal lattice structures, and their CUC is a rhombohedral parallelepiped. The primitive vectors of these RHL structures can be written as

$$\vec{a}_1 = b\hat{e}_x + c\hat{e}_y + c\hat{e}_z; \quad \vec{a}_2 = c\hat{e}_x + b\hat{e}_y + c\hat{e}_z;$$
$$\vec{a}_3 = c\hat{e}_x + c\hat{e}_y + b\hat{e}_z; \tag{3.9}$$

As a trigonal lattice, the lattice constant of α-arsenic is $a = \sqrt{b^2 + 2c^2} = 4.13$Å, and the bond angle is $\theta = \cos^{-1}(\vec{a}_1 \cdot \vec{a}_2/a^2) = 54''10'$. The ratio $b/c = 0.0877$ is quite small, so the primitive vectors of the RHL α-arsenic are only slightly deviated from that of the FCC structure, with a symmetry breaking from point group O to D_3, as shown in Figure 3.15(b). In a RHL crystal, the basis of the two atoms in the PUC usually stay at $\vec{\delta}_1 = 0$ and $\vec{\delta}_2 = u(\vec{a}_1 + \vec{a}_2 + \vec{a}_3)$. When parameters $b = 0$ and $u = \frac{1}{4}$, the rhombohedral structure turns to a diamond structure with a lattice constant $2c$.

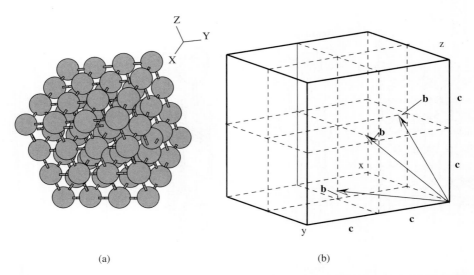

Figure 3.15 Crystal structure of α-As with RHL structure (a) 3D view (b) primitive vectors $\vec{a}_1, \vec{a}_2, \vec{a}_3$

The arsenic crystal has a rhombohedral structure due to the covalent bonds between arsenic atoms. Following the 8–N rule, an arsenic atom in the VA family can form three covalent bonds. These covalent bonds connect the atoms to form a non-flat layer, and the layers are then bound together by van der Waals forces. Graphite structure is similar except that the covalent bonds connect carbon atoms with sp^2 hybridization to form a flat layer.

3.3.2 Structures of Compounds: Pauling's Rules

Most compounds in nature are minerals and are formed mainly by ionic bonds. Natural intermetallic compounds are rare, because metals are easily oxidized. The structures and phases of intermetallic compounds are traditionally taught in materials science or physical metallurgy courses. In this section, the structure of compounds is analyzed using Pauling's rules.

In 1925, Linus Pauling obtained his Ph.D. in Chemistry from the California Institute of Technology under the supervision of R. Dickinson in the field of X-ray crystallography. As a Ph.D. student, he had already demonstrated his ability to "guess" atomic structures and explain the X-ray diffraction data. In 1926, he went to Germany and worked with Arnold Sommerfeld, where he gained a good grounding in quantum mechanics. In 1928, he published six principles, which were later summarized as five "Pauling's rules," to determine the structure of complex crystals. Pauling had disagreements with W. L. Bragg over the use of Bragg's X-ray diffraction data and analysis of silicates. Nevertheless,

Table 3.6 Pauling Rules for minerals: the key points

	Core concept	Main impact factors
First rule	coordination polyhedron	cation-anion radius ratio
Second rule	electrostatic valency	$S = \sum_i S_i = \sum_i (w_i v_i)$
Third rule	sharing edges or faces	total energy of a crystal
Fourth rule	no common edges or faces	electro-valency of cation
Fifth rule	principle of parsimony	kinds of coordination polyhedra

L. C. Pauling is the founding father of modern structural chemistry, in which quantum mechanics plays a significant role.

In structural chemistry, Pauling treated cations and anions as hard spheres with a definite radius, therefore, Pauling's model is also known as the hard-sphere model. It should be emphasized that the hard sphere assumption is contrary to the quantum mechanical electron-cloud picture, therefore the ionic "radius" is not unique. In this hard-sphere model, the cations and anions must have contacts with each other, and the electrostatic energy of the whole system must be minimized.

Pauling's first rule listed in Table 3.7 uses the cation-anion radius ratio to explain the structure of the coordination polyhedron. From the view of chemists, ionic crystals may be considered as sets of linked polyhedra, where the building block—the coordination polyhedron—is totally different from the concept of the PUC. The cation-anion distance is regarded as the sum of the ionic radii, which is the minimum distance R_0 of the ionic potential discussed in Chapter 2. Pauling chose the cation, but not the anion, as the center of the coordination polyhedron. The species of anions, such as oxygen ions, oxyhydrogen ions, and chlorine ions, are quite limited. The roles of cations and anions are different: anions, with their relatively larger radii, form the "bones" of a crystal; while the smaller cation with the highest valence charge is the center of

Table 3.7 Pauling's first rule: common geometry of coordination polyhedra in solids

Coordination polyhedron	Coordination number	Cation-anion radius ratio	Examples (coincident ones)
Cube octahedron	12	$[1.000, \infty]$	HCP
Cube	8	$[0.732, 1.000]$	CsCl
Octahedron	6	$[0.414, 0.732]$	NaCl, FeO, MgO, CaO
Tetrahedron	4	$[0.225, 0.414]$	ZnS, (SiO_4), (AlO_4)
Triangle	3	$[0.155, 0.225]$	(CO_3)

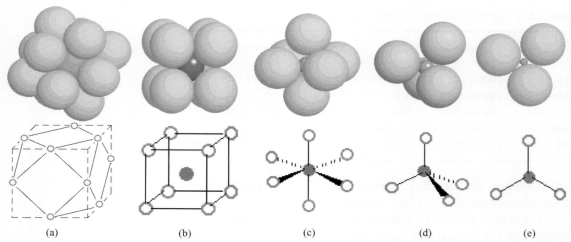

Figure 3.16 Coordination polyhedra with a cation in the center: (a) cube octahedron; (b) cube; (c) octahedron; (d) tetrahedron; (e) triangle

a coordination polyhedron, and these cations are separated by the anions to minimize the total electrostatic energy.

The most significant class of minerals is the oxides, where natural minerals are often formed by linked polyhedra of carbonate, silicate, or aluminate groups. The carbonate group (CO_3) forms a triangle in space, as shown in Figure 3.16(e), since $r_{C^{4+}}/r_{O^{2-}}$ is within the range [0.155, 0.225]. The silicate group (SiO_4) forms a tetragonal polyhedron, as shown in Figure 3.16(d), because $r_{Si^{4+}}/r_{O^{2-}} = 0.30$ is within the range [0.225, 0.415]. The aluminum-oxygen bonds can form a tetragonal aluminate group (AlO_4), because $r_{Al^{3+}}/r_{O^{2-}} = 0.36$; a octahedral aluminate group (AlO_6) is also possible with $r_{Al^{3+}}/r_{O^{2-}} = 0.385$ but it does not follow Pauling's first rule.

The radii of hydroxyl, chlorine, flourine, and sulfur ions are $r_{OH^-} = 1.36$Å, $r_{F^-} = 1.33$Å, $r_{Cl^-} = 1.81$Å, and $r_{S^{2-}} = 1.84$Å, respectively. In sodium chloride, there is a ($NaCl_6$) group with $r_{Na^+}/r_{Cl^-} = 0.54$; in cesium chloride, there is a ($CsCl_8$) group with $r_{Cs^+}/r_{Cl^-} = 0.79$. Both structures follow Pauling's first rule.

Pauling' first rule is called a "rule" but not a "law" because the correspondences listed in Table 3.7 are not always true for all kinds of ionic crystals. A cation with a large radius and low valence charge is less likely to follow Pauling's first rule. For example, both sodium oxide (Na_2O) and potassium oxide (K_2O) have an anti-fluorite structure with Na^+ or K^+ locating at the center of a tetrahedron. However the radius ratio $r_{Na^+}/r_{O^{2-}}$ or $r_{K^+}/r_{O^{2-}}$ is not within the range [0.225, 0.415], which means that Pauling's first rule does not fit for sodium oxide or potassium oxide. In fact, the potassium ion can have between six and 12 oxygen ion coordinations.

Table 3.8 Cation-anion radius ratio in oxides (oxygen ion radius $r_{O^{2-}} = 1.40\text{Å}$)

Metal ions	B^{3+}	Be^{2+}	Si^{4+}	Li^+	Al^{3+}	Ti^{4+}	Fe^{3+}
radius ratio	0.21	0.23	0.30	0.36	0.36	0.44	0.465
Metal ions	Mg^{2+}	Fe^{2+}	Mn^{2+}	Na^+	Ca^{2+}	K^+	Cs^+
radius ratio	0.47	0.53	0.57	0.69	0.80	0.95	1.02

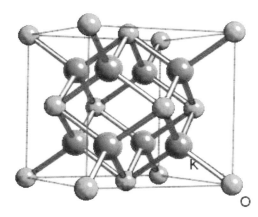

Figure 3.17 Structure of K_2O: a structure incompatible with Pauling's first rule

Pauling's second rule is useful for determining the positions of cations other than those at the center of a coordination polyhedron. It is called the electrostatic valency principle because, in a stable crystal structure, the total strength of the valency bonds that connect an anion to all the neighboring cations is equal to the charge of the anion:

$$S_{\text{anion}} = \sum_i^{NN} S_i = \sum_i^{NN} (w_i/v_i) = \sum_\alpha m_\alpha(w_\alpha/v_\alpha) \qquad (3.10)$$

where S_{anion} is the valence charge of the anion, and for the α'th kind of neighboring cations, w_α is the valence charge, v_α is its (anion) coordination number, and m_α is the number of anions around the cation. W. L. Bragg summarized Pauling's second rule thus: there are many electric field lines starting off from the i'th cation, and the total strength of field lines is the valence charge w_i. Therefore the field lines reaching one anion in the cation's coordination polyhedron with a coordination number v_i are simply w_i/v_i, which is called the electrostatic valency S_i, a measurement of the strength of the bond. Obviously the total strength of all the field lines ending up at the anion would add up to the valence charge S_{anion} of the anion. The electrostatic valency principle is also a

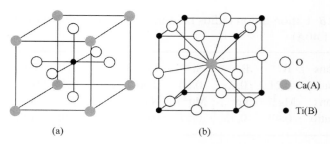

Figure 3.18 Calcium titanate (ABO$_3$) structure: (a) Ti^{4+} at center; (b) Ca^{2+} at center

"rule," which is phenomenological and can not correctly describe the structure of about 16% of natural minerals.

Calcium titanate is a typical ceramic crystal with an ABO$_3$ structure. Ti^{4+} is the cation with the highest valence charge, and according to the Pauling's first rule, Ti^{4+} is the center of the coordination polyhedron. The radius ratio $r_{Ti^{4+}}/r_{O^{2-}}$ is within the range [0.415, 0.732]; therefore, the titanium ion and its six oxygen ion neighbors form an octahedron, as shown in Figure 3.18(a). So where do the calcium ions go? Following Pauling's second rule, the valence charge of O^{2-} can be expressed as:

$$2 = 2 \times (4/6) + m \times (2/v_{Ca}). \tag{3.11}$$

An oxygen ion O^{2-} has two nearest neighboring Ti^{4+} ions. If the calcium ions were at the corners of the conventional cubic cell, an O^{2-} would have $m = 4$ nearest neighboring Ca^{2+} ions; on the contrary, the coordination number of Ca^{2+} ion with respect to O^{2-} is $v_{Ca} = 12$. Then Eq. (3.11) is perfectly satisfied and it means that the calcium titanate structure satisfies both Pauling's first and second rules.

Pauling's third and fourth rules discuss how coordination polyhedrons are piled up in space to form a crystal. Pauling's third rule says that when neighboring coordination polyhedra share edges or faces, the stability of the crystal structure decreases due to a higher Coulomb repulsion. Pauling's fourth rule is complementary to the third rule: in a crystal containing different cations, the coordination polyhedra with a high valency cation at the center and a small coordination number tend not to share polyhedral elements such as edges and faces. The third and fourth rules are especially significant when the tetrahedral or octahedral coordination polyhedra are piled up and the cation at the center is not perfectly screened. In Figure 3.18, it can be seen that in calcium titanate, the (TiO$_6$) groups only share acmes, because Ti^{4+} has a high valence charge; in sodium chloride, the NaCl$_6$ coordination polyhedra share edges, because Na$^+$ has a lower electrovalency.

Pauling's fifth rule is called the principle of parsimony: the number of essentially different kinds of constituents in a crystal, such as the coordination polyhedra, tends to be small. In other words, in the CUC of a compound non-Bravais lattice, the classes of anion and cation lattice sites tend to be minimized; the ions with the same electrovalency might share the same site and they are considered as constituents of that site.

To summarize the previous sections in this chapter, the symmetry of lattices is used to classify crystal structures. Pauling's rules are introduced to describe compound structures, and the corresponding Bravais lattice of a compound can be processed using the first to fifth rules. For more detailed descriptions of crystals, such as the positions of basis sites in complex lattices, the following International Tables for Crystallography are recommended: Volume A: Space-group symmetry; and Volume B: Reciprocal space (which is discussed in the next section).

3.4 Reciprocal Lattices and Brillouin Zones

The wave is a fundamental form of matter. Energy and information can be transferred using waves. There are three classes of wave in nature: (1) the mechanical wave described by Newtonian or analytical mechanics; (2) the electromagnetic wave characterized by Maxwell's equations; (3) the matter wave proposed by L. V. de Broglie and represented by Schrodinger's equation.

All three classes of wave in crystals play significant roles in the analysis of physical properties: (1) the mechanical wave: atomic vibration is a mechanical wave and is the key to explaining the thermal and acoustic properties of solids; (2) the electromagnetic wave: the X-ray is an electromagnetic wave and is used to analyze the structure of solids by diffraction methods; (3) the matter wave: the electron matter wave is the central concept in the electronic theory of solids, and the neutron matter wave is used in the measurement of the phonon spectrum.

The wave in a solid has to propagate in a crystal, where atoms are arranged in a lattice structure. Therefore, these waves must be related to the symmetry of the lattice, including the translation, rotation, and reflection symmetries. In this section, the reciprocal lattice and the Brillouin zone are discussed to illustrate some common geometrical properties of waves in solids.

3.4.1 Reciprocal Lattices

In 1913, P. P. Ewald at the University of Munich introduced the concepts "reciprocal space" and "reciprocal lattice" to explain the X-ray diffraction patterns in a single crystal. Nowadays, the reciprocal lattice is used in the study of all kinds of waves in solids, and has thus become a common concept in solid state physics.

Mathematically speaking, the Bravais lattice and its reciprocal lattice form a pair of Fourier spaces. Any wavefunction can be expressed by the Fourier expansion:

$$\Psi(\vec{r}, t) = \iiint d^3\vec{k}\, \tilde{\psi}_{\vec{k}} \exp(i\vec{k}\cdot\vec{r} - i\omega t) \qquad (3.12)$$

In other words, the real space is the space of $\vec{r}$, but the reciprocal space is a space of wavevector $\vec{k}$. If the wavefunction also has the lattice periodicity $\Psi(\vec{r}) = \Psi(\vec{r} + \vec{R})$, $\Psi(\vec{r})$ can be expanded by a discrete Fourier series:

$$\Psi(\vec{r}) = \sum_{\vec{G}} \tilde{\psi}_{\vec{G}} e^{i\vec{G}\cdot\vec{r}} \rightarrow e^{i\vec{G}\cdot\vec{R}} \equiv 1$$
$$\rightarrow \vec{G} = \sum_{\alpha} h_\alpha \vec{a}_\alpha^*; \quad \vec{a}_\alpha \cdot \vec{a}_\beta^* = 2\pi \delta_{\alpha\beta}, \qquad (3.13)$$

where $\vec{G}$ is the wavevector in a reciprocal lattice and can form a reciprocal lattice by a series of translations with a set of reciprocal primitive vectors $\{\vec{a}_j^*\}$. The orthonormal relationship between primitive vector $\{\vec{a}_i\}$ and reciprocal primitive vector $\{\vec{a}_j^*\}$ reflects the mutual Fourier space relationship between the Bravais lattice and the reciprocal lattice.

Geometrically speaking, the reciprocal lattice in solid state physics has the Bravais lattice structure. If a crystal in real space has a complex lattice structure, the corresponding Bravais lattice can be found with the help of a basis, and the reciprocal lattice structure can be derived from the Bravais lattice. In a 3D lattice, the expression of the reciprocal primitive vectors can be derived from Eq. (3.13):

$$\begin{cases} \vec{a}_1^* = \frac{2\pi}{\Omega_c}(\vec{a}_2 \times \vec{a}_3) \\ \vec{a}_2^* = \frac{2\pi}{\Omega_c}(\vec{a}_3 \times \vec{a}_1) \\ \vec{a}_3^* = \frac{2\pi}{\Omega_c}(\vec{a}_1 \times \vec{a}_2) \end{cases}$$
$$\Omega_c = \vec{a}_1 \cdot (\vec{a}_2 \times \vec{a}_3); \quad \Omega_c^* = \vec{a}_1^* \cdot (\vec{a}_2^* \times \vec{a}_3^*) = \frac{(2\pi)^3}{\Omega_c}. \qquad (3.14)$$

The above equation shows the relationship between the volume of the reciprocal PUC Ω_c^* and the volume of the PUC in real space Ω_c. The calculation of the volume of the reciprocal PUC is an exercise at the end of this chapter. In Eq. (3.14), $\{\vec{a}^*\}$ is related to the vector cross product of $\{\vec{a}\}$; therefore the reciprocal lattice and the lattice in real space must belong to the same crystal system. In most cases, the reciprocal lattice is the Bravais lattice of the real crystal. There are exceptions, for example the body-centered lattice would be reciprocal to the face-centered lattice, and vice versa. The real FCC to reciprocal BCC lattice transformation is

$$\begin{cases} \vec{a}_1 = \frac{a}{2}(\hat{e}_y + \hat{e}_z) \\ \vec{a}_2 = \frac{a}{2}(\hat{e}_z + \hat{e}_x) \\ \vec{a}_3 = \frac{a}{2}(\hat{e}_x + \hat{e}_y) \end{cases} \Rightarrow \begin{cases} \vec{a}_1^* = \frac{a^*}{2}(-\hat{e}_x + \hat{e}_y + \hat{e}_z) \\ \vec{a}_2^* = \frac{a^*}{2}(+\hat{e}_x - \hat{e}_y + \hat{e}_z) \\ \vec{a}_3^* = \frac{a^*}{2}(+\hat{e}_x + \hat{e}_y - \hat{e}_z) \end{cases}$$

$$\Omega_c^* = \frac{(2\pi)^3}{\frac{1}{4}a^3} = \frac{1}{2}(a^*)^3. \tag{3.15}$$

In Table 3.9, the lattice constants and typical angles of the reciprocal conventional unit vectors are listed for seven 3D crystal systems. It should be emphasized that, for a FCC or BCC crystal, the reciprocal lattice constant is $a^* = 4\pi/a$, but not the general rule $a^* = 2\pi/a$ in Table 3.9.

The reciprocal CUC also has the same $\langle 100 \rangle$ orientation as that in real space for the ORC, TET, and cubic crystal systems. The reciprocal CUC for other crystal systems has different orientations from the CUC in real space. The reciprocal HEX CUC rotates by 30° with respect to the CUC in real space, as shown in Figure 3.19(b), and the angle γ^* between $\vec{a}_1^*$ and $\vec{a}_2^*$ becomes $\gamma^* = 180° - \gamma$.

Table 3.9 Reciprocal conventional unit cell sizes for seven crystal systems

Crystal systems	a^*	b^*	c^*	Angle
ORC,TET,CUB	$2\pi/a$	$2\pi/b$	$2\pi/c$	
HEX	$2\pi/(a \sin \gamma)$	$2\pi/(b \sin \gamma)$	$2\pi/c$	$\gamma^* = \pi - \frac{2\pi}{3}$
MCL	$2\pi/(a \sin \beta)$	$2\pi/b$	$2\pi/(a \sin \beta)$	$\beta^* = \pi - \beta$
RHL,Triclinic	$2\pi bc \sin \alpha / \Omega_c$	$2\pi ca \sin \beta / \Omega_c$	$2\pi ab \sin \gamma / \Omega_c$	

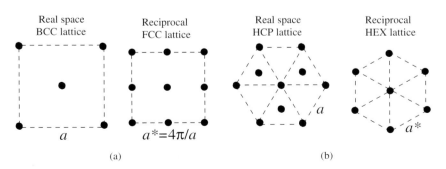

Figure 3.19 Conventional unit cell in real space and reciprocal space: (a) cubic crystal system: BCC → FCC; (b) hexagonal crystal system: HCP → HEX

A reciprocal lattice can be "recorded" using the X-ray diffraction method. In 1964, M. Buerger designed a single crystal diffraction camera, which can record a 2D picture of the reciprocal lattice, such as $\vec{G}_{hkl} = h\vec{a}_1^* + k\vec{a}_2^* + l\vec{a}_3^*$ with a fixed integer h. The mechanism of single crystal diffraction is discussed in the next section. Nowadays, data for thousands of reciprocal lattice sites can be obtained within a day using an automatic single crystal diffraction spectrograph.

3.4.2 Brillouin Zones

The Brillouin zone is an useful concept for describing the energy spectrum of a quasi-particle in a crystal, such as the electron or phonon spectrum in solids. In 1913, Léon Brillouin had only just graduated from the École Normale Supérieure in France when, together with Arnold Sommerfeld, he published a paper on electron scattering. In 1926, G. Wentzel, H. A. Kramers, and L. Brillouin independently published papers proposing an approximation theory (later called the WKB model), which dealt with the difficulty in solving the Schrodinger's equation with an electric potential. In 1930, when Léon Brillouin was a professor at the Sorbonne in France, he found that the electron energy $E(\vec{k})$ had a discontinuous gap when the plane wavevector $\vec{k}$ of an electron crossed the orthogonal plane of a reciprocal vector $\vec{G}$. To provide a better description of energy bands, Brillouin proposed the idea of dividing the reciprocal space of $\vec{k}$ into a series of Brillouin zones.

The energy spectra $E(\vec{k})$ of quasi-particles in solids, such as band electrons, phonons, and magnetrons, have the periodicity of the reciprocal lattice:

$$E(\vec{k}) = E(\vec{k} + \vec{G}) \tag{3.16}$$

therefore, it is only necessary to express $E(\vec{k})$ in a Brillouin zone, which is a PUC of the reciprocal lattice with the central symmetry of $\vec{k} \to -\vec{k}$. In fact, after 1936, when F. Seitz and E. Wigner first studied the symmetry of electron energy bands using point groups, the concept of Brillouin zones gradually matured.

When Brillouin zones are defined, a common origin should be chosen for all reciprocal vectors $\vec{G}_{hkl}(\forall h, k, l)$, and all orthogonal planes of $\{\vec{G}_{hkl}\}$ are called "Bragg planes." The Bragg planes divide reciprocal space into an infinite number of zones. The n'th Brillouin zone is defined as the assembly of zones which are connected to the origin by a route passing through $n-1$ Bragg planes (the route should not pass through a line or a dot neighboring more than two zones to avoid miscounting). The First Brillouin Zone (FBZ) is especially significant in solid state theory since most of the spectra $E(\vec{k})$ are expressed at that zone.

All Brillouin zones have central symmetry $\vec{k} \to -\vec{k}$, which is important for the spectrum description of quasi-particles. The FBZ is

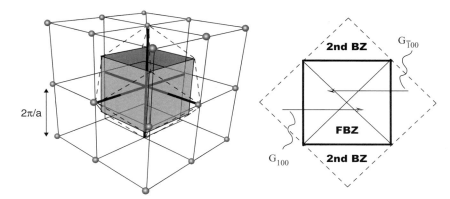

Figure 3.20 First and second Brillouin zone in a SC reciprocal lattice: the arrows with thick lines are the reciprocal position vectors $\vec{G} = \pm\vec{a}_1^*, \pm\vec{a}_2^*, \pm\vec{a}_3^*$; the translucent cube is the FBZ; the pyramids of dashed lines build up the second Brillouin zone

a set of points in reciprocal space that can be reached from the origin without crossing any Bragg plane. A higher order Brillouin zone contains several zones, as shown in Figure 3.20. Any Brillouin zone is a PUC of the reciprocal lattice, therefore the volume of any Brillouin zone equals $\Omega^* = \vec{a}_1^* \cdot (\vec{a}_2^* \times \vec{a}_3^*)$. A higher order Brillouin zone can be transferred into the FBZ by a series of translational symmetry operations $\vec{G}_{hkl}$ in the reciprocal lattice. For example, the second Brillouin zone in Figure 3.20 can be "moved" into the FBZ by six reciprocal vector translations $\vec{G}_{\langle 100 \rangle} = \mp\vec{a}_1^*, \mp\vec{a}_2^*, \mp\vec{a}_3^*$, respectively. The "moved" pieces of the second Brillouin zone will just fill the FBZ, because both of them are reciprocal PUCs.

Geometrically, the FBZ of a crystal is simply the Wigner–Seitz cell of the reciprocal lattice. The BCC and FCC lattices are reciprocal, therefore the FBZ of an FCC crystal is just the W-S cell of the BCC lattice, and the FBZ of a BCC crystal is just the W-S cell of the FCC lattice. The FBZs of the FCC, BCC, and HCP crystals are a truncated octahedron, dodecahedron, and hexagonal prism respectively, as shown in Figure 3.21. The FBZ of an HCP lattice—a hexagonal prism—has exactly the same orientation as the CUC in real space.

A real crystal has a finite size, therefore it is not a "true" Bravais lattice or non-Bravais lattice. In solid state theory, proper boundary conditions have to be assumed to connect the real crystal to an infinite lattice. In 1911, a student of David Hilbert, Herman Weyl, proved that the macroscopic properties were independent of the choice of boundary conditions. In 1912, Max Born and another student of Hilbert, Theodore von Karman, were studying lattice vibration theory, which led them to propose the Born-Karman boundary condition. This stated that the

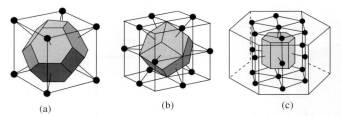

Figure 3.21 Reciprocal lattice (black dots are reciprocal sites) and FBZ: (a) Wigner-Seitz cell of BCC lattice, FBZ of FCC crystal; (b) Wigner-Seitz cell of FCC lattice, FBZ of BCC crystal; (c) FBZ of HCP crystal, the most outside light hexagonal prism shows the orientation of CUC in real space

wave in a solid had periodicity at the boundary of the real crystal, from which an expression for the wavevector $\vec{k}$ could be obtained. The Born-Karman boundary condition is now widely used in all areas of solid state physics. It should be noted that, in a nano-sized crystal with only tens or hundreds of atoms, the Born-Karman boundary condition might not be an appropriate choice, and further studies are needed.

Here a crystal with a set of primitive vectors $(\vec{a}_1, \vec{a}_2, \vec{a}_3)$ is used to examine the Born-Karman condition in more detail. We will assume that the whole crystal is a parallelepiped spanned by $(L_1\vec{a}_1, L_2\vec{a}_2, L_3\vec{a}_3)$, where L_1, L_2, L_3 are the number of PUCs along the three directions $(\vec{a}_1, \vec{a}_2, \vec{a}_3)$ respectively. $N_L = L_1 \times L_2 \times L_3$ is the total number of primitive unit cells in the whole crystal. If the wavefunction of a quasi-particle in a solid satisfies the Born-Karman condition

$$\Psi(\vec{r}) = \Psi(\vec{r} + L_j\vec{a}_j), \quad j = 1, 2, 3, \tag{3.17}$$

it implies that any wavevector in reciprocal space meets the condition $\exp(i\vec{k} \cdot L_j\vec{a}_j) = 1$ at $j = 1, 2, 3$ dimension. The expression $\vec{k}$ under the Born-Karman condition can thus be derived using the orthogonal relationship between $\vec{a}_i$ and $\vec{a}_j^*$ in Eq. (3.13):

$$\vec{k} = \frac{l_1}{L_1}\vec{a}_1^* + \frac{l_2}{L_2}\vec{a}_2^* + \frac{l_3}{L_3}\vec{a}_3^* \tag{3.18}$$

where l_1, l_2, l_3 are arbitrary integers. In other words, the wavevector $\vec{k}$ in a finite-sized crystal is discretized in the reciprocal space with the minimum unit spanned by $(\vec{a}_1^*/L_1, \vec{a}_2^*/L_2, \vec{a}_3^*/L_3)$. The discretization volume Ω^*/N_L of $\vec{k}$ is tiny compared with the volume of the PUC, because L_1, L_2, L_3 are large integers; therefore, $\vec{k}$ can be viewed as quasi-continuous in the reciprocal space.

In the FBZ, $\vec{k}$ should stay within the range of $(-\vec{a}_j^*/2, \vec{a}_j^*/2]$ ($j = 1, 2, 3$) along the three reciprocal primitive vectors. Therefore, l_1, l_2, l_3 in Eq. (3.18) should be limited to $l_j \in (-L_j/2, L_j/2]$ ($j = 1, 2, 3$).

The total number of $\vec{k}$ in the FBZ is simply $L_1 \times L_2 \times L_3$, which equals the total number of PUCs N_L in the real crystal. This is an important conclusion.

In Chapter Four, lattice vibration is discussed. In a branch of the spectrum, there are N_L types of phonon corresponding to N_L possible values of $\vec{k}$ in an FBZ. In Chapter Five, electron band theory is introduced. In an energy band, there are $2N_L$ electron quantum states corresponding to N_L possible values of $\vec{k}$ in an FBZ, and two possible values $\pm \hbar/2$ of electron spins. These are important results derived from the Born-Karman boundary condition.

3.5 Measurement of Crystal Structure by Diffraction

In former sections, mathematical theories of crystal structure were introduced. Physics is an experimental science, and the measurement of crystal structure is necessary to prove the correctness of the structural theory. Historically, various theoreticians and experimentalists in Germany and England contributed to the development of the diffraction method for the measurement of crystal structure.

In 1902, Max von Laue went to the University of Berlin to work under Professor Max Planck. There he attended lectures by O. Lummer on interference spectroscopy, the influence of which was shown in von Laue's dissertation on interference phenomena in plane-parallel plates. After he obtained his doctorate from Berlin in 1903, von Laue spent two years at the University of Gottingen. In 1905, he was offered the post of assistant to Max Planck at the Institute for Theoretical Physics in Berlin. In 1909 he went to the University of Munich, where he lectured on optics, thermodynamics, and the theory of relativity.

Max von Laue's best known work, for which he received the Nobel Prize in Physics in 1914, was his discovery of the X-ray diffraction of crystals at the University of Munich. This arose from discussions about problems relating to light diffraction through a periodic arrangement of particles. Von Laue then hypothesized that much shorter electromagnetic rays, such as X-rays, might also cause diffraction phenomena in crystals. Some of his colleagues, such as A. Sommerfeld and W. Wien, disagreed. However, W. Friedrich, one of Sommerfeld's assistants, and P. Knipping, after some intial failures, eventually found the X-ray crystal diffraction patterns in Figure 3.22(a). Von Laue worked out the mathematical formulation, and the Laue formula was published in 1912. This X-ray diffraction experiment established both the electromagnetic wave nature of X-rays and the periodic lattice structure of crystals. Laue's discovery also paved the way for later work by the Braggs.

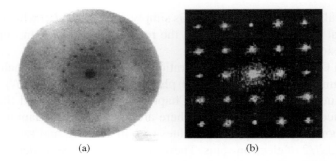

Figure 3.22 Early X-ray diffraction pictures: (a) by Laue; (b) by Debye

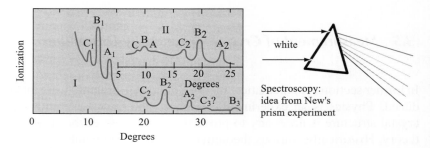

Figure 3.23 Early X-ray spectroscopy (based on Bragg and Bragg, 1913)

William Henry Bragg was born at Westward, England in 1862. He overcame childhood hardship, and went to Cambridge to study mathematics on a scholarship in 1881. In 1885, he studied physics at the Cavendish Laboratory, and at the end of that year, he was elected to the Professorship of Mathematics and Physics at the University of Adelaide, South Australia. He worked in Australia for 24 years, and married Gwendoline Todd, the daughter of Sir Charles Todd, F.R.S. Their son, William Lawrence Bragg, was born in 1890, and received his early education at St. Peter's College, Adelaide.

In 1909, William Henry Bragg went back to England, where he became Cavendish Professor of Physics at Leeds. That same year, William Lawrence Bragg entered Trinity College, Cambridge. In 1912, Lawrence began working under Prof. J. J. Thomson. That autumn, Lawrence formulated his own theoretical explanation of the von Laue X-ray diffraction phenomenon—Bragg's law—and published his first paper on the subject in the Proceedings of the Cambridge Philosophical Society in November.

In January 1913, William Henry Bragg built the first X-ray spectrograph and measured the X-ray line spectra of various elements. In 1913, in accordance with W. Barlow's hypothesis on sodium chloride structure and Bragg's law, they accurately measured the lattice constant of NaCl and the wavelength of X-rays using the X-ray spectrograph. The work of the Braggs in 1913–1914 founded a new branch of science of greatest importance and significance: the X-ray diffraction of crystals.

If the fundamental discovery of the wave nature of X-rays, evidenced by their diffraction in crystals, came from von Laue and his collaborators, it is equally true that the development of the use of X-rays to systematically reveal the structure of crystals was entirely due to the Braggs. This was recognized by the award of a Nobel Prize jointly to father and son in 1915.

William Henry Bragg was the president of Royal Society from 1935 to 1940; William Lawrence Bragg held the same position from 1954 to 1966. At Cambridge, students and visiting scholars from various countries used X-ray spectroscopy to analyze the structure of crystals in condensed matter, including complicated crystal structures with more than ten parameters; phase transition and defect structure in alloys; and biological structures such as hemoglobin, proteins, and muscles. Under William Lawrence Bragg's leadership, 12 scientists won Nobel Prizes in the field of structural biology.

Max von Laue, William Henry Bragg, and William Lawrence Bragg all lived to about 80 years of age, and all three made important contributions to physics and education.

3.5.1 X-ray, Electron, and Neutron Diffraction

Since 1913, X-ray diffraction has become the most commonly used method for analyzing crystal structure. However, there are two other diffraction methods which are also very important in solid state physics: electron microscopy and neutron diffraction. All three methods have many common characteristics in terms of their physical principles. In this sub-section, the scientific origins of each of the methods are introduced, followed by an analysis of their advantages and disadvantages.

The discovery of X-rays in 1895 was the beginning of a revolutionary change in our understanding of the physical world. While Dr. Rector W. K. Roentgen was working at the University of Wurzburg, he noticed that the barium platinum cyanide (BaPtCN) screen was fluorescing as he was generating cathode rays (electron beams) in glass vacuum tube. He subsequently brought his wife, Bertha Roentgen, into the laboratory and took an X-ray picture of her hand using a sodium barium cyanide (NaBaCN) developer, as shown in Figure 3.24(b). He called the unknown radiation "X-rays" because "X" usually stands for an unknown quantity in mathematics. In 1901, Dr. Roentgen was awarded the Nobel Prize in Physics.

The industrial X-ray vacuum tube was invented by William Coolidge in 1913, and it was one of the most important events in the development of radiology. A diagram of the first X-ray vacuum tube mass-produced by Siemens AG is shown in Figure 3.24(a). There is so

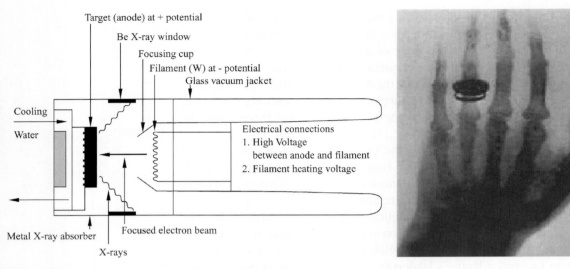

Figure 3.24 Structure of X-ray vacuum tubes and an X-ray picture of Bertha Roentgen

little gas inside the Coolidge X-ray tube that the gas will not interfere with the production of X-rays. The electrons emitted by the heated filament reach the metal target at high speed, and the X-ray spectrum in Figure 3.26(b) is produced.

In 1905, Albert Einstein proposed the hypothesis of light quanta—later called "photons"—to explain photoemission phenomena. This achievement resulted in Einstein being awarded his Nobel Prize in 1921. Photon energy is defined by Planck's formula for the electromagnetic wave energy quantum $\varepsilon = h\nu$. Einstein believed that X-rays were electromagnetic waves with short wavelengths and a strongly particulate nature. Photoemission and X-ray generation are invertible processes and can be expressed by Einstein's photoemission formula as

$$h\nu = E_{\text{out}} + W - E_{\text{i}}; \quad E_{\text{in}} = eV = h\nu + E_{\text{out}} \tag{3.19}$$

In the photoemission process, the energy of the incident photon $h\nu$ equals the energy of the emitted electron E_{out} plus the work function of the metal surface W minus the initial electron energy E_{i} in a solid; in the inverse process, i.e., X-ray generation, the energy of the incident photon E_{in} (controlled by the acceleration voltage V) equals the energy of the X-ray photon $h\nu$ plus the energy of the emitted electron E_{out}.

The X-ray spectrum was defined by two phenomena: (1) the continuous spectra from bremsstrahlung (radiation from an accelerated charge) and (2) the line spectra from characteristic radiation with respect to the quantum transition of inner-shell electrons in metal atoms. The continuous X-ray spectrum has the minimum wavelength $\lambda_{\min}$, which corresponds to the extreme case $E_{out} = 0$:

$$eV = h\nu_{max} \rightarrow \lambda_{min} = \frac{12.4\text{Å}}{V/\text{kVolts}} \qquad (3.20)$$

The atomic distance in a crystal is around 1Å; therefore, a characteristic X-ray with a wavelength of around 1Å is suitable for diffraction. The acceleration voltage V should be high enough to assure the appearance of the chosen wavelength.

In 1913, Henry Moseley, who worked in Rutherford's group at the University of Manchester, heard about the X-ray diffraction method developed by the Braggs. He decided to study the X-ray spectrum using this method. Moseley went to the University of Leeds and learnt how to measure X-ray wavelengths using the crystal diffraction method. Soon afterwards, he found the K_α, K_β lines shown in Figure 3.25(b), as well as the more complex L lines. In the autumn of 1913, Moseley concluded the formula for the K_α and L_α lines based on the X-ray spectrum of ten elements (from calcium to zinc):

$$\lambda_{K_\alpha}^{-1} = R_H \left(\frac{1}{1^2} - \frac{1}{2^2} \right)(Z-1)^2; \quad \lambda_{L_\alpha}^{-1} = R_H \left(\frac{1}{2^2} - \frac{1}{3^2} \right)(Z-7.4)^2. \qquad (3.21)$$

where Z is the atomic number. The Moseley formula fits with experiments so well that the Bohr model is still qualitatively valid for inner

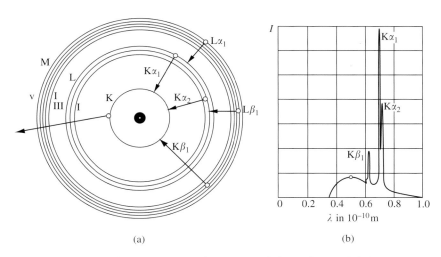

Figure 3.25 (a) Characteristic radiations and the Bohr model (b) X-ray spectrum of an X-ray tube with a molybdenum target; the intensity ratios of the indicating lines are $K_{\alpha 1} : K_{\alpha 2} : K_{\beta 1} = 10:5:2$

Table 3.10 Characteristic wavelength of X-ray line spectra w.r.t. common targets

Target	V(kV)	$K_{\alpha 1}:\lambda$(Å)	$K_{\alpha 2}:\lambda$(Å)	$K_{\beta 1}:\lambda$(Å)	$K_\alpha:\lambda_{\text{Moseley}}$(Å)
Mo	20	0.70926	0.71354	0.63223	0.72319
Cu	9	1.54050	1.54434	1.39217	1.55061
Co	7.7	1.78890	1.79279	1.62073	1.79834
Fe	7.1	1.93597	1.93991	1.75654	1.94509

energy levels in multi-electron atoms. More importantly, it proved that the atomic number in the periodic table is just the total positive charge in the nucleus. The characteristic X-ray wavelengths of four common metal targets, together with an estimation λ_{Moseley} calculated using Moseley's formula, are listed in Table 3.10.

The principle of wave-particle duality was extended to all fundamental particles by de Broglie in 1924, which fully opened the door to Schrodinger's quantum mechanics. Prince Louis-Victor de Broglie's first degree was in history in 1910; then, as his liking for science prevailed, he obtained a science degree in 1913. In 1924, at the Faculty of Sciences at Paris University, he delivered a one-and-a-half page thesis on quantum theory, in which he proposed the hypothesis that a fundamental particle was intrinsically a wave, which was the inverse of A. Einstein's hypothesis that an electromagnetic wave was intrinsically a photon. The concept of the "de Broglie matter wave" was highly praised by A. Einstein. In 1929, de Broglie won the Nobel Prize in Physics for his discovery of the wave nature of electrons.

De Broglie's wave-particle duality is simply a restatement of Einstein's photoemission formula. A free fundamental particle with definite energy ε and momentum $\vec{p}$, such as the electron, photon, or nucleon (fundamental at that time), can be described by a plane wavefunction with a fixed frequency ν and wavelength λ. It follows the de Broglie relationship as:

$$E = h\nu = \hbar\omega; \quad p = h/\lambda = \hbar k \tag{3.22}$$

The matter wave is also called the de Broglie wave with de Broglie wavelength λ.

The de Broglie wave theory of electrons was first proved by C. J. Davisson. Davisson was born in Illinois, U.S.A., and in 1902 he was granted a scholarship by the University of Chicago due to his proficiency in mathematics and physics. In 1911 he obtained his Ph.D. at Princeton, after which he worked at the Carnegie Institute of Technology. In 1924, he and his collaborator, Dr. C. H. Kunsman, accidentally discovered low

3.5 Measurement of Crystal Structure by Diffraction

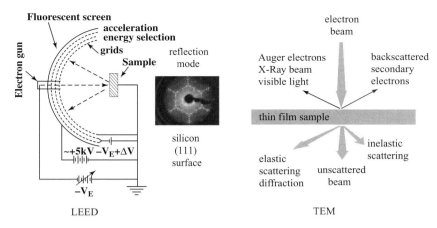

Figure 3.26 Schematics of low energy electron diffraction (LEED) and tunnelling electron microscope (TEM) apparatus

energy electron diffraction (LEED) while they were studying secondary electron emission from nickel crystals using a 100 eV electron incident beam. In 1927, Davisson and another collaborator, Dr. L. H. Germer, found that the maxima of electron diffractions occured at the positions predicted by Bragg's law of X-ray diffraction.

In 1928, G. P. Thomson and his student, A. Reid, tried to prove de Broglie's concept using a new electron diffraction method. A narrow electron beam was transmitted through a polycrystal thin film with a μm-order thickness. The scattered beam was received on a photographic plate normal to the beam which, when developed, showed a diffraction pattern of rings, which corresponded well with the rings obtained from similar experiments with X-rays. The de Broglie wavelength of the electron matter wave is simply h/mv:

$$\lambda = \frac{h}{p} = \frac{2\pi}{\sqrt{2m_e E/\hbar^2}} = \frac{2\pi a_B}{\sqrt{E/13.6\text{eV}}} = \frac{12.25\text{Å}}{\sqrt{E/\text{eV}}} \quad (3.23)$$

If the electron beam has energy of 20 to 60,000 volts, the de Broglie wavelength will be 2.7–0.05Å. G. P. Thomson's experiment agreed with the de Broglie's relationship within an accuracy of 1%. C. J. Davisson and G. P. Thomson shared the 1937 Nobel Prize in Physics for their work.

In 1931, Dr. Max Knoll and his student, Ernst Ruska, from the Technical University of Berlin, created the first Transmission Electron Microscope (TEM). The principle they used was the same as that in G. P. Thomson's experiment, where an electron beam was used to penetrate a thin film. The resolution obtained using the TEM technique quickly surpassed that of optical microscopy, and in 1937 the first commercial TEM was produced by the Siemens-Halske Company.

Figure 3.27 Sir James Chadwick's tool box and his neutron detector

The earliest known work describing the concept of a Scanning Electron Microscope (SEM) was also from Max Knoll in 1935. Subsequently, in 1938, M. von Ardenne constructed a scanning transmission electron microscope (STEM) by adding scan coils to a TEM, and his first STEM micrograph was of a zinc oxide (ZnO) crystal. In 1942, the first SEM with a 50 nm resolution was used to examine the surface of a solid specimen by Valdimir Kosma Zworykin and his group, who were working for RCA in the United States. In 1986, Ernst Ruska shared the Nobel Prize for his work in TEM with Gerd Binnig and Heinrich Rohrer, inventors of the scanning tunneling microscope (STM).

The structure of the nucleus had not been discovered when de Broglie first proposed his matter wave theory. In 1932, Sir James Chadwick made a fundamental breakthrough in the realm of nuclear science: he proved the existence of neutrons—elementary particles devoid of any electrical charge. Chadwick graduated from the University of Manchester in 1911 and spent the next two years under Professor (later Lord) Rutherford as a master student. During the First World War, Chadwick was in Germany as an intern student, and he was detained as a civilian prisoner of war. He was allowed to read books and talk to other physicists, but he could not conduct experiments. In 1919, he went back to England and joined Rutherford's group in Cambridge to study the properties and structures of atomic nuclei. Rutherford was then the leader of the Cavendish Laboratory, and steered numerous Nobel Prize winners towards their great achievements, including G. P. Thomson and N. Bohr. Rutherford had predicted the existence of the neutron in 1920, but he was not able prove it.

In 1930, German physicists Walther Bothe and Herbert Becker, noticed something odd. When they shot alpha rays at beryllium, it emitted a neutral radiation that could penetrate 200 millimeters of lead. In contrast, it takes less than one millimeter of lead to stop a proton. Bothe and Becker assumed the neutral radiation was high-energy gamma rays (photons). Meanwhile, Marie Curie's daughter, Irene Joliot-Curie, and her husband, Frederic, put a block of paraffin wax in front of beryllium rays. They observed high-speed protons coming from the paraffin. They knew that gamma rays could eject electrons from metals, so they also thought the neutral beryllium rays were gamma rays.

When Chadwick heard about Joliot-Curies' experiment, he realized that the radiation could not be gamma rays. To eject protons at such a high velocity, the beryllium rays would have needed an energy of 50 MeV. Chadwick had another explanation for the beryllium rays. He thought they were neutrons, and he set up an experiment to test his hypothesis. To prove that the particles were indeed neutrons, their mass had to be measured. He could not weigh them directly, so he designed a collision experiment to do the measurement. Firstly, he bombarded boron with alpha particles. Like beryllium, the boron emitted neutral rays. Chadwick then placed a hydrogen target—water—in the path of the rays. When the rays struck the target, protons flew out. Chadwick measured the velocity of the protons. Using the laws of conservation of momentum and energy, he calculated the mass of the neutral particle. It was 1.0067 times the mass of the proton, and so, in 1932, twelve years after Rutherfold's prediction, his assistant James Chadwick had finally proved the existence of the long-sought neutron. In a way, Chadwick opened the door to the development of U-235 fission and the creation of the atomic bomb. For the epoch-making discovery of the neutron, Chadwick was awarded the Nobel Prize in Physics in 1935; F. Joliot and I. Joliot-Curie won the Nobel Prize in Chemistry in the same year; and and in 1954, W. Bothe shared the Nobel Prize with Max Born for his investigations of cosmic rays.

During the Second World War, many great nuclear physicists, including Enrico Fermi and another 20 Nobel laureates, joined the Manhattan Project at the University of Chicago. After the Second World War, most scientists recognized the horror of nuclear war. Instead, the peaceful applications of nuclear technology were promoted, such as

Figure 3.28 An early (1950) neutron diffractometer with flexible wavelength control used by E. O. Wollan and C. G. Shull (standing) at Oak Ridge National Laboratory

nuclear power, nuclear magnetic resonance (NMR), and nuclear diffraction methods. In 1946, Clifford G. Shull started work at the Clinton Laboratories, (now the Oak Ridge National Laboratory), under the director of E. P. Wigner. Shull teamed up with Ernest Wollan, and during the next nine years, they explored ways of using the neutrons produced by nuclear reactors to determine the atomic structure of materials. In Shull's opinion, the most important problem he worked on at the time was to determine the positions of hydrogen atoms in materials. The hydrogen atom is ubiquitous in all biological materials, as well as many other inorganic materials, but it is too light to be detected by X-ray diffraction or electron microscopy. With neutrons, the hydrogen-containing structures could be clearly seen using the elastic scattering process.

In 1950, Bertram N. Brockhouse joined the Chalk River Nuclear Laboratory in Ontario, Canada. In 1951, influenced by Shull and Wollan's work, Brockhouse and his students developed the inelastic scattering 3-axis neutrons spectrometer, which could detect the change of both the direction and energy of neutrons when they collided with atoms. Brockhouse studied how atomic structures in liquids change with time. He also measured energies of phonons (atomic vibrations) and magnons (spin waves), which are discussed further in Chapters Four and Seven. In 1994, Shull and Brockhouse shared the Nobel Prize in Physics for their contributions to neutron diffraction.

Neutron diffraction is based on the wave nature of neutrons. The neutrons in the diffraction process are non-realistic particles with energy E and momentum $\vec{p}$; and they form a neutron de Broglie wave with the wavelength:

$$\lambda = \frac{h}{p} = \frac{2\pi}{\sqrt{2m_n E/\hbar^2}} = \frac{2\pi a_B}{\sqrt{E/13.6\text{eV}}}\sqrt{\frac{m_e}{m_n}} = \frac{0.286\text{Å}}{\sqrt{E/\text{eV}}} \quad (3.24)$$

where m_n is the neutron mass, 1836 times the electron mass m_e. A neutron with the energy $E = 0.082\,eV$ corresponds to the de Broglie wave with the wavelength $\lambda = 1\text{Å}$ at $T_n = E/k_B = 949K = 676°C$. This neutron is called a "hot neutron" and it is suitable for crystal diffraction.

The mechanism of neutron diffraction is the strong interaction between neutrons and nuclei, which is different from the electromagnetic interaction mechanism in X-ray and electron diffraction. A list comparing the advantages and disadvantages of the three diffraction methods is given in Table 3.11.

Neutron diffraction can detect light elements because the interaction is a short-distance one, and is still quite strong between a single proton and a neutron. The scattering cross-section of a neutron is slightly dependent on the number of hadrons in a nucleus; therefore, the isotopes can be detected. The energy of a hot neutron is still quite small. The neutron also has a magnetic moment $\mu_N = 5.44 \times 10^{-4} \mu_B$; therefore, the

3.5 Measurement of Crystal Structure by Diffraction

Table 3.11 Advantages and disadvantages of the three diffraction methods

Diffraction	Advantages	Disadvantages
X-ray	simplest equipment; broad application	not suitable for very thin film or quasi-particles in solids
Electron	real and reciprocal image; thin film and surfaces; many different microscopes	hard to detect light elements; hard to detect some quasi-particles or collective motions in solids
Neutron	good for detecting light elements; can distinguish isotopes; can detect quasi-particles	large equipment (nuclear reactor); neutron beams hard to detect; low luminosity, long exposure time

inelastic scattering phenomena related to phonons or magnetons in a solid can easily be observed.

For the same de Broglie wavelength, the energy of the photon, electron, and neutron approximately follow the ratio $10^6 : 10^3 : 1$, because the photon is a pure relativistic particle with zero mass, and the electron mass is three times smaller than the neutron mass. According to wave-particle duality, the diffraction process in wave terms is scattering in particle terms. The photon energy with respect to an X-ray with the wavelength $\lambda = 1\text{Å}$ is quite large, therefore elastic scattering is more likely to occur. The electron energy in different electron microscopes varies greatly, therefore both elastic and inelastic electron diffraction are of interest.

The electromagnetic interaction is stronger in electron diffraction (electron-electron scattering) than in X-ray diffraction (photon-electron scattering). An electron beam can only penetrate 100–1000 nm into solids, whereas X-rays can penetrate several microns. Therefore, electron diffraction is the better tool for very thin films, whereas X-ray diffraction is good for determining the bulk property of a solid sample. The tunneling electron microscope is a powerful tool as it can see the microscopic image, the diffraction pattern in the reciprocal space, and the chemical composition of the same area in a sample.

3.5.2 Diffraction Theory

The elastic scattering theories of X-ray, electron, and neutron diffraction methods are similar except for the details of their interactions, and they are discussed together in this sub-section.

Diffraction theory starts with Bragg's law, published in 1912. Earlier that year, based on the Friedrich-Knipping-Laue effect of X-ray

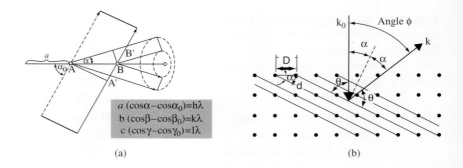

Figure 3.29 (a) Laue's formula and his diffraction theory; (b) Bragg's law: θ is the Bragg angle, $\vec{k}_0$ & $\vec{k}$ are the incoming & outgoing wave vector, and d is the lattice plane separation distance

diffraction, Max von Laue had constructed a fairly complicated diffraction theory. Bragg's Law greatly simplified von Laue's description of X-ray interference to the form:

$$2d \sin\theta = n\lambda; \quad D \sin\phi = D \sin 2\theta = n\lambda \qquad (3.25)$$

The Braggs used X-rays in reflection geometry with respect to crystal planes to analyze X-ray diffraction spectra. Based on the diffraction peaks of a fixed line in the X-ray line spectra, the Braggs identified the FCC structure of the sodium chloride crystal, which corrected von Laue's misidentification of it as a simple cube. The lattice constant of sodium chloride and the wavelength of the line could also be determined.

In reality, a solid consists of a huge number of atoms, and one atom can have many electrons moving around the nucleus. In the following derivation, a solid is treated as a 3D grid with "moving" electrons, and we can still prove that Bragg's law is valid, without using simple reflection geometry.

In the elastic scattering process, the incoming and outgoing particles in the diffraction beam have the same energy (in the particle picture) and the same wave number $k_0 = k = 2\pi/\lambda$ (in the wave picture). The wavevector difference $\vec{s} = \vec{k} - \vec{k}_0$ is a significant quantity in diffraction theory and it equals the reciprocal vector $\vec{G}$ with respect to a set of lattice planes at a diffraction peak.

The incoming diffraction beam with a wavevector $\vec{k}_0 = \vec{p}/\hbar$ and an angular frequency $\omega = \varepsilon/\hbar$ can be described by the de Broglie wave in the plane form:

$$u(\vec{r}, t) = A \exp[i(\vec{k}_0 \cdot \vec{r} - \omega t)] \qquad (3.26)$$

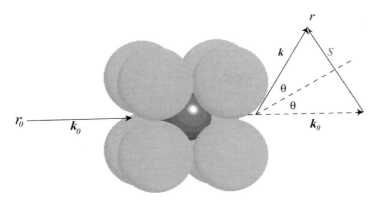

Figure 3.30 Schematics of multi-electron scattering or the diffraction process

The single-particle (electron or nucleon) scattering wave is the building block of the diffraction beam. After the scattering on the l'th particle in a solid, the diffraction part of the incident wave (except the unscattered wave) has the spherical wave form:

$$u'_l(\vec{r}, t) = f \frac{A}{|\vec{r} - \vec{r}_l|} \exp[i(\delta_l^0 + k|\vec{r} - \vec{r}_l| - \omega t)]; \quad \delta_l^0 = \vec{k}_0 \cdot (\vec{r}_l - \vec{r}_0)$$

(3.27)

where δ_l^0 is the phase of the l'th particle in a solid with respect to the incoming plane wave from the "infinitely far" position $\vec{r}_0$; f is the scattering length, which is related to intrinsic properties of the electromagnetic or strong interaction; and $D_l = |\vec{r} - \vec{r}_l|$ is the distance from the l'th particle to the "infinitely far" observation point $\vec{r}$.

Before summing up the scattered waves from all particles in a solid, a proper expression for the phase difference can be found by analyzing the bi-particle scattering wave:

$$u'_1 + u'_2 \cong f \frac{A}{D} e^{i(\delta_1^0 + kD_1 - \omega t)}[1 + e^{i\delta_{21}}]$$

(3.28)

$$\delta_{21} = (\vec{k}_0 \cdot (\vec{r}_2 - \vec{r}_0) + k|\vec{r} - \vec{r}_2|) - (\vec{k}_0 \cdot (\vec{r}_1 - \vec{r}_0) + k|\vec{r} - \vec{r}_1|)$$
$$\cong (\vec{k}_0 - \vec{k}) \cdot (\vec{r}_2 - \vec{r}_1) = -\vec{s} \cdot \vec{r}_2 + \vec{s} \cdot \vec{r}_1$$

(3.29)

where approximations $\frac{A}{|\vec{r} - \vec{r}_l|} \cong \frac{A}{D}$ and $k|\vec{r} - \vec{r}_l| \cong \vec{k} \cdot (\vec{r} - \vec{r}_l)$, with $l = 1, 2$, are used in the former derivation. D is the distance from the part of the sample interacting with the beam to the spherical screen receiving the diffraction pattern.

Based on the bi-particle scattering wave, the total diffraction of the de Broglie wave from all electrons or hadrons in a solid can be written as:

$$u'(\vec{r}, t) = \sum_n u'_n(\vec{r}, t) \cong f \frac{A}{D} e^{i(\delta_1 + kD - \omega t)} \sum_n \exp(-i\vec{s} \cdot \vec{r}_n) \quad (3.30)$$

where the constant phase $\delta_1 = \delta_1^0 + \vec{s} \cdot \vec{r}_1$. Eq. (3.30) has a neatly symmetrical form.

On the spherical screen receiving the diffraction signals, the total diffraction wave and the diffraction intensity of a solid sample are

$$u'(\vec{r}, t) = f_{\text{cr}} \frac{A}{D} e^{i(kD - \omega t)}; \quad I = |f_{\text{cr}}|^2 = \left| f \sum_n e^{-i\vec{s} \cdot \vec{r}_n} \right|^2 \quad (3.31)$$

where the sum covers all particles in a solid (electrons for X-ray or electron diffraction, or hadrons for neutron diffraction). The former Eq. (3.31) can be used for the analysis of all solids since the former derivation is irrelevant to solid structures.

In a crystal with $N = N_L n_a$ atoms, the crystal scattering factor f_{cr} can be calculated by decomposing position vector $\vec{r}_n = \vec{R}_l + \vec{\delta}_j + \vec{r}$ into three levels: CUC (sum over $l = 1, \ldots, N_L$), atoms inside a CUC (sum over $j = 1, \ldots, n_a$), and sub-atomic particles interacting with the diffraction beam (integral over $\vec{r}$):

$$f_{\text{cr}} = \sum_l \sum_j f_{\text{a},j} \exp(-i\vec{s} \cdot (\vec{R}_l + \vec{\delta}_j)) = FS \quad (3.32)$$

$$S = \sum_l \exp(-i\vec{s} \cdot \vec{R}_l) \quad (3.33)$$

$$F = \sum_j f_{\text{a},j} \exp(-i\vec{s} \cdot \vec{\delta}_j) \quad (3.34)$$

$$f_{\text{a},j} = f \iiint_j d^3\vec{r} |\psi(\vec{r})|^2 e^{-i\vec{s} \cdot \vec{r}} = 4\pi f \int_0^\infty dr\, r^2 \rho(r) \frac{\sin sr}{sr} \quad (3.35)$$

S is the lattice structure factor, which reflects the symmetry of a crystal system; F is the geometrical scattering factor, which stands for different Bravais or non-Bravais lattices in the crystal system; and $f_{\text{a},j}$ is the atomic scattering factor, which shows the strength of electromagnetic or strong interactions. In the X-ray or electron diffraction process, $\rho(r) = |\psi(\vec{r})|^2$ is the total probability density of electrons in an atom, the atomic scattering factor $f_{\text{a},j}$ is slightly dependent on the Bragg angle θ via the quantity $s = 2k_0 \sin\theta$, and $f_{\text{a},j}$ maximizes as fZ_j at the unscattered angle $\theta = 0$, where Z_j is the total number of electrons in the j'th atom in a CUC. In neutron diffraction, the calculation of $f_{\text{a},j}$ of short-range

Table 3.12 Conditions for diffraction points in the elastic-scattering process

Factors	Diffraction points appear under the condition
$S = \sum_l \exp(-i\vec{s} \cdot \vec{R}_l)$	$\vec{s} = \vec{G}'_{hkl} \Rightarrow 2k_0 \sin\theta = G'_{hkl} = 2\pi/d_{hkl}$
$F = \sum_j f_{a,j} \exp\left(-i\vec{s} \cdot \vec{\delta}_j\right)$	$\sum_j f_{a,j} \exp\left(-i2\pi(hu_1^j + ku_2^j + lu_3^j)\right) \neq 0$

strong interactions is more complicated, and the result is that $f_{a,j}$ would only vary slightly versus the number of hadrons in a nucleus. This is the reason why both X-ray and electron diffraction are better for detecting heavier atoms, but neutron diffraction can "see" light atoms.

The diffraction pattern on the receiving screen is determined by the distribution of the intensity $I = |f_{cr}|^2 = |FS|^2$. When both S and F are nonzero, the single crystal (elastic scattering) diffraction point will appear, as listed in Table 3.12.

In Table 3.12, $\vec{R}_l = l_1\vec{a}'_1 + l_2\vec{a}'_2 + l_3\vec{a}'_3$ is the CUC position vector. By following Eq. (3.18) for periodic boundary conditions, the wavevector difference $\vec{s} = \vec{k} - \vec{k}_0 = s_1\vec{a}'^*_1 + s_2\vec{a}'^*_2 + s_3\vec{a}'^*_3$ is quasi-continuous in the reciprocal space: $s_\alpha = h_\alpha/L_\alpha$ (h_α the integer; L_α the total number of CUC in the α-direction). The lattice structure factor S can be calculated by the sum rule:

$$S = \sum_l e^{-i\vec{s}\cdot\vec{R}_l} = \sum_{l_1=1}^{L_1} e^{-i2\pi s_1 l_1} \sum_{l_2=1}^{L_2} e^{-i2\pi s_2 l_2} \sum_{l_3=1}^{L_3} e^{-i2\pi s_3 l_3} = N_L \delta_{\vec{s},\vec{G}'} \quad (3.36)$$

The geometric progression $\sum e^{-is_\alpha l_\alpha} = (1 - e^{-is_\alpha L_\alpha})/(1 - e^{-is_\alpha}) = 0$ because $s_\alpha L_\alpha$ is an integer. Thus only when $s_\alpha = m_\alpha$ (m_α the integer) or $\vec{s} = \vec{k} - \vec{k}_0$ equals a CUC reciprocal vector $\vec{G}'_{hkl} = h\vec{a}'^*_1 + k\vec{a}'^*_2 + l\vec{a}'^*_3$, S is nonzero and just equals N_L.

Among all possible positions $\vec{k} = \vec{k}_0 - \vec{G}'_{hkl}$ of diffraction points, the appearance of diffraction points has to be further determined by the kinds and positions of atoms at $\vec{\delta}_j = u_1^j\vec{a}'_1 + u_2^j\vec{a}'_2 + u_3^j\vec{a}'_3$ in a CUC, as represented by the $F_{hkl} \neq 0$ condition in Table 3.12. All possible (u_1^j, u_2^j, u_3^j) in a Bravais lattice are listed in Table 3.3. The geometrical scattering factors for BCC and FCC lattices are

$$\text{BCC } F_{hkl} = f_a \sum_{j=1}^{2} e^{-i\pi(h+k+l)(j-1)} \neq 0 \Rightarrow h+k+l = \text{even integer};$$

(3.37)

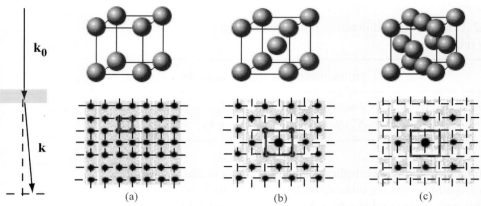

Figure 3.31 When $k_0 = k \gg a^*$, lattice structure in real space (upper) and diffraction points in reciprocal space (lower), conditions of diffraction points are $\vec{s} = h\vec{a}_1'^* + k\vec{a}_2'^* + l\vec{a}_3'^*$, on the receiving plane $l \simeq 0$: (a) SC, $F_{hkl} = 1$ for $\forall h, k$; (b) BCC, $F_{hkl} = 1$ for $h + k = 2m$; (c) FCC, $F_{hkl} = 1$ for $h = 2m, k = 2n$

$$\text{FCC } F_{hkl} = f_a \sum_{j=1}^{4} e^{-i\pi(hu_1^j + ku_2^j + lu_3^j)} \neq 0 \Rightarrow h, k, l \text{ all even or all odd}$$

(3.38)

The former condition is stated with respect to the simple cubic reciprocal position vector $\vec{G}'_{hkl} = \frac{2\pi}{a}(h\hat{e}_x + k\hat{e}_y + l\hat{e}_z)$. Therefore the reciprocal lattice constant of a BCC/FCC crystal is $a^* = 2(2\pi/a)$, as illustrated in Figure 3.31.

The CUC reciprocal lattice vector $\vec{G}'_{hkl}$ is in the normal direction of the lattice planes (hkl) with respect to the CUC (which is the conventional labeling for lattice planes), because

$$\vec{G}'_{hkl} \cdot \left(\frac{\vec{a}'_1}{h} - \frac{\vec{a}'_2}{k}\right) = 0, \quad \vec{G}'_{hkl} \cdot \left(\frac{\vec{a}'_2}{k} - \frac{\vec{a}'_3}{l}\right) = 0,$$
$$\vec{G}'_{hkl} \cdot \left(\frac{\vec{a}'_3}{l} - \frac{\vec{a}'_1}{h}\right) = 0.$$

(3.39)

The distance between two neighboring lattice planes equals $d_{hkl} = (\vec{a}'_1/h) \cdot \vec{G}'_{hkl}/G'_{hkl} = 2\pi/G'_{hkl}$ for the (hkl) crystal planes. Therefore, by the $\vec{s} = \vec{G}'_{hkl}$ condition in the multi-particle scattering theory, Bragg's law is derived again with the form $2d_{hkl} \sin\theta_{hkl} = \lambda$! It should be noted that the Bragg angles with respect to $F_{hkl} = 0$ cannot appear. The Laue formula can also be derived from $\vec{s} = \vec{G}'_{hkl}$.

The appearance condition of diffraction points can be explained clearly by the Ewald structure, which was formulated by P. P. Ewald

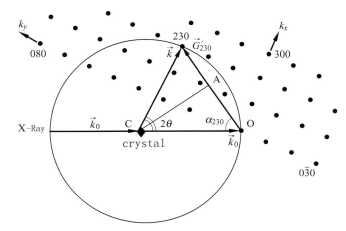

Figure 3.32 Ewald structure and single crystal diffraction

from the University of Munich in 1913. The condition $\vec{s} = \vec{k} - \vec{k}_0 = \vec{G}'_{hkl}$ of diffraction point appearance can be visualized by the Ewald structure or Ewald ball. The radius of the Ewald ball is $k_0 = 2\pi/\lambda$. There is a 2θ angle between the outgoing wavevector $\vec{k}$ and the incoming wavevector $\vec{k}_0$. Most significantly, when the single crystal sample at the center C rotates, the reciprocal lattice originating at the surface O should rotate simultaneously around O. The receiving screen of diffraction is spherical. If a reciprocal lattice site $\vec{G}'_{hkl}$ (with $F_{hkl} \neq 0$) happens to be on the Ewald sphere, a diffraction point will appear.

There are three main X-ray diffraction methods: (1) The Laue method, in which the incoming diffraction wave has a white spectrum, and the sample can be rotated in three dimensions. The symmetry of the crystal structure can be found, but the lattice constant cannot be determined. (2) The rotating crystal method, which was also devised by Laue in 1912. The incident diffraction wave has a fixed wavelength and direction, and the single crystal sample should be rotated three-dimensionally to obtain a complete picture of the crystal structure. (3) The powder method, which was established by the Dutch physicist, Peter Debye, and the Swiss researcher, Paul Scherrer, from the University of Gottingen in 1916. The powder sample or the poly-crystalline sample does not have to be rotated, and the complete crystal structure can be analyzed from the diffraction ring-pattern. The diffraction ring-patterns obtained by the powder method are explained by Table 3.12. A diffraction ring appears when both Bragg's law $2d_{hkl} \sin \theta_{hkl} = \lambda$ ($s = G'_{hkl}$) and the geometrical scattering factor $F_{hkl} \neq 0$ are satisfied.

The pattern of diffraction points measured by these three methods will obey the symmetry group of the respective crystal, as listed in Table 3.4; for example, the diffraction points of an SC/FCC/BCC lattice

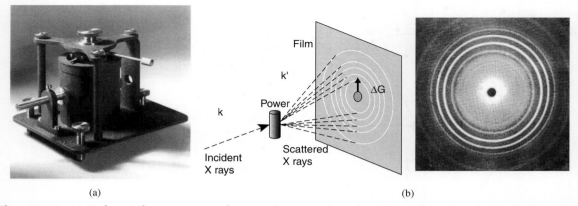

Figure 3.33 (a) Debye-Scherrer camera for powder XRD; (b) Schematics of the powder method and the diffraction ring-pattern

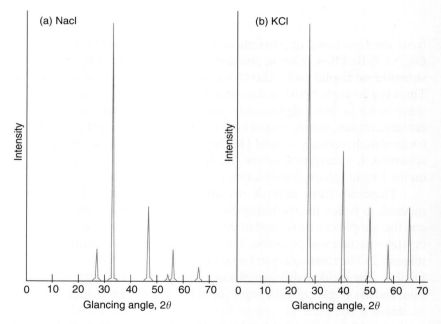

Figure 3.34 Powder X-ray diffraction pattern of NaCl and KCl

must have two, three, or four fold symmetry; the diffraction points of a HEX lattice must have two, three, or six fold symmetry, and so on.

When the de Broglie wavelength of the diffraction beam λ is fixed and comparable to the lattice constant a, i.e., when the Ewald ball radius $\vec{k}_0$ is of the same order as the reciprocal lattice constant a^*, only a small percentage of the diffraction points appear on the spherical receiving screen, as seen in Figure 3.32. Therefore it is hard to obtain complete diffraction information for a crystal by the rotating crystal method, so in this case, the powder method would be more suitable for the crystal structure analysis. When $\lambda \ll a$ (which is true in high energy XRD and TEM), i.e., when $\vec{k}_0 \gg a^*$, the Ewald sphere is most "flat" near point "O", and a large number of inner diffraction points can be obtained. For high energy diffraction equipment, the rotating crystal method is more practical for structural analysis.

3.6 Disordered Solid Structure

Disordered solid structure is an important field in modern solid state physics. There are many types of disordered natural and artificial solids. Some of them, such as glass, have been used by humans for hundreds of years. However the study of the atomic and molecular structures of disordered solids is still a relatively new research area.

The classification of solid disorder must be defined with respect to a reference. The proper reference is obviously an ideal crystal structure. In S. R. Elliott's book, *Physics of Amorphous Materials*, there is a clear classification of solid disorder into four classes, as illustrated in Figure 3.35.

The first class of disorder in solids is topological disorder, which is the atomic arrangement in amorphous materials. Amorphous materials are prepared by rapid quenching of melting liquid, so that the disorder of the atomic arrangement in the liquid is retained and solidified in the amorphous material. There are no translational, rotational, and mirror image symmetries of lattices in amorphous materials. Quartz glass is a typical amorphous solid, and its atomic structure is shown in Figure 3.35(a). Quartz crystal structure is similar to the diamond structure, except that there is an oxygen ion in the middle of each Si-Si bond of the silicon crystal. Quartz glass has a long-range disordered structure, however the short-range atomic structure, or chemical bonding, is quite analogous to quartz crystal, and is a good example of topological disorder.

The second class of disorder in solid is spin disorder, which is illustrated in Figure 3.35(b). The atomic spins still stay on lattice sites, but the spins themselves are randomly oriented. Typical solids with spin disorder are the paramagnets or diamagnets. When exchange interactions among neighboring atoms are stronger than the thermal fluctuations, a disorder-order phase transition of atomic spins occurs, and a paramagnet becomes a ferromagnet.

The third class of disorder in solids is substitutional disorder, which often appears in metallic alloys, as illustrated in Figure 3.35(c). In alloys, the positions of different elements can be substituted. The most important chemical bond in alloys is the metallic bond, and the source of the

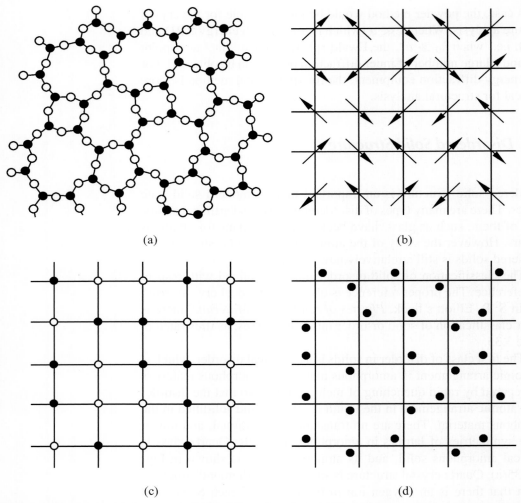

Figure 3.35 Disorder in solids: (a) topological disorder; (b) spin disorder; (c) substitutional disorder; (d) vibrational disorder (based on Elliott, 1990)

metallic bond is free electron gas. If the crystal structure is similar after substitution, the substitution has little effect on the metallic bonds, therefore substitutional disorder in intermetallic compounds is quite possible.

The fourth class of disorder in solids is vibrational disorder, as illustrated in Figure 3.35(d), which occurs in all types of solid. Vibrational disorder is caused by the continuous thermal motion of atoms, and it is obviously more severe with higher temperatures. However, the degree of disorder of atomic vibration is quite small compared to the other three classes of disorder in solids, because the vibration amplitude is tiny compared to the lattice constant.

3.6 Disordered Solid Structure

In reality, one or more classes of disorders occur in solids. The physical properties of solids are highly affected by their disorders, such as the resistance of metals, the susceptibility, the specific heat, and so forth. In this section, the structures of amorphous materials, quasi-crystals, and liquid crystals are discussed briefly.

3.6.1 Amorphous Materials

The structure of an amorphous solid is similar to that of a liquid, in which the atoms are distributed almost randomly. There are different ways of describing solids: a crystal can be described by the primitive vectors and the basis; a polycrystal needs both the description for the crystal phase and information about the grain boundaries; while an amorphous solid can only be described by statistical illustration of the positions of the atoms, as shown in Figure 3.36.

The structure of amorphous materials or liquids can be described by the average number of atoms $\Delta N(R) = n(R)4\pi R^2 \Delta R$ in a spherical shell from R to $R + \Delta R$ around an arbitrary atom. In the atomic density $n(R)$, the singularity $\delta(R)$ with respect to the selected central atom at $R = 0$ has been excluded. If $n(R)$ is zero when $R < 2.2\text{Å}$, then the cutoff $d = 2.2\text{Å}$ is a little larger than the hard sphere atom diameter. $n(R)$ is an oscillating function of R and it approaches a constant n_0 at $R = \infty$. n_0 is just the average density of the whole amorphous solid.

The pair distribution function $g(R) = n(R)/n_0 = \Delta N(R)/(4\pi R^2 \Delta R n_0)$ is the physical quantity characterizing the structure of amorphous materials or liquids. $g(R)$ can be measured using the synchrotron X-ray diffraction method or the high-energy X-ray (HXR) method with a radial diffraction function (RDF) analysis. The analysis of the diffraction mechanism of amorphous materials can start with the

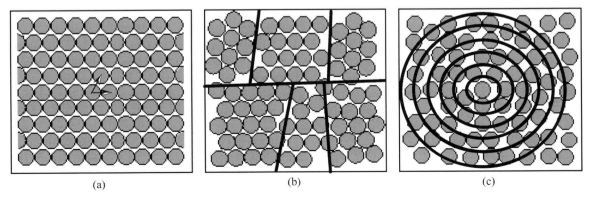

Figure 3.36 Description of solids: (a) single crystal with primitive vectors; (b) polycrystal with grain boundaries; (c) amorphous solid with statistics $\Delta N(R)$

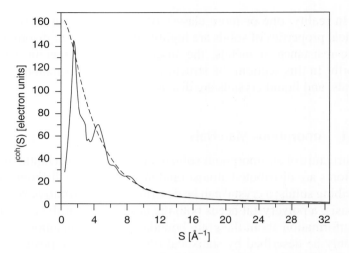

Figure 3.37 Synchrotron XRD or high XRD intensity of amorphous materials (solid line). The energy of the X-ray is 130 keV. The dashed curve shows the isotropic part of the intensity (self-scattering + Compton scattering) (based on Schlenz, Neuefeind, and Rings, 2003)

correlation function of the atomic density $\rho(\vec{r})$:

$$\langle \rho(\vec{r})\rho(\vec{r}')\rangle = n_0^2 g(|\vec{r}-\vec{r}'|) + n_0 \delta(\vec{r}-\vec{r}') \qquad (3.40)$$

which is a better definition for the pair distribution function $g(r)$.

The HXR diffraction intensity of an amorphous material with only one type of atom can be expressed as a function of the incoming-outgoing wave vector difference $\vec{s} = \vec{k} - \vec{k}_0$ following the general Eq. (3.31):

$$I = |f_{\text{lq}}|^2 = \left|\sum_n f_a e^{-i\vec{s}\cdot\vec{r}_n}\right|^2 = f_a^2 \sum_n \sum_j e^{-i\vec{s}\cdot(\vec{r}_n-\vec{r}_j)} \qquad (3.41)$$

In amorphous materials, the position vector of an atom $\vec{R}_l$ is distributed randomly throughout the whole body, and the Born-Karman boundary condition for $\vec{k}$ or $\vec{s}$ in the reciprocal space is no longer valid; therefore it is quite difficult to obtain the sum in Eq. (3.41).

The total amorphous structure factor can be defined as the scaled intensity in an amorphous solid with N atoms:

$$\begin{aligned} S_{\text{lq}}(s) &= \frac{I}{Nf_a^2} = \frac{V}{N}\iiint d^3\vec{R}\langle \rho(\vec{r})\rho(\vec{r}')\rangle e^{-i\vec{s}\cdot\vec{R}} \\ &= 1 + \iiint d^3\vec{R} n_0 g(R) e^{-i\vec{s}\cdot\vec{R}} \end{aligned} \qquad (3.42)$$

where the correlation function $\langle \rho(\vec{r})\rho(\vec{r}')\rangle$ is only a function of $R = |\vec{r} - \vec{r}'|$. The pair distribution function can be obtained by an inverse Fourier transformation:

$$g(R) = \frac{1}{n_0} \iiint \frac{d^3\vec{s}}{(2\pi)^3}[S_{lq}(s) - 1]e^{i\vec{s}\cdot\vec{R}}$$
$$= \frac{1}{2\pi^2 n_0 R} \int_0^\infty ds\, s[S_{lq}(s) - 1]\sin(sR) \qquad (3.43)$$

Based on the measured total structure factor $S_{lq}(s)$ shown in Figure 3.37, the pair distribution function $g(R)$ can finally be found by Eq. (3.43).

In Figure 3.38, the typical pair distribution functions of a gas, liquid, amorphous solid, and crystal are shown. In the gas, $g(R)$ is statistically constant outside the diameter of an atom or a molecule. In the liquid and the amorphous solid, $g(R)$ has similar declining oscillating behavior, but the frequency of the oscillations in the liquid is smaller due to the larger motions of the atoms or molecules. In the crystal, the sharp peaks of $g(R)$ correspond to the first, second, third, etc., nearest neighbors with a long-range order. In summary, the pair distribution function $g(R)$ is a universal function for describing solid structures.

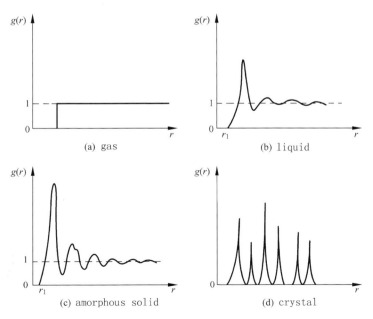

Figure 3.38 Typical pair distribution functions $g(R)$ of (a) a gas, (b) a liquid, (c) an amorphous solid, and (d) a crystal (based on Encyclopedia of China: Physics, 1987)

3.6.2 Quasicrystals

In Section Two of this chapter, it was proved that in a crystal, only one-, two-, three-, four-, and six-fold rotational symmetries could possibly exist. Quasicrystals are solids with five-fold rotational symmetry, and they are of fundamental interest in crystallography as they complete the definition of solids with long-range order.

Quasicrystals were discovered by Dan Shechtman when he was using transmission electron microscopy and other electron diffraction methods to investigate the phases formed by rapidly-quenched aluminum alloys in 1984. Professor Shechtman was then on leave from his permanent position at the Israel Institute of Technology, and was working at the National Bureau of Standards (now known as NIST, the National Institute for Standards and Technology). The alloy of aluminum and manganese discovered by Shechtman is produced by the super-fast cooling of the molten metals at a speed of 10^6K per second. Figure 3.39 shows the TEM diffraction pictures with two-fold, six-fold, and ten-fold symmetry of the same $Al_{86}Mn_{14}$ alloy.

The quasicrystal structure can occur in many binary or trinary alloys produced by super-fast cooling. In a quasicrystal the translational symmetry of lattice no longer exists, therefore the description of quasicrystal structure becomes a complex mathematical problem because the concept of the PUC in a crystal can no longer be used.

In the language of geometry, quasicrystals are described as having quasi-periodic structure, which can be configured using the "tiling" method. In 1961, mathematician and logician Hao Wang, inspired by an old mathematical conundrum—"The Four-Color Map Problem"—proposed the theory of aperiodic tiling. While trying to determine whether the tiling problem was computable, Wang was able to show that a given finite set of prototypes (squares, rectangles, parallelograms, and so forth) of different sizes could be configured to tile a 2D plane, where each prototype has four different colors on the four sides, and the

Figure 3.39 The 2-fold, 6-fold, and 10-fold symmetry shown in TEM diffraction pictures of $Al_{86}Mn_{14}$ quasicrystal alloy (based on Shechtman, Blech, Gratias, and Cahn, 1984)

contact plane of two neighboring prototypes has the same color. Three years later, Robert Berger, a Ph.D. student in Harvard, proved in his thesis that the general tiling rule proposed by Hao Wang did not in fact exist; however, he achieved a tiling of a 2D plane which consisted of 20,426 Wang tiles.

In 1974, Sir Roger Penrose, an English mathematical physicist at the University of Oxford, made another breakthrough in tiling. He used two different "unit cells" to tile a plane using the "matching rule." The two unit cells of Penrose tiling are shown in Figure 3.40(a). Rhombus A has vertex angles 72° and 108°; rhombus B has vertex angles 36° and 144°. These angles are obviously multiples of $\frac{2\pi}{10}$, illustrating the icosahedral symmetry. Also, all the edges have equal lengths in Rhombus A and B.

The matching rule in Penrose tiling is an extension of Wang's tiling theory. Two, but not four, "colors" are introduced to the edges of the

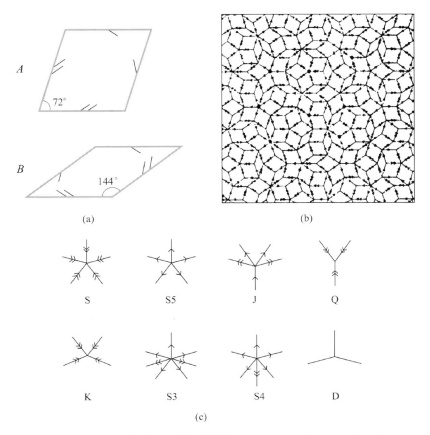

Figure 3.40 Penrose Tiling with 5-fold symmetry: (a) two unit cells; (b) pattern of the Penrose Tiling; (c) 8 kinds of vertices of tiles (based on Liu and Fu, 1999)

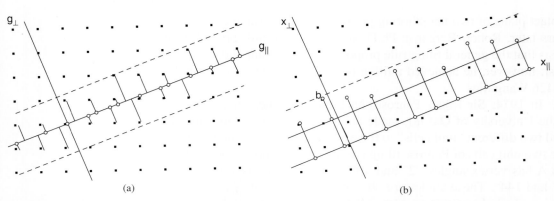

Figure 3.41 High dimensional projection method for quasicrystal structure and diffraction pattern analysis: (a) in the k-space; (b) in the x-space (based on Elser, 1985)

two unit cells of Penrose tiling: red is marked by double ticks and blue is marked by single ticks. Only edges with the same color are allowed to join. The core pattern of Penrose tiling, shown in Figure 3.40(b), is formed by five Rhombi A with five-fold rotational symmetry, which resemble the rose window in a Gothic church. Outside the rose-like core pattern, five Rhombi B can just fit into the spaces between two Rhombi A "leaves", which follow the matching rule $144° + 108° + 108° = 360°$. All sites in the Penrose lattice can be classified into eight kinds of vertices, which are listed in Figure 3.40(c).

It is not possible to explain the diffraction patterns of the quasicrystal shown in Figure 3.39 using Penrose tiling because it is difficult to analyze Penrose lattice sites using linear algebra. In 1981, a Dutch mathematician N. G. de Bruijn used an algebraic method to prove that the Penrose lattice could be formed by a high dimensional projection from the shadowed area (the area between two dotted lines in Figure 3.41) of a five-dimensional space to a 2D sub-space.

In 1985, one year after Shechtman's observation, Veit Elser from the AT&T Bell Laboratory explained quasicrystal diffraction using a high dimensional projection method with a six-dimensional space ($x_\parallel$ stands for the dimensions at real 3D space and $x_\perp$ represents the other three dimensions) and a six-dimensional reciprocal space ($g_\parallel$ stands for the real reciprocal dimensions of the quasicrystal and $g_\perp$ represents the other three reciprocal dimensions). The quasicrystal's structure factor of diffraction can be constructed in three-dimensional real space as follows:

$$S_{\text{qc}}(g_\parallel) = \sum_{n=1}^{N} e^{ix_\parallel^n g_\parallel} = -\sum_{n=1}^{N} e^{ix_\perp^n g_\perp}; \quad e^{ix^n \cdot g} = e^{ix_\parallel^n g_\parallel + ix_\perp^n g_\perp} = 1$$

(3.44)

where N is the total number of atoms; $g_\perp = \frac{\pi}{a} \sum n_i \hat{e}_\perp^i$ (n_i the integer, a the edge length of Penrose tiling) is the "pseudo-Bragg-vector," and the diffraction pattern is represented by $g_\parallel = \frac{\pi}{a} \sum n_i \hat{e}_\parallel^i$. The six primitive vectors $\vec{e}^i = (\hat{e}_\parallel^i, \hat{e}_\perp^i)$ of the reciprocal six-dimensions are defined as

$$\hat{e}_\parallel^i \cdot \hat{e}_\parallel^j = \delta_{ij} \pm (1 - \delta_{ij}) \frac{1}{\sqrt{5}}; \quad \hat{e}_\perp^i \cdot \hat{e}_\perp^j = \delta_{ij} \mp (1 - \delta_{ij}) \frac{1}{\sqrt{5}} \quad (3.45)$$

where $i, j = 1, 2, \ldots, 6$; and it should be noted that $\frac{1}{\sqrt{5}}$ is a typical irrational number related to the angle $2\pi/10$, which leads to the five-fold symmetry of diffraction patterns.

3.6.3 Liquid Crystals

In the early stages of solid state physics, the research focused mainly on crystal structure. Disordered structures or multi-scaled structures were ignored due to their complexity. Nowadays, soft condensed matter, including liquid crystals and other macromolecular materials, is an important research subject in science and engineering.

Liquid crystal is a state of matter between the totally disordered liquid and the strictly ordered crystal. In 1888, the Austrian chemist, Friedrich Reinitzer, was working at the Institute of Plant Physiology at the University of Prague when he noticed a strange phenomenon. Reinitzer was trying to ascertain the correct formula for the molecular weight and the melting point of a cholesterol substrate. He was struck by the fact that the substance seemed to have two melting points. At 145.5°C, the solid crystal melted into a cloudy liquid, and remained in that state until it reached 178.5°C, when the cloudiness suddenly disappeared, giving way to a clear transparent liquid. This two-melting-point phenomenon still existed after further purification of the material.

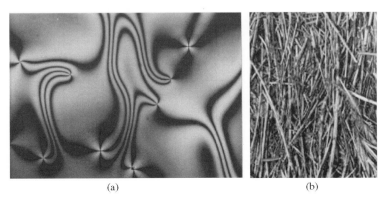

(a)   (b)

Figure 3.42 Liquid crystal: (a) macroscopic view; (b) molecular scale

Puzzled by his discovery, Reinitzer turned for help to the German physicist, Otto Lehmann, who was an expert in crystal optics. In 1889, Lehmann became convinced that the cloudy liquid had a unique kind of order. Eventually he realized that the cloudy liquid was a new state of matter, and named it "liquid crystal." In a normal liquid the optical properties are isotropic. In a liquid crystal, the optical properties are highly anisotropic even if the substance itself is fluid.

The underlying physics of liquid crystals was explored by a French theoretical physicist, Pierre-Gilles de Gennes, who made great contributions to the theories of second order phase transition in complex systems such as liquid crystals, magnetic materials, and superconductors. After he graduated from the École Normale in 1955, he worked at the Atomic Energy Center on neutron scattering and magnetism. In 1959, he became a postdoctor with C. Kittel at Berkeley. In the 1960s, de Gennes was one of the leading scientists in the field of superconductivity and liquid crystal systems. His theory of second order phase transition expanded Lev Landau's theory in the similar field. Pierre-Gilles de Gennes was rewarded with the Nobel Prize in Physics in 1991 for his theories on the complex form of matter.

The pioneer in liquid crystal display (LCD) technology, George H. Heilmeier, obtained his B.S., M.A., M.S.E., and Ph.D. degrees in Electrical Engineering from the University of Pennsylvania. Soon after graduation, he joined RCA Laboratories in Princeton, and he became

Figure 3.43 Early liquid crystal device (LCD) invented by George H. Heilmeier

the Head of the Solid State Device Research Group in 1966. In 1963, he and Richard Williams published a paper in *Nature* suggesting the use of liquid crystal materials for display screens in products such as televisions. However the group realized that it would take many years to develop LCD TVs, so they concentrated on digital time displays for clocks and watches. Heilmeier's group found that a voltage of only 1–10 V could change the color of an LCD when the liquid crystal was mixed with dyes. This discovery became the basis for a wide range of future LCD applications.

About one in 200 organic materials are liquid crystals. There are two classes of liquid crystal: thermotropic and lyotropic, as shown in Figure 3.44. Thermotropic liquid crystals show anisotropic optical properties within a certain range of temperatures, which enables them to be used in LCDs. The molecular structures of thermotropic liquid crystals also usually contain benzene rings. Lyotropic liquid crystals only display

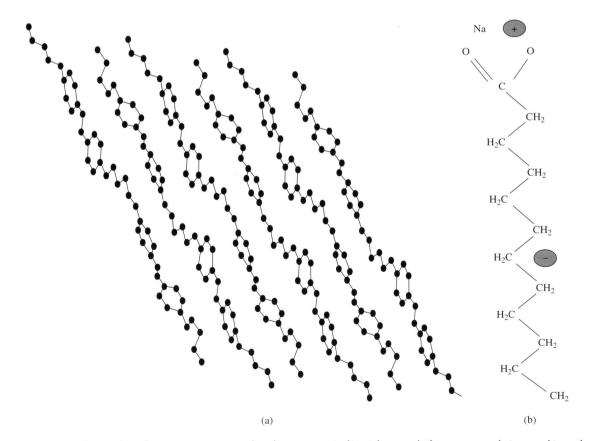

Figure 3.44 The molecular structure in (a) the thermotropic liquid crystal (from www.ch.ic.ac.uk) and (b) one type of lyotropic liquid crystal

anisotropic optical properties when dissolved in water or organic solvents, which in living organisms can be found in the eye, brain, muscle, neural tissues, and some germ cells.

The lyotropic liquid crystal is composed of amphipathic compounds in a polar solvent. The two ends of a lyotropic liquid crystal molecule are asymmetric; for example, sodium fatty acid has a hydrophilic end (attracted to water), i.e., the sodium acetate (–COONa) and a lipophilic end (attracted to oil), i.e., the hydrocarbon chain. When the lyotropic liquid crystal molecules are in a polar solvent such as water, the hydrophilic ends are attracted to water and arranged in order; therefore, the flower-like (low molecular density) or biological membrane (high molecular density) phase might appear, as seen in Figure 3.45.

Thermotropic liquid crystals have three phases, namely the nematic phase, smectic phase, and cholesteric phase, as shown in Figure 3.46. The number of benzene rings in a thermotropic liquid crystal molecule is usually two or three, although it can be zero, one, or four. The length and diameter of the rod-like molecule is several nanometers and several angstroms respectively. The aspect ratio of the molecule should be larger than 4:1 to form a liquid crystal phase.

Nematic liquid crystals are most commonly used in LCDs. The nematic phase N is similar to the liquid phase and is the simplest of

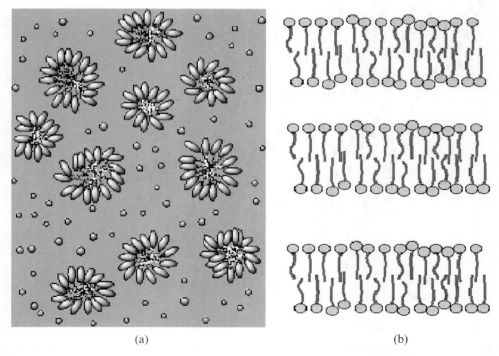

(a)   (b)

Figure 3.45 Lyotropic liquid crystal: (a) flower-like; (b) biological membrane phase

3.6 Disordered Solid Structure 99

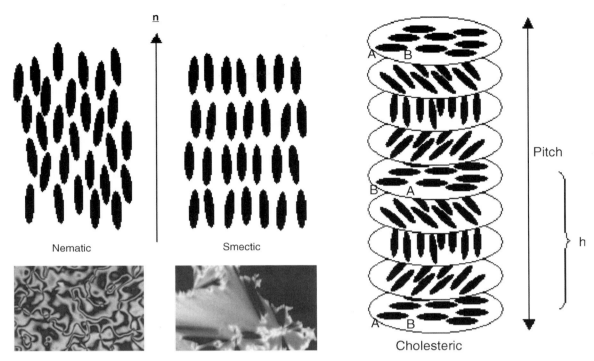

Figure 3.46 Thermotropic liquid crystal: nematic, smectic, and cholesteric phase (from www.ch.ic.ac.uk)

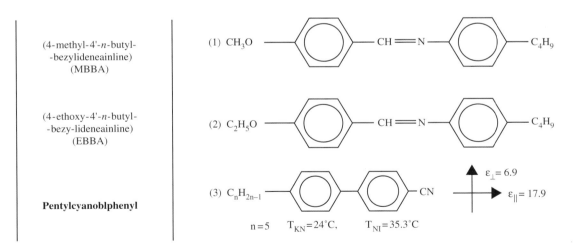

Figure 3.47 Molecular structures of the nematic liquid crystal

the thermotropic liquid crystal phases. Most of the molecules float around as in a liquid phase, but are still ordered along the director $\hat{n}$. In Figure 3.47, three examples of nematic liquid crystal molecules are illustrated. The nematic molecule has two benzene rings, which can be classified according to the chemical cluster in the middle of the two rings: –CH=N–, –N=N–, –N(O)=N–, –CO–O– or simply a direct link. The most useful nematic liquid crystals are those with the central cluster –CH=N–. MBBA and EBBA, shown in Figure 3.47, are used commercially, and mixing the two enables the production of LCDs with a wide temperature range.

The cholesteric phase N* is also called the helix phase, and is sometimes treated as a special case of the nematic phase. All the molecules arrange themselves into a strongly twisted structure, as illustrated in Figure 3.46, in which the pitch distance h is several hundred nanometers. The name "cholesteric" refers back to the substances with which Reinitzer made his discovery. Cholesteric liquid crystal molecules can be classified into two types: one is the cholesterol hydrin derivative, and the other is the chiral matter with asymmetric carbon atoms, as shown in Figure 3.48. The definition of chirality is "something that cannot be superimposed on its own mirror image, like a hand." For example, in the cholesteric molecule $COO(CH_2)_8CH=CH(CH_2)_7–CH_3$, a carbon atom can have four chemistry clusters forming tetragonal covalent bonds; therefore, the molecule has chiral asymmetry.

The smectic phase is similar to the solid phase, in that the liquid crystals are ordered in layers. The liquid crystal molecules normally float around freely within the layer, but they cannot move freely between the layers. The smectic molecules tend to arrange themselves in the same direction, as illustrated in Figure 3.46. The smectic phase is divided into several sub-phases: as the temperature decreases, the smectic phases A, C, and B appear in sequence. When the molecules in a layer arrange in disorder, and all molecules are perpendicular to the layer surface, it is called smectic phase A; when the molecules in a layer arrange in disorder, but the molecules are tilted towards the layer surface, it is called smectic phase C; when the molecules in a layer arrange in a triangular lattice, and all molecules are perpendicular to the layer surface, it is called smectic phase B. In the smectic phase, the nearest neighboring molecules have stronger side-by-side interactions than those at the ends, so layers are formed; therefore longer molecules are preferred, as shown in Figure 3.49.

The nematic, cholesteric, and smectic phases of thermotropic liquid crystals can be detected using X-ray diffraction. This is another research area, so here we will only briefly discuss the mechanism of soft matter diffraction. The diffraction intensity I of a liquid crystal can be calculated in two steps: (1) the scattering length f_m of a molecule and (2) the structure factor of all molecules. The accurate calculation of f_m is obviously very complex, and is dependent on the wavefunction of the macromolecule. The calculation of the structure factor is also a little difficult without the sum rule in crystals.

3.6 Disordered Solid Structure 101

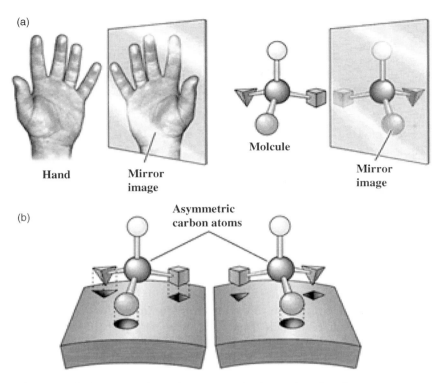

Figure 3.48 Asymmetric carbon atoms in the molecule of some cholesteric liquid crystals

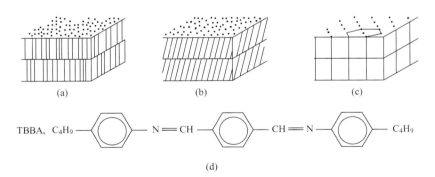

Figure 3.49 (a) Smectic A; (b) Smectic C; (c) Smectic B; (d) A molecule: TBBA

A simplified model can be built for f_m of a long liquid crystal molecule with the aspect ratio $l : d = m : 1$, where m identical atomic clusters aligned in a chain are used to simulate the true complex molecule. If each atomic cluster has a scattering length f_a, the total molecular scattering length is

$$f_m = f_a \sum_{j=1}^{m} e^{-i\vec{s}\cdot\vec{r}_j} = f_a \sum_{j=1}^{m} e^{-i(\vec{s}\cdot\hat{n})dj} = f_a \frac{\sin(l\vec{s}\cdot\hat{n}/2)}{\sin(d\vec{s}\cdot\hat{n}/2)} e^{i\theta_0} \quad (3.46)$$

where $\vec{s} = \vec{k} - \vec{k}_0$ is the difference between the incoming and outgoing wavevector, $\hat{n}$ is the director, and $l = md$ is the length of the molecule. The phase constant is $\exp(i\theta_0) = \exp(-i(m+1)d\vec{s}\cdot\hat{n}/2)$.

In the nematic phase, the centers of molecules are randomly distributed in 3D. If there are N molecules interacting with the X-ray beam, the total diffraction intensity is simply a direct sum of each molecule:

$$I = \left| f_m \sum_l e^{-i\vec{s}\cdot\vec{r}_l} \right|^2 = N|f_m|^2 = N|f_a|^2 \left| \frac{\sin(l\vec{s}\cdot\hat{n}/2)}{\sin(d\vec{s}\cdot\hat{n}/2)} \right|^2 \quad (3.47)$$

The total intensity is maximized as $I = N|mf_a|^2$ when $\vec{s} \perp \hat{n}$ and minimized with $\vec{s} \parallel \hat{n}$, i.e., the reflection from the liquid crystal surface is strongest when the molecule is parallel, and weakest when the molecule is perpendicular to the surface. The former characteristics explain the optical anisotropy.

In the nematic phase, if the director $\hat{n}$ of the liquid crystal molecules is perpendicular to the liquid crystal layer and parallel to the incident beam $\vec{k}_0$, a ring diffraction pattern will appear, and the diffraction peaks can be calculated directly by Eq. (3.47), as illustrated in Figure 3.50(a). If the nematic director $\hat{n}$ is parallel to the surface of the LCD layer and perpendicular to $\vec{k}_0$, two banana-like intensity peaks will occur at two poles of a circle on the receiving screen, where the line connecting the two poles is in the $(\vec{k}_0, \vec{k})$ plane and perpendicular to the director $\hat{n}$, as seen in Figure 3.50(b) and (c).

If the liquid crystal is in the smectic phase, sharper diffraction peaks will occur due to the layered structure of the molecules, as shown in Figure 3.50(d). In the smectic A phase, if the layer-layer distance is D, the bright ring locates at

$$D\cos(2\theta) = \lambda; \quad \Rightarrow \quad 2\theta = \arccos(\lambda/D) \quad (3.48)$$

If the liquid crystal is in the smectic B phase, and the Laue method is used with $\vec{k}_0$ perpendicular to the layers, hexagonal diffraction points can be detected.

The most important application of liquid crystals is the liquid crystal screen, used in televisions and computers. In Figure 3.51, the cell mechanism in a conventional liquid crystal screen design is illustrated, where there is a package of two crossed polarizers with a liquid crystal in between. The LCD cell is dark when an external electric field forces all

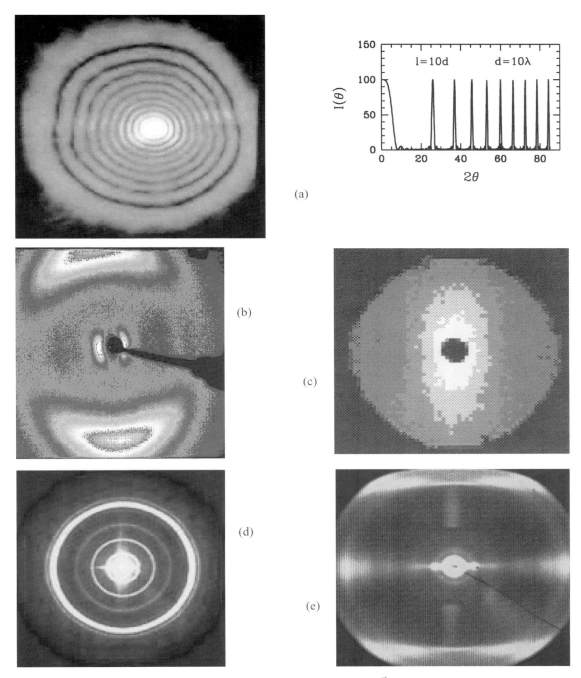

Figure 3.50 Diffraction patterns of liquid crystals, incoming beam $\vec{k}_0$ is perpendicular to the lc-layer: (a) nematic phase, $\hat{n} \perp$ plane, experiment and simple calculation (from http://people.na.infn.it); (b) nematic phase, $\hat{n} \parallel$ plane, X-ray diffraction (from http://lhedl.jinr.ru); (c) nematic phase, $\hat{n} \parallel$ plane, neutron diffraction (from www.ch.ic.ac.uk); (d) smectic phase, $\hat{n} \perp$ plane, the bright ring is w.r.t the inter-plane distance 5.45 nm (from www.lrsm.upenn.edu); (e) cholesteric phase (from www.lrsm.upenn.edu)

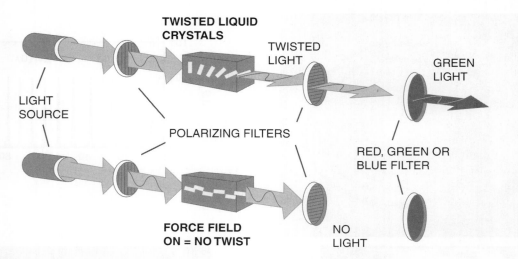

Figure 3.51 Bright and dark LCD cells w.r.t. the green light (from http://nobelprize.org)

liquid crystal molecules parallel to the direction of the light ray, and these molecules then have no influence on the state of polarization. Thus, the package of crossed polarizers allows no light through. The cell is bright when the external field is turned off. The molecules are twisted by 90° from the vertical to the horizontal from one end to another, realized by surface chemical hydrophile or lipophile treatment, and the polarization of light is tuned from the vertical to the horizontal. Thus, the package of crossed polarizers allows the light through.

This brings us to the end of a long chapter in which the structures of ordered and disordered solids have been discussed with reference to the theoretical and experimental diffraction methods. There are other branches of solid structure theory which are not covered in this book, such as the crystallography of alloys and compounds, and defects and their motion, which the reader may wish to explore further elsewhere.

Summary

This chapter has focused on the structure of solids, including crystal structures and the structures of amorphous solids, quasi-crystals and liquid crystals. Significant experiments for measuring solid structures have also been discussed, including X-ray, electron and neutron diffraction. To summarize, the main areas covered are:

1. Crystallography: the historical achievements of Seeber, Bravais, Hessel, Schoenflies, Barlaw, and Feodorov in crystallography were

introduced. A crystal can be described by a periodic Bravais lattice and a basis. All possible Bravais lattices are classified by the point groups, which are defined by the rotational and reflectional symmetries of the lattices. The structures of elements were introduced, including the most common structures: BCC, FCC, HCP, DIA, and the non-Bravais lattice structures.
2. Pauling's rules: the rules introduced by Linus Pauling can help us understand the lattice structures of compounds. The structures of coordination polyhedra, such as the silicate group (SiO_4), can be explained by the radius of ions. The compound crystal is thus constructed by the proper arrangements of the coordination polyhedra together with other ions.
3. Reciprocal space: the reciprocal lattice was introduced to describe diffraction, the structure of which geometrically correlates to Bravais lattices in real space. The Brillouin zone is a special unit cell of the reciprocal lattice, for the description of the wave vectors for a variety of waves in solid state physics.
4. Diffraction: the discovery of X-ray diffraction was based on the contributions of Roentgen, Laue, the Braggs, and Moseley. Electron and neutron diffraction were invented by Davisson, G. P. Thomson, Shull, and Brockhouse, following the matter wave concept formulated by Einstein and de Broglie, and the discovery of electrons and neutrons by J. J. Thomson and Chadwick.
5. Diffraction theory: crystal structure can be determined by diffraction points or rings, which are related to reciprocal lattices. The Ewald ball is useful for determining the diffraction points from the reciprocal lattice sites.
6. Disordered solid structure: the structure of amorphous solids can be described by a statistical pair-distribution function of atomic positions, which can be measured by synchrotron X-ray diffraction. Quasi-crystal structure can be explained by the mathematical high-dimension-projection theory. Liquid crystals are a self-alignment of organic molecules, and their nematic, smectic, and cholesteric phases can be determined by XRD.

References

Ashcroft, N. W., and Mermin, N. D. (1976). *Solid State Physics*, Holt, Rinehart and Winston, New York.
Bragg, W. L., and Bragg, W. H. (1913). *Proceedings of the Royal Society of London*, Vol. 88A, 413.
Chaikin, P. M., and Lubensky, R. C. (1995). *Principles of Condensed Matter Physics*, Cambridge University Press, Cambridge.
Elliott, S. R. (1990). *Physics of Amorphous Materials*, Wiley, New York.

Elser, V. (1985). "Indexing Problems in Quasicrystal Diffraction," *Phys. Rev. B*, Vol. 32, 4892–4898.

Encyclopedia of China: Physics (1987) (in Chinese). Encyclopedia of China Publishing House, Beijing.

Feng, D., and Jin, G. (2003). *Condensed Matter Physics* (in Chinese), Higher Education Press, Beijing.

Huang, K., and Han, R. (1988). *Solid State Physics* (in Chinese), Higher Education Press, Beijing.

Kittel, C. (1986). *Introduction to Solid State Physics*, Wiley, New York.

Landau, D. L., and Lifshitz, E. M. (1987). *Course of Theoretical Physics, Vol. 3, Quantum Mechanics*, Pergamon Press, New York.

Liu, Y. Y., and Fu, X. J. (1999). *Quasi-crystals* (in Chinese), Shanghai Scientific and Technical Publishers, Shanghai.

Omar, M. A. (1975). *Elementary Solid State Physics: Principles and Applications*, Addison-Wesley, New York.

Pauling, L. (1960). *The Nature of the Chemical Bond*, Cornell University Press, Ithaca, New York.

Schlenz, H., Neuefeind, J., and Rings, S. (2003). "High Energy X-ray Diffraction Study of Amorphous (Si0.71Ge0.29)O2 H," *Phys. Rev. B*, Vol. 32, 4892–4898.

Shechtman, D., Blech, I., Gratias, D., and Cahn, J. W. (1984). "Metallic Phase with Long-range Orientational Order and No Translational Symmetry," *Phys. Rev. Lett.*, Vol. 53, 1951–1953.

Yan, H. (2000). *Supermolecular Liquid Crystals* (in Chinese), Science Press, Beijing.

"X-Rays Imaging Detectors (1D and 2D) for Bio-Medical Application." Photo Gallery, Laboratory of High Energies (LHE), Russia. 2 September 2005. <http://lhedl.jinr.ru/g_biomed.html>

"History and Properties of Liquid Crystals." Nobel Prize Organization, City of Stockholm, Sweden. 14 November 2005. <http://nobelprize.org/educational_games/physics/liquid_crystals/history/index.html>

"Optics in Soft-Matter Group." Dipartimento di Scienze Fisiche, Universitá di Napoli. City of Napoli, Italy. 10 October 2005. <http://people.na.infn.it/~marrucci/softmattergroup/>

"Thermotropic Liquid Crystals." Prof. J. M. Seddon, Dept. Of Chemistry, Imperial College London, City of London, UK. 25 August 2005. <http://www.ch.ic.ac.uk/liquid_crystal/pages/thermo.html>

"Multi-Angle X-ray Scattering Facility." Materials Research Science and Engineering Center, University of Pennsylvania, City of Pennsylvania, USA. 2 September 2005. <http://www.lrsm.upenn.edu/lrsm/facMAXS.html>

"WebElements Periodic Table." University of Sheffield and WebElements Ltd, City of Sheffield, UK. 1 May 2005. <http://www.webelements.com/>

Exercises

3.1 Prove that the distance between two neighboring lattice planes equals $d_{hkl} = (h^2/a^2 + k^2/b^2 + l^2/c^2)^{-1/2}$ for the (*hkl*) crystal planes in orthorhombic, tetragonal, or cubic crystal systems. Find d_{111} for silicon (lattice constant $a = 5.43$Å).

3.2 Draw the crystal direction [213] and the crystal plane (213) in the cubic crystal system.

3.3 If the simple cubic lattice $\vec{R} = (n\hat{e}_x + m\hat{e}_y + h\hat{e}_z)a$ is chosen as a basis, what is the lattice w.r.t. n, m, l—all even integers or all odd integers? Is this a Bravais lattice?

3.4 Is the base-centered cubic lattice a Bravais lattice? Prove your conclusion.

3.5 Draw the lattice site arrangement in (100) and (110) crystal planes in BCC and FCC lattices, respectively.

3.6 Find the crystal direction in BCC lattices w.r.t. the cross lines of (111) and (100) planes, and the crystal direction w.r.t. the cross lines of (111) and (110) planes.

3.7 Draw all the rotational axes w.r.t. the T-groups in the silicon crystal.

3.8 Explain why D_6, D_{3h}, and D_{6h} are all symmetry groups of the hexagonal lattices.

3.9 Find the ratio of the PUC for BCC and FCC lattices.

3.10 Prove that the Wigner-Seitz cell of the BCC lattice has the same volume as its PUC.

3.11 Find the total number of lattice sites and the related distance of the first, second, third, ..., sixth nearest neighbors in SC, BCC, and FCC lattices, respectively.

3.12 Find the maximum packing density of SC, BCC, and FCC lattices, respectively.

3.13 Calculate the aspect ratio of the lattice constants c/a for the HCP CUC.

3.14 The crystal of Na (atomic number 23) has a BCC structure with a lattice constant $a = 4.23\text{Å}$. Find the density of sodium.

3.15 The density of the Ga (atomic number 69.72) crystal is 5.91 g/cm^3, the lattice constants are $a = 4.51\text{Å}$, $b = 4.515\text{Å}$, and $c = 7.64\text{Å}$. How many atoms are there in the CUC? And in the PUC?

3.16 Can Pauling's rules explain the structures of ZnO and ZnS crystals? Explain the reasons.

3.17 Can Pauling's rules explain the difference between the structures of CaCO$_3$ and CaTiO$_3$ crystals? Explain why.

3.18 Prove that the BCC and FCC lattices are the reciprocal lattices of one another.

3.19 Calculate the volume of the reciprocal PUC Ω^*, and find its relationship with the volume of the PUC Ω in real space.

3.20 A monoclinic crystal has a PUC of $a = 4\text{Å}$, $b = 6\text{Å}$, $c = 8\text{Å}$, $\alpha = \beta = 90°$, and $\gamma = 120°$. Find the reciprocal primitive vectors. Calculate the volume of the reciprocal PUC, and the (110) lattice plane separation.

3.21 In a FCC lattice with the lattice constant a, there are $N_L = L_1 \times L_2 \times L_3$ cubic cells. Find the accurate description of the wave vector $\vec{k}$, and the averaged volume of each grid of $\vec{k}$.

3.22 Derive the Laue formula from the diffraction theory given in this chapter.

3.23 The probability density of the 1s electrons in an hydrogen atom is $\rho(r) = \frac{2}{\pi a_B^3} e^{-2r/a_B}$, where a_B is the Bohr radius. Calculate the atomic scattering factor f_a of the hydrogen atom.

3.24 Prove that the 1D lattice structure factor S has the first non-zero maximum at $S^2 = 0.04 N^2$, where N is the total number of 1D unit cells.

3.25 Prove that the geometrical scattering factor F_{hkl} is non-zero only when $h + k + l = 2m$ for a BCC lattice. Is the result the same, when the sum of j within a CUC takes over 8corner + 1center sites, or just takes over the two independent sites?

3.26 Prove that the geometrical scattering factor F_{hkl} is non-zero only when the integers h, k, l are all even or all odd.

3.27 Calculate the geometrical scattering factor F_{hkl} of the diamond structure. What is the difference between the F_{hkl} of a FCC lattice and a diamond structure?

3.28 In a CsCl lattice, if $f_{Cs} = 3 f_{Cl}$ is assumed, what is the geometrical scattering factor F_{hkl}? Can you distinguish the CsCl structure from the BCC or FCC lattices by F_{hkl}?

3.29 Using the results of Exercise 11 in this chapter, find the first six Bragg angles (with $K_{\alpha 1}$ of Cu target $\lambda = 1.54$Å as the diffraction beam) of the NaCl lattice with the lattice constant $a = 5.64$Å. What is the difference between Bragg angles of a NaCl lattice and a pure FCC lattice?

3.30 When a single crystal HCP lattice is tested in a TEM (where the wavelength λ is fixed and it is much smaller than the lattice constant), how many kinds of symmetry of a 2D lattice could be found in the diffraction patterns? What is the typical scale (expressed by lattice constants c, a, and λ) in the patterns?

3.31 The thermal expansion of a crystal will cause a shift of the Bragg angle. Prove that $\delta\theta = -\frac{\gamma}{3} \tan\theta$, where γ is the bulk modulus.

3.32 In a spherical volume $V_0 = 4\pi R_0^3$ of an amorphous material, there are N_0 atoms which could fluctuate. The averaged atomic density is $n_0 = \langle N_0/V_0 \rangle$. Prove the relationship between the atomic number fluctuation and the pair distribution function $g(r)$:

$$\frac{\langle N_0^2 \rangle - n_0^2 V_0^2}{n_0 V_0} = 1 + n_0 \int_0^{R_0} 4\pi r^2 dr [g(r) - 1] \qquad (3.49)$$

3.33 If the amorphous structure factor is approximately described by the function $S_{lq}(s) - 1 = [\sin(\beta s)/(\beta s)] \exp(-\alpha s)$, what is the relationship between the first peak R_0 of the pair distribution function $g(R)$ and the parameters α, β?

3.34 Is it possible for the diffraction pattern of a quasicrystal to have two-fold, six-fold, or ten-fold rotational symmetry? Does that mean the projection of the quasicrystal lattice to a certain plane could become a 2D Bravais lattice?

3.35 In a nematic phase with $\hat{n} \perp$ liquid crystal layer, the molecule has an aspect ratio of $l : d = 5 : 1$. When the X-ray wavelength $\lambda = d/10$ ($\vec{k}_0 \perp$ liquid crystal layer), draw the scaled intensity $I(\theta)$, and find the first two Bragg angles w.r.t. bright diffraction rings.

Thermal Properties of Solids and Lattice Dynamics

4

ABSTRACT

- Einstein Phonon Model of Specific Heat
- Debye Phonon Model of Specific Heat
- Lattice Vibration and Neutron Diffraction

Man has been aware of thermal phenomena for thousands of years; indeed it is likely that the concepts of "hot" and "cold" were recognized even before fire was first discovered. However, there has long been a controversy over the essence of heat, with two theories competing: the caloric theory, and the mechanical theory of heat or the conservation of energy.

The caloric theory was derived from the Ancient Greek four-elements theory. This theory treated heat as a matter, which could not be created or annihilated, and which flowed from fire to other matter in the heating process. However the caloric theory could not explain the heat generated by friction.

The study of thermal physics dates from Galileo's era, arising from the attempts to produce a thermometer. In the 19th century, a new concept of the nature of heat gradually evolved, which was heavily influenced by Newtonian mechanics. In 1842, a German doctor, Julius Robert Mayer, proposed the idea that heat was a kind of energy, which could be converted to or from mechanical energy. In 1840, the English physicist, James Prescott Joule, proved that heat could be converted from electrical energy, and derived Joule's law $P = I^2 R$. In 1847, Joule gave the most accurate measurement of the heat-energy conversion constant: 1 cal = 4.18 Joule. Joule's work established the experimental foundations for the conservation law of energy—the first law of thermodynamics. It is one of the few laws that is true in both classical and quantum physics.

In 1850, a Prussian, Rudolf Clausius, concluded that heat was not a caloric flow, but rather a representation of motion. During the period 1850 to 1854, the second law of thermodynamics was developed by

R. Clausius and an Englishman, William Thomson (Lord Kelvin). The second law of thermodynamics reconciled Frenchman Sadi Carnot's work on gas engine efficiency (1824) with Joule's dynamic theory of heat.

Thermodynamics evolved into statistical physics in 1887 when Ludwig Boltzmann presented his greatest achievement, the formula $S = k_B \ln \Omega$. The Boltzmann principle established a relationship between macroscopic entropy and the probability of micro-states, which opened the door to quantum physics. In 1900, Max Planck found that it was necessary to introduce the quantized energy level $h\nu$ for the electromagnetic wave oscillator to explain the black-body radiation law using the Boltzmann principle. Moreover, Planck found the Planck constant $h = 6.5 \times 10^{-27}$ erg·s and obtained the Boltzmann constant $k_B = 1.37 \times 10^{-16}$ erg/K when he compared his radiation formula with Wien's displacement law.

In the 1860s, the specific heat of gas was explained reasonably well by the kinetic theory of gases, which was based on the Maxwell speed distribution and the Boltzmann factor $e^{-\beta\varepsilon}$. The Dulong-Petit law, published in 1820, which stated that the solid specific heat is 5.96 cal/K/mole, could also be explained by the kinetic theory of gases: $3R = 6(R/2)$ (six degrees of freedom for a vibrating atom). In 1875, H. F. Weber found that the specific heats of diamond and graphite were much lower than the values predicted by the Dulong-Petit law, as shown in Figure 4.1. A. Einstein became aware of this fact when, as a student, he attended a lecture by Weber. In 1907, based on Max Plank's statistics, Einstein explained the specific heat of solids versus temperature quantitatively.

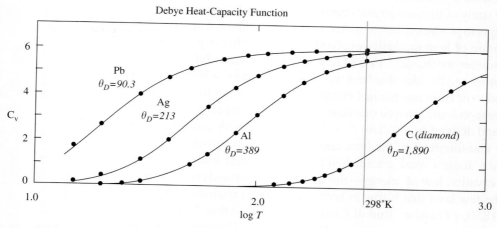

Figure 4.1 Dulong-Petit law and the corrections (Lewis, Randall, 1961)

However, the Einstein model of phonons did not explain specific heat at very low temperatures. In 1911, P. Debye modified the Einstein model by assuming that the elastic vibration in a solid had a similar dispersion relation $\omega = vk$ to the photon (electromagnetic wave) dispersion $\omega = ck$, thus formulating the famous T^3-law of specific heat at very low temperatures, which was later proved by Walther Nernst.

A mathematical expression for sound propagation was first given by Isaac Newton, and a differential equation of the displacement y of air at x was given as:

$$\rho_0 \frac{\partial^2 y}{\partial t^2} = -\frac{\partial \Delta P}{\partial x} = \kappa \rho_0 \frac{\partial^2 y}{\partial x^2} \qquad (4.1)$$

where the wave speed $v = \sqrt{\kappa}$, and κ is the ratio of pressure fluctuation ΔP over density fluctuation $\Delta \rho$. Much of the progress in sound propagation theory rested on the developments in analytical mechanics during the 18th century by Euler (1707 to 1783), Lagrange (1736 to 1813), and d'Alembert (1717 to 1783).

During the period 1912 to 1913, inspired by the Debye model of phonons, M. Born and T. von Karman calculated the dispersion relation $\omega(k)$ of atomic vibrations (phonon spectrum) in solids more accurately using analytical mechanics. In 1951, Brockhouse's inelastic-scattering neutron diffraction proved the phonon spectrum found by Born and Karman. The first book, *Dynamical Theory of Crystal Lattice*, was written by Born and a Chinese physicist, Huang Kun. Lattice dynamics is the basic theory which explains the acoustic properties of solids.

4.1 Albert Einstein and his Phonon Model

Albert Einstein was born at Ulm, Germany, in 1879. He began his schooling in Munich, and continued it in Milan, Italy and then Aarau, Switzerland. In 1896, he entered the Swiss Federal Polytechnic School in Zurich. In 1901, after he gained his diploma, Einstein was unable to find a teaching post, the job he was trained for. In 1903, he accepted a

Figure 4.2 Albert Einstein: the symbol of wisdom

position as a technical assistant at the Swiss Patent Office, and married his Serbian undergraduate classmate, Mileva Maric.

In 1905, Einstein's "miraculous year," he wrote five great papers in his spare time which changed the face of physics. Einstein realized the inadequacies of Newtonian mechanics. In June 1905, in the epochal paper "Electrodynamics of Moving Bodies," he discarded the concept of "ether" in Maxwell-Lorentz electrodynamics, and established his special theory of relativity by introducing the same limit of speed c in both mechanics and electrodynamics. In September, he published a short paper explaining the energy source of radiation from radium, which featured the famous mass-energy relation: $E = mc^2$. This legendary equation was of great importance to the future development of nuclear physics.

In March and April 1905, Einstein wrote two papers on molecular kinetic theory. He tried to prove atomism using Brownian motion, stating that a pollen particle moving randomly in a liquid had a root-mean-square displacement in the x-direction:

$$\langle \delta_x^2 \rangle = \frac{k_B T}{3\pi \eta a} \Delta t = 2D \Delta t; \quad D = \frac{k_B T}{6\pi \eta a} = \frac{k_B T}{m} \tau \qquad (4.2)$$

where D is the diffusion coefficient, Δt is the time interval, η is the liquid viscosity coefficient, and a/m is the ratio of the radius to the mass of the pollen particle. This was the thesis for his doctor's degree in 1905, and it is also Einstein's most cited paper.

In March 1905, Einstein published another paper "On a Heuristic Point of View Concerning the Production and Transformation of Light." In it, he said, "The wave theory of light has proved itself superbly in describing purely optical phenomena and will probably never be replaced by another theory." He subsequently went on to state the completely opposite hypothesis that "the energy of light is discontinuously distributed," based on Philipp Lenard's photoemission experiment in 1902. Due to the controversy over his relativity theory, his work on the quantum theory of light and photons was the only achievement for which, in 1921, he was awarded a Nobel Prize.

Einstein's ideas were not readily accepted by other physicists, but he continued to apply his quantum physics to different fields. In 1907, Einstein introduced the energy quanta of atomic vibrations—phonons—to explain the specific heat of solids. In 1912, he used the photon concept to establish the "quantum equivalence law" of photochemistry. According to his law, a photon is absorbed by an atom or a molecule, which then enables a photochemical reaction to take place. Einstein's photochemistry law was finally proved experimentally by Emil Warburg and James Franck in the 1930s. In 1917, Einstein published a paper entitled "On the Quantum Theory of Radiation." Einstein's idea was that, in addition to the spontaneous, random emission of photons from excited atoms, there is an emission forced by radiation, and its emission probability depends on the strength of the background radiation component at the related frequency. This stimulated emission is the theoretical basis of maser and laser developed in 1960s.

From 1913 to 1916, Einstein expanded his special theory to include the effects of gravitation on space and time. This theory, referred to as the General Theory of Relativity, proposed that matter causes space to curve. At the end of June 1915, Einstein spent a week at Göttingen, where he lectured on his (incorrect) 1914 version of general relativity. David Hilbert attended his lectures. Following their conversation, both Einstein and Hilbert discovered the same final form of the gravitational field equations at almost the same time in November 1915. In 1916, Einstein proposed the concept of the gravitational wave, although it was not finally proved by experiments until 1979.

In 1917, Einstein proposed the first static model of the universe using the gravitational field equations, where the universe is viewed as a 3D closed curved surface in 4D space-time. In the 1920s, Einstein converted to the new cosmology after a series of controversial meetings with the Dutch astronomer, Willem De Sitter. This conversion came about after it was shown by de Sitter, Alexander Friedmann, and Georges Lemaitre that Einstein's original static solutions would not work in practice, because the slightest deviation from perfect uniformity would cause the universe to either expand or contract as a whole. Following the discovery of cosmological redshifts by Edwin Hubble in 1929, the model that universes expand in time, or the Big Bang theory, becomes the standard model in cosmology.

Figure 4.3 General relativity and cosmology

By 1919, Einstein was internationally famous. Among his non-scientific works, *About Zionism* (1930), *Why War?* (1933), *My Philosophy* (1934), and *Out of My Later Years* (1950) are perhaps the most well-known. Einstein believed that the true mission of physicists was to explore the most general and basic laws, from which the natural phenomena could be deduced.

In the following section, Einstein's phonon model in 1907 for the thermal properties of solids, one of his contributions to quantum physics, is discussed.

The basic assumptions of Einstein's phonon model reflected his scientific philosophy. Einstein's basic law of thermal atomic motions in solids was influenced by Planck's radiation law, and was a continuation of his own photon concept namely:

1. All atoms in solids vibrate at a fixed frequency ν. The oscillating energy is discontinuously distributed: $E = nh\nu$, and $\varepsilon = h\nu$ is the energy of the quasi-particle, or "phonon," in a crystal.
2. The basic principles of statistical physics are utilized, especially those developed by Boltzman.

At the temperature T, in the $3N$-fold degeneracy energy level $\varepsilon = h\nu$, the average number of phonons $\bar{n}$ can be calculated by minimizing the difference between the Helmholtz free energy and the Gibbs free energy $F - G = E - TS - \mu n$ of n identical phonons (where T, V are fixed in solids and μ is simply 0 in the phonon gas):

$$0 = \frac{d}{dn}\left\{nh\nu - k_B T \ln\frac{(n+3N)!}{n!(3N)!}\right\}\bigg|_{n=\bar{n}}$$

$$= h\nu - k_B T \ln\frac{\bar{n}+3N}{\bar{n}}$$

$$\bar{n} = 3N\frac{1}{e^{\beta h\nu}-1} = 3N f_{B-E}(\varepsilon); \quad \beta = 1/k_B T \qquad (4.3)$$

$f_{B-E}(\varepsilon) = (e^{\beta h\nu} - 1)^{-1}$ is just the Bose-Einstein statistics, which was formally brought up by an Indian physicist, Bose, for all fundamental particles with integer spins, and was introduced to the German physics society by A. Einstein in 1924.

At the temperature T, the equilibrium total energy $\bar{E} = \bar{n}h\nu$ and the specific heat $C = d\bar{E}/dT$ of the phonon gas or atomic vibrations are simply:

$$\bar{E} = 3N\left[\frac{1}{2}h\nu + \frac{h\nu}{e^{\beta h\nu}-1}\right]; \qquad (4.4)$$

$$C_{\text{mole}}(T) = 3N_A k_B \left(\frac{\Theta_E}{T}\right)^2 \frac{e^{\Theta_E/T}}{(e^{\Theta_E/T}-1)^2}. \qquad (4.5)$$

It should be emphasized that the ground-state or zero-point energy $\frac{1}{2}h\nu$ of an oscillator did not appear in the Einstein model in 1907, but was first brought up by Max Planck in 1910. The temperature coefficient $\Theta_E = \hbar\omega/k_B$ is called the Einstein temperature and it is a characteristic constant of a certain solid.

The high-temperature and low-temperature expansions of the specific heat in the Einstein model are helpful to understand the varying tendency versus temperature:

$$C_{\text{mole}}(T) \sim 3R\left[1 - \left(\frac{\Theta_E}{T}\right)^2\right] \quad T \gg \Theta_E \quad (4.6)$$

$$C_{\text{mole}}(T) \sim 3R\left(\frac{\Theta_E}{T}\right)^2 e^{-\Theta_E/T} \quad T \ll \Theta_E \quad (4.7)$$

The high-temperature expansion agrees quite well with experiments, and obviously tends to the Dulong-Petit law $C_{\text{mole}} = 3R$ ($c_p \simeq c_v$ in solids). The low-temperature expansion of the Einstein model was not complete and the derived results deviated from experiments, due to the first basic assumption in the Einstein model: the over simplicity of phonon spectrum $\varepsilon = h\nu =$ const.

4.2 Debye Model of Specific Heat

Peter Debye was born in 1884 in Maastricht in the Netherlands. He obtained his degree in electrical technology at the Aachen Institute of Technology in 1905. After that, he worked for two years as an assistant in Technical Mechanics at the Aachen Technological Institute, and then at Munich University, where, in 1908, he obtained his Ph.D. in Physics under the supervision of Prof. Arnold Sommerfeld.

In 1911, Debye became Professor of Theoretical Physics at Zurich University, succeeding Albert Einstein. He read the notes left by Einstein about phonons and specific heat. Based on his own work from 1910, in which he developed Planck's radiation law, Debye modified the Einstein model and proposed his own phonon model.

The basic assumptions of the Debye phonon model utilized the concept of the phonon introduced by Einstein, but replaced the constant frequency ν in the Einstein model with a photon-like dispersion in the Planck radiation theory:

1. The dispersion relation is $\omega = vk$ for atomic vibrations, where v is the sound wave velocity in the solid. The phonon energy quantum is $\varepsilon = \hbar\omega = \hbar v k$.

2. Utilize the quantum statistics, i.e., the statistics of phonons in Eq. (4.3) in the Einstein model. In the statistical calculation of the total phonon energy, the frequency distribution $g(\omega)$ cuts off at a Debye frequency ω_D, which is determined by the total degree of freedom $3N$ of atomic vibrations.

The concept of "density of states" (DOS) $\rho(k)$ was developed by Peter Debye when he worked with Max Planck on the formula of electromagnetic radiation based on the statistical theory. Debye then extended the concept of DOS to the Einstein model of phonons. $\rho(k)$ only depends on the dimension and boundary conditions, not the details of the quasi-particle:

$$\bar{E}_{\text{em}} = 2 \sum_{\vec{k}} \frac{\hbar \omega}{e^{\beta \hbar \omega} - 1} = 2 \int dn \, \frac{\hbar \omega}{e^{\beta \hbar \omega} - 1}$$

$$= 2 \int dk \, \rho(k) \frac{\hbar c k}{e^{\beta \hbar c k} - 1}; \quad (4.8)$$

$$dn = \frac{4\pi k^2 dk}{\Omega^*/N_L} = \frac{V}{2\pi^2} k^2 dk; \quad \rho(k) = \frac{dn}{dk} = \frac{V}{2\pi^2} k^2 \quad (4.9)$$

where 2 in Eq. (4.8) stands for the ± 1 spin degeneracy of a photon; dn is the total degeneracy in a volume $k \to k + dk$ in the reciprocal space; with the Born-Karman boundary condition, $\Omega^*/N_L = (2\pi/La)^3 = 8\pi^3/V$ is the average volume in the reciprocal space occupied by a wavevector $|\vec{k}\rangle$ (quantum state), as discussed in Section 4.3.2. DOS is an important tool in solid state physics.

The energy density of states (EDOS), or the frequency distribution $g(\omega)$ in the Debye model, can be found from the DOS $\rho(k)$ based on the dispersion relation $\omega(k)$ of phonons. The EDOS of other quasi-particles in a solid, such as band electrons or magnons, can also be found based on their energy spectra $E(k)$. The photon has a two-fold spin degeneracy because the electromagnetic wave is a pure transverse wave; similarly, the phonon has a three-fold degeneracy for each $\vec{k}$, because an elastic wave in a solid medium should have one longitudinal mode and two transverse modes, whose dispersion relations are $\omega_1 = v_l k$ and $\omega_{2,3} = v_t k$ respectively, so:

$$\bar{E} = \sum_{m=1}^{3} \int dk \, \rho(k) \frac{\hbar \omega}{e^{\beta \hbar \omega_m(k)} - 1} = \int d\omega \, g(\omega) \frac{\hbar \omega}{e^{\beta \hbar \omega} - 1};$$

$$g(\omega) = \sum_{m=1}^{3} \rho(k) \frac{dk}{d\omega_m} = \sum_{m=1}^{3} \frac{V}{2\pi^2} k^2 \frac{1}{v_m} = \frac{V}{2\pi^2} \sum_{m=1}^{3} \frac{1}{v_m^3} \omega^2$$

$$= \frac{V}{2\pi^2} \left(\frac{1}{v_l^3} + \frac{2}{v_t^3} \right) \omega^2 = \frac{V}{2\pi^2} \frac{3}{v_s^3} \omega^2$$

(4.10)

where v_s is the average sound velocity. The Debye frequency distribution $g(\omega) \propto \omega^2$ would cause divergence at high-frequency, similar to the Rayleigh-Jeans catastrophe $P_{\text{rad}} \propto \lambda^{-4} T$ of electromagnetic radiation. To solve this problem, Debye introduced a cut-off frequency, namely the Debye frequency ω_D, to satisfy the criteria of total degrees of freedom for atomic vibrations:

$$3N = \int_0^{\omega_D} d\omega\, g(\omega) = \frac{V}{2\pi^2} \frac{1}{v_s^3} \omega_D^3 \quad \rightarrow \quad g(\omega) = \frac{9N}{\omega_D^3} \omega^2 \quad (4.11)$$

The physical meaning of ω_D is the maximum frequency of atomic vibrations. With the frequency distribution known, the equilibrium total energy and the specific heat $C_V = d\bar{E}/dT$ of phonons in a crystal can easily be solved:

$$\bar{E} = \int_0^{\omega_D} d\omega\, g(\omega) \frac{\hbar\omega}{e^{\beta\hbar\omega} - 1}$$

$$= 9N\hbar\omega_D \int_0^{\beta\hbar\omega_D} dx \frac{x^3}{e^x - 1} \quad (4.12)$$

$$C_{\text{mole}} = \int_0^{\omega_D} d\omega\, g(\omega) \frac{\frac{\hbar^2\omega^2}{k_B T^2} e^{\beta\hbar\omega}}{(e^{\beta\hbar\omega} - 1)^2}$$

$$= 9R \left(\frac{T}{\Theta_D}\right)^3 \int_0^{\Theta_D/T} dx \frac{x^4 e^x}{(e^x - 1)^2}, \quad (4.13)$$

where $\Theta_D = \hbar\omega_D/k_B$ is defined as the Debye temperature, which is a better characteristic constant than Θ_E to distinguish the thermal properties in different solids.

The Debye model explained the specific heat of simple crystals very well. The high-temperature and low-temperature expansions of the specific heat are:

$$C_{\text{mole}}(T) \sim 3R\left[1 - \frac{3}{5}\left(\frac{\Theta_D}{T}\right)^2\right] \quad T \gg \Theta_D \quad (4.14)$$

$$C_{\text{mole}}(T) \sim 9R\left(\frac{T}{\Theta_D}\right)^3 \frac{4\pi^4}{15} = \frac{12\pi^4}{5} R \left(\frac{T}{\Theta_D}\right)^3 \quad T \ll \Theta_D \quad (4.15)$$

The high-temperature expansion in the Debye model decreases slightly more slowly than the temperature in the Einstein model, as shown in Figure 4.4(a); the low-temperature expansion illustrates Debye's T^3-law, which was later proved by Walther Nernst's low-temperature experiments.

When specific heat is expressed as a function of $C_V/3Nk_B$ versus T/Θ_D, the measured data of all solids will locate around the solid line in

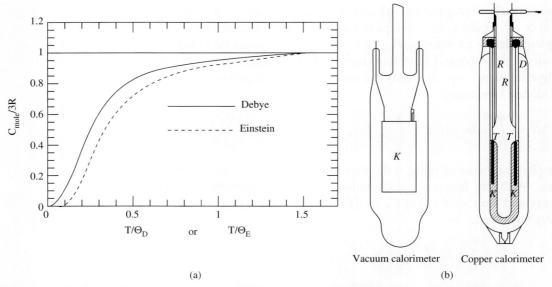

Figure 4.4 Debye specific heat curve: (a) comparison between the Einstein model and the Debye model; (b) low-temperature equipment of Nernst in 1911

Figure 4.4(a); which illustrates the universality of the Debye model. The Debye temperatures are listed for some simple elements and compounds in Table 4.1. Following Eq. (4.11), the relationships between Θ_D and the basic properties of solids are:

$$\Theta_D = \frac{\hbar}{k_B}\omega_D = \frac{\hbar}{k_B}\left(\frac{6\pi^2 N}{V}v_s^3\right)^{1/3} \sim \frac{\hbar}{k_B}\sqrt{\frac{Ea}{M_{atom}}}, \quad (4.16)$$

where $v_s = \sqrt{E/\rho}$ is the average sound velocity, E is the elastic constant; a is the lattice constant; and M_{atom} is the average atomic mass. In the same family, when the atomic number increases, M_{atom} rises, E decreases, and a increases slightly, resulting in a lower Debye temperature. In the same period, Θ_D is mainly determined by the structure or the elastic constant E. The Debye temperature calculated by Eq. (4.16) fits very well with the data in Table 4.1 for most crystals if the sound speed at room temperature is used.

The deviations from the Debye model of specific heat come from several sources. The largest deviations are caused by other quasi-particles in solids: (1) In metals, at very low temperatures the specific heat follows bT but not aT^3, as shown in Figure 4.5(b). This is caused by the conducting electrons, which will be discussed in the next chapter. (2) In ferromagnetic insulators, at very low temperatures, the specific heat follows $cT^{3/2}$. This is caused by the spin wave, which will be

Table 4.1 Debye temperature (room temperature fit) of simple crystals

Matter	$\Theta_D(K)$		$\Theta_D(K)$		$\Theta_D(K)$		$\Theta_D(K)$		$\Theta_D(K)$
Na	150	Be	1000	B	1250	C	1860	NaCl	281
K	100	Mg	318	Al	390	Si	625	KCl	227
Cu	315	Ca	230	Ga	240	Ge	360	KBr	177
Ag	215	Zn	250	In	129	Sn	170	AgCl	183
Au	170	Cd	172	Tl	96	Pb	88	AgBr	144

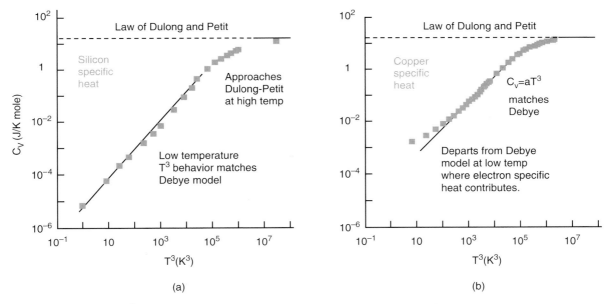

Figure 4.5 Debye T^3-law of specific heat at low temperature: (a) silicon; (b) copper (from http://hyperphysics.phy-astr.gsu.edu)

discussed in Chapter Seven. To summarize, phonons, electrons, and magnons are the three sources of specific heat in solids.

Other deviations are caused by the complicity of atomic vibrations, or experimental conditions: (3) In experiments, the parameters (P, T) are usually fixed, therefore the measured specific heat is C_P, which is a little different from C_V calculated in this section: $C_P = C_V + TV(\alpha_V^2/\beta)$ (the volume expansion coefficient $\alpha_V = \frac{1}{V}(\partial V/\partial T)_P$, the depression coefficient $\beta = -\frac{1}{V}(\partial V/\partial P)_T$). The difference could be up to 10% at room temperature; (4) the fitted Debye temperature is not a constant versus temperature. Macroscopically, at very low temperatures, E must be larger, and the fitted Θ_D would be higher, as shown

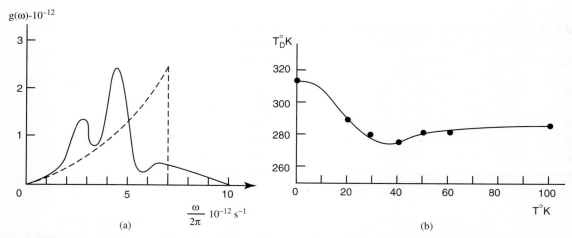

Figure 4.6 (a) Comparison of the measured frequency distribution in NaCl and the Debye approximation (based on Kellermann, 1941); (b) Debye temperature of NaCl at different temperatures (based on Kellermann, 1941)

in Figure 4.6(b); microscopically, the frequency distribution $g(\omega)$ measured by neutron diffraction is much more complex than Debye's ω^2 form, as shown in Figure 4.6(a), which will be discussed in the following section. (5) Some solids, such as sodium chloride and silver, have smaller specific heat than the values obtained from the Dulong-Petit law at high temperatures. Born and Brody's work (1921) explained it in terms of the Δx^3 expansion in the potential, which was beyond the scope of the lattice harmonic theory, covered later in this chapter.

Peter Debye was successful in both physics and chemistry. In 1915, he used the X-ray diffraction method to measure the bond lengths & angles of molecules such as N_2, O_2, F_2, CO_2, and H_2O. Therefore the unit of the electric dipole of molecules became known as the "debye," as seen in Chapter Two. In 1916, Debye worked with Sherrer to develop the powder X-ray diffraction method. In 1936, Debye was awarded the Nobel Prize in Chemistry for his contributions in determining the electrical and geometrical structures of molecules using X-ray diffraction. He moved to the United States in 1940, and became a Professor of Chemistry and the head of the Chemistry Department at Cornell University.

4.3 Lattice Dynamics and Neutron Diffraction

The atomic vibrations of a crystal determine a wide range of macroscopic behaviors, such as the specific heat and sound speed versus phonon spectra, the resistivity versus electron-phonon interactions, infrared absorption, and Raman scattering versus photon-phonon interactions.

The atomic vibration wave and the phonon reveal the wave-particle duality of a quasi-particle representing the thermal motion in solids. Historically, the concept of the phonon was described by A. Einstein in 1907 as the energy quanta of thermal motions in solids, and the lattice dynamics for the calculation of phonon spectra or lattice vibration wavefunctions was published by Max Born and Theodore von Karman in 1913.

4.3.1 Lattice Harmonic Theory

The atomic vibration wave is a microscopic mechanical wave, which can be described using a complex wavefunction on a discretizing lattice. The lattice harmonic theory is based on Hooke's Law, where only the $\Delta U \sim \Delta x^2$ expansion of potential is considered. The basic assumptions are:

1. The equilibrium position of an atom stays at the lattice site $\vec{R}_{lm} = \vec{R}_l + \vec{\delta}_m$ ($l = 1, 2, \ldots, N_L$ labels a Bravais lattice, in n-dimensional systems and l should be a set of integers $\{l_\beta\}$ for $\vec{R}_l = l_1 \vec{a}_1 + \cdots + l_n \vec{a}_n$ vectors; $m = 1, 2, \ldots, n_a$ is the displacement of a basis site $\vec{\delta}_m$ inside the PUC). The atomic motion from the equilibrium position is defined as $\vec{u}_{lm} = \vec{r}_{lm} - \vec{R}_{lm}$, where $\vec{r}_{lm}$ is the nucleus position of the (l, m) atom.
2. The displacement $|\vec{u}|$ is much less than the lattice constant a, b, c, therefore the harmonic approximation can be applied: the empirical potentials are expanded into the second order of the displacement in the Hamiltonian.
3. In principle, the potential among atoms is quantum mechanical electromagnetic interactions among electrons. The values of the potential are usually empirical and can sometimes be found by *ab initio* calculations.

The potential $U(\vec{r}_{lm} - \vec{r}_{l'm'})$ between the (l, m) atom and the (l', m') atom in a crystal can be expanded in the Taylor series:

$$U(\vec{r}_{lm} - \vec{r}_{l'm'})$$
$$= \phi_0 + \frac{1}{2!}(u_{lm\alpha} - u_{l'm'\alpha})U_{\alpha\alpha'}(\vec{R}_{lm} - \vec{R}_{l'm'})(u_{lm\alpha'} - u_{l'm'\alpha'})$$

(4.17)

where Einstein's notation is used for 3D components $\alpha, \alpha' = 1, 2, 3$. The first-order expansion disappears because, at equilibrium positions, atoms are free from force. The second-order derivative of the potential $U_{\alpha\alpha'}(\vec{R}_{lm} - \vec{R}_{l'm'}) = \partial_\alpha \partial_{\alpha'} U|_{u=0}$ has characteristics of the elastic coefficients K, and it can be estimated from the intrinsics of the chemical bonds. When the atom-atom distance variation $|\vec{u}_{lm} - \vec{u}_{l'm'}|$ is of the order of 1Å, the variation of the atomic potential is about 1 eV, as discussed in Chapter Two. Therefore the microscopic elastic constant K and the macroscopic elastic constant E can be estimated as:

$$K \simeq \frac{2\Delta U}{\Delta x^2} \simeq \frac{2 \times 1\text{eV} \times 1.6 \times 10^{-19} \text{ J/eV}}{(10^{-10} \text{ m})^2} = 32 \text{ N/m} \quad (4.18)$$

$$E = \frac{F/A}{\Delta x/a} = \frac{K}{a} \simeq \frac{32 \text{N/m}}{3 \times 10^{-10} \text{ m}} = 10^{11} \text{ N/m}^2 \quad (4.19)$$

which agrees with the order of Young's module 100 GPa (Hooke's law $F = K\Delta x$; the area of a sample A, and the lattice constant of a crystal a).

The motion equation in the lattice harmonic theory can be found from the Hamiltonian, which is a standard method in analytical mechanics. Analytical mechanics was developed from Newtonian mechanics by some of the great European mathematicians of the eighteenth and nineteenth centuries, including D. Bernoulli, J. L. Lagrange, P. de Laplace, and W. R. Hamilton. The physics itself was not changed, but systems with large numbers of particles could be investigated. The total Hamiltonian $\mathcal{H}$ of atomic vibrations in a crystal can be expressed by the function of displacements $\vec{u}_{lm}$:

$$\mathcal{H} = \sum_{lm\alpha} \frac{1}{2} M_m \dot{u}_{lm\alpha}^2 + \frac{1}{2} \sum_{lm\alpha} \sum_{l'm'\alpha'} \phi_{\alpha\alpha'}(lm; l'm') u_{lm\alpha} u_{l'm'\alpha'} \quad (4.20)$$

$$\phi_{\alpha\alpha'}(lm; lm) = \sum_{l'm' \neq lm} U_{\alpha\alpha'}(\vec{R}_{lm} - \vec{R}_{l'm'})$$

$$\phi_{\alpha\alpha'}(lm; l'm') = -U_{\alpha\alpha'}(\vec{R}_{lm} - \vec{R}_{l'm'}) \quad (l \neq l' \text{ or } m \neq m') \quad (4.21)$$

where the equilibrium binding energy of the crystal is set to zero at the hamiltonian; $\phi_{\alpha\alpha'}(lm; l'm')$ is a symmetric potential matrix; and the atomic mass M_m only depends on the index m inside the PUC. The equation of motion is:

$$M_m \ddot{u}_{lm\alpha} = -\sum_{l'm'\alpha'} \phi_{\alpha\alpha'}(lm; l'm') u_{l'm'\alpha'} \quad (l = 1 - N_L, m = 1 - n_a)$$

$$(4.22)$$

To solve a set of $3N_L n_a$ correlated equations, the Fourier method, developed in 1811, can be used. The standard trial function is a plane wave on a lattice with a frequency $\omega = 2\pi/T$ and a wave vector $\vec{k} = \hat{k}(2\pi/\lambda)$:

$$u_{lm\alpha}(t) = \frac{1}{\sqrt{M_m}} \tilde{u}_{m\alpha}(\vec{k}) e^{i\vec{k}\cdot\vec{R}_l - i\omega t} \quad (l = 1 - N_L, m = 1 - n_a)$$

$$(4.23)$$

where $A_{m\alpha} = \tilde{u}_{m\alpha}/\sqrt{M_m}$ is the amplitude of the (l, m) atom in the α-direction. By substituting Eq. (4.23) into Eq. (4.22), $3N_L n_a$ motion equations can be simplified into $3n_a$ eigen-equations in the Fourier space by defining a $3n_a \times 3n_a$ dynamic matrix $D_{m\alpha}^{m'\alpha'}(\vec{k})$ with ($m, m' = 1, 2, \ldots, n_a$; $\alpha, \alpha' = 1, 2, 3$):

$$\omega^2 \tilde{u}_{m\alpha}(\vec{k}) = \sum_{m'\alpha'} D_{m\alpha}^{m'\alpha'}(\vec{k}) \tilde{u}_{m'\alpha'}(\vec{k}) \tag{4.24}$$

$$D_{m\alpha}^{m'\alpha'}(\vec{k}) = \frac{1}{\sqrt{M_m M_{m'}}} \sum_{l'} \phi_{\alpha\alpha'}(lm; l'm') \, e^{i\vec{k}\cdot(\vec{R}_{l'}-\vec{R}_l)} \tag{4.25}$$

The $3n_a \times 3n_a$ dynamic matrix is independent of l and l' because the lattice has a translational symmetry. By Eq. (4.21), $\phi_{\alpha\alpha'}(lm; l'm')$ only depends on $\vec{R}_l - \vec{R}_{l'}$.

The eigen-equations Eq. (4.24) can be solved using a method similar to the matrix method of the Heisenberg representation in quantum mechanics. The dispersion relation of the phonons can be found by the determinant:

$$\left| D_{m\alpha}^{m'\alpha'}(\vec{k}) - \omega^2 \delta_{mm'} \delta_{\alpha\alpha'} \right| = 0 \tag{4.26}$$

With the eigenvalues $\omega_j^2(\vec{k})$ ($j = 1, 2, \ldots 3n_a$) solved, the corresponding eigenvectors $\tilde{u}_{m\alpha}^j = e_{m\alpha}(j, \vec{k})$ and eigenfunctions $\vec{u}_{j\vec{k}}(\vec{r}, t) = \vec{u}_{lm}^j(t)$ can be found as well:

$$\vec{u}_{lm}^j(t) = \frac{\vec{e}_m(j, \vec{k})}{\sqrt{M_m}} e^{i\vec{k}\cdot\vec{R}_l - i\omega_j(\vec{k})t} = A_m(j, \vec{k}) e^{i\vec{k}\cdot\vec{R}_l - i\omega_j(\vec{k})t} \tag{4.27}$$

Following the Born-Karman condition, the wavevector $\vec{k}$ has N_L values in the first Brillouin zone (FBZ). Therefore there are $3n_a \times N_L$ kinds of atomic vibrations $\vec{u}_{j\vec{k}}(\vec{r}, t)$, which are called the "normal modes" of lattice vibration in a crystal.

Following the wave-particle duality principle, there are $3n_a N_L$ types of phonon corresponding to the $3n_a N_L$ vibrational degrees of freedom in a crystal. Therefore the quantum mechanical version of the atomic vibration Hamiltonian is:

$$\mathcal{H} = \sum_{j=1}^{3n_a} \sum_{\vec{k}}^{N_L} \hbar \omega_j(\vec{k}) \left(n_{j\vec{k}} + \frac{1}{2} \right) \Rightarrow \bar{E} = E_0 + \int d\omega \, g(\omega) \frac{\hbar\omega}{e^{\beta\hbar\omega} - 1}$$

(4.28)

At 0 K, the quantum number $n_{j\vec{k}}$ of any phonon is zero; at temperature T, the equilibrium number $\bar{n}_{j\vec{k}}$ follows the Bose-Einstein statistics

$(\exp[\beta\hbar\omega_j(\vec{k})] - 1)^{-1}$. The ground state energy $\frac{1}{2}\hbar\omega_j(\vec{k})$ is non-zero due to the Heisenberg uncertainty principle.

Both the phonon and photon are bosons; however, the phonon is a quasi-particle, and the photon is a fundamental particle. A free fundamental particle must satisfy the Einstein energy-momentum relation $\varepsilon^2 = c^2 p^2 + m^2 c^4$, so intrinsic properties like charge and spin can be defined. A quasi-particle, such as the phonon, has a much more complex dispersion $\varepsilon = \hbar\omega_j(\vec{k})$, which does not satisfy the Einstein energy-momentum relation.

4.3.2 Optical Branches and Acoustic Branches

The phonon wave of discrete atoms, as illustrated in Figure 4.7, is very complex. The wave is three-dimensional, and it depends on the magnitude of wave vector and the detailed orientation of $\vec{k}$ with respect to the unit cell. For example, the sound speed is dependent on the orientation of $\vec{k}$ in the conventional unit cell.

The phonon spectrum $\hbar\omega_j(\vec{k})$ or dispersion $\omega_j(\vec{k})$ includes $j = 1, 2, \ldots, 3n_a$ branches. Due to the relationship $\vec{G} \cdot \vec{R} = 2\pi n$, the dynamic matrix in Eq. (4.25) satisfies $D_{m\alpha}^{m'\alpha'}(\vec{k}) = D_{m\alpha}^{m'\alpha'}(\vec{k} + \vec{G})$, and is therefore a periodic function at the reciprocal lattice. ω_j^2 is an eigen-value of the D-matrix, hence all branches of dispersion $\omega_j(\vec{k})$ are periodic functions at the reciprocal lattice, and can be expressed in the first Brillouin zone (FBZ). The phonon dispersion branches with $\omega \simeq vk$ near the

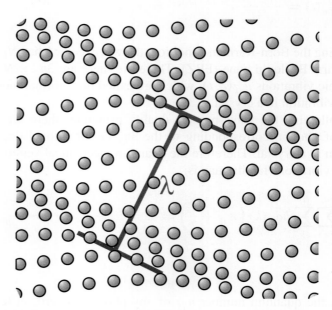

Figure 4.7 Phonon wave in the atomic scale of a crystal

center Γ of FBZ are called acoustic branches; others are called optical branches.

In solid state physics, the quasi-particle (electron, phonon, and magnon) spectrum $\varepsilon(\vec{k})$ forms a 3D surface in a 4D $\varepsilon - \vec{k}$ space, which is impossible to show visually in the 3D space we live in. In a similar manner to phonons, the spectra of other quasi-particles also have periodicity $\varepsilon(\vec{k}) = \varepsilon(\vec{k} + \vec{G})$ in the reciprocal space. The conventional expression of $\varepsilon(\vec{k})$ is to draw a series of "curves" $\varepsilon(k = \hat{n} \cdot \Delta \vec{k})$ along the symmetric director $\hat{n}$ in the FBZ, such as $\Gamma - H - P - \Gamma - N - H$ for BCC crystals, $\Gamma - X - K - \Gamma - L$ for FCC crystals, and $\Gamma - M - L - \Gamma - K$ for HCP crystals, as illustrated in Figure 4.8. The chosen symmetric direction is usually one of the most important [100], [110], [111] directions: the [100] or equivalent direction is labeled as Δ, where the wave vector $\Delta \vec{k} = \vec{k} = (k, 0, 0)$; the [110] or equivalent direction is labeled as Σ, where $\vec{k} = \sqrt{1/2}(k, k, 0)$; and the [111] or equivalent direction is labeled as Λ, where $\vec{k} = \sqrt{1/3}(k, k, k)$.

The silicon crystal is an important functional material, the phonon spectrum of which is important in various applications. The silicon crystal has a DIA structure, which has two independent sites in a PUC; therefore there must be $3n_a = 6$ branches in its phonon spectrum, i.e., for each $\vec{k}$, the maximum number of eigen-frequencies is $3n_a$, and these eigen-frequencies can degenerate due to the symmetry of the lattice. In Figure 4.9, there are six branches in the [110] direction (Σ), but only four branches in the [100] (Δ) and [111] (Λ) directions, because the two transverse (optical and acoustic) phonon dispersions degenerate in the [100] and [111] directions. In the silicon phonon spectrum, the maximum frequency $\omega_{j\vec{k}}/2\pi$ is in the order of 10 THz, which is consistent

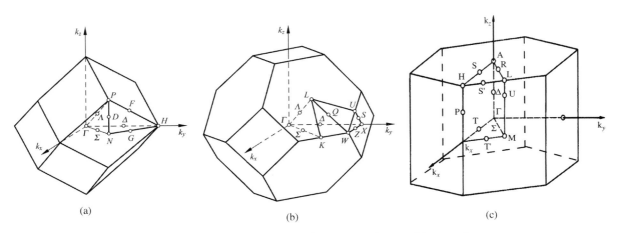

Figure 4.8 FBZ and the special points for (a) BCC, (b) FCC, and (c) HCP crystals

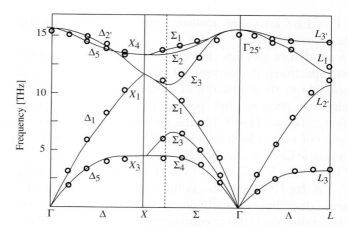

Figure 4.9 Phonon dispersion of silicon crystals (from http://www.personal.psu.edu/pce3/Research%20Topics.htm)

with the expectation of the Debye model:

$$\omega_D = k_B \Theta_D / \hbar = \frac{1.38 \times 10^{-23} \times 625}{1.05 \times 10^{-34}} = 2\pi \times 13 \text{ THz} \quad (4.29)$$

The calculated Debye frequency is a little less than the maximum frequency in the true phonon spectrum, and this can be verified by the comparison of the theoretical/experimental $g(\omega)$, as shown in Figure 4.6(a). The sound speed in silicon can also be estimated from the dispersions in Figure 4.9: $v \simeq \omega_D/(a^*/2) \simeq 10^4$ m/s.

The optical and acoustic branches in real 3D crystals are quite complex, therefore a simpler model is required to calculate the typical optical and acoustic phonon dispersions. Here we consider the longitudinal lattice vibration wave propagating in the [100] direction of a sodium chloride crystal, where the lattice can be dissolved into an assembly of independent 1D chains. The lattice constant, or the periodicity of the diatomic 1D sodium chloride chain, is labeled a. The masses of Cl^- and Na^+ are M and m respectively. The elastic coefficient $K = U_{11}(R_0)$ is assumed for the ionic bond potential. The potential matrix $\phi(ln; l'n')$ ($n, n' = 1, 2$) can be found following the equation of motion in Eq. (4.22), or directly obtained by Eq. (4.21), and the non-zero matrix elements are:

$$\begin{aligned}
M\ddot{u}(l, 1) &= -K[u(l, 1) - u(l - 1, 2)] - K[u(l, 1) - u(l, 2)] \\
m\ddot{u}(l, 1) &= -K[u(l, 2) - u(l, 1)] - K[u(l, 2) - u(l + 1, 1)]; \\
\phi(l, 1; l, 1) &= 2K; \quad \phi(l, 1; l, 2) = -K; \quad \phi(l, 1; l - 1, 2) = -K \\
\phi(l, 2; l, 2) &= 2K; \quad \phi(l, 2; l, 1) = -K; \quad \phi(l, 2; l + 1, 1) = -K.
\end{aligned}$$

$$(4.30)$$

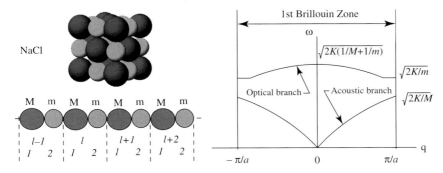

Figure 4.10 Optical branch and acoustic branch of [100] longitudinal waves in NaCl

The 1D 2 × 2 dynamic matrix can be calculated from Eq. (4.25):

$$\tilde{D} = \begin{pmatrix} \frac{2K}{M} & \frac{1}{\sqrt{Mm}}(-K - Ke^{-ika}) \\ \frac{1}{\sqrt{Mm}}(-K - Ke^{ika}) & \frac{2K}{m} \end{pmatrix} \quad (4.31)$$

The eigenvalues of the dynamic matrix can easily be solved:

$$\omega_{\pm}^2 = 2K \frac{(M+m) \pm \sqrt{M^2 + m^2 + 2Mm\cos ka}}{2Mm} \quad (4.32)$$

where $\omega_+(k)/\omega_-(k)$ are the optical/acoustic branches, as shown in Figure 4.10.

The long-wavelength and short-wavelength expansions of optical and acoustic phonon spectra and eigenvectors are:

$$\omega_+(ka \leq 1) \simeq \sqrt{2K(1/M + 1/m)}; \quad \begin{pmatrix} A_M \\ A_m \end{pmatrix} \propto \begin{pmatrix} m \\ -M \end{pmatrix}$$

$$\omega_-(ka \leq 1) \simeq \sqrt{2K/(M+m)}\, ka/2; \quad \begin{pmatrix} A_M \\ A_m \end{pmatrix} \propto \begin{pmatrix} 1 \\ 1 \end{pmatrix}$$

$$\omega_+(ka = \pi) \simeq \sqrt{2K/m}; \quad \begin{pmatrix} A_M \\ A_m \end{pmatrix} \propto \begin{pmatrix} 1 \\ 0 \end{pmatrix}$$

$$\omega_-(ka = \pi) \simeq \sqrt{2K/M}; \quad \begin{pmatrix} A_M \\ A_m \end{pmatrix} \propto \begin{pmatrix} 0 \\ 1 \end{pmatrix} \quad (4.33)$$

Following Eq. (4.27), the relationship between the vector amplitude of an atom in a crystal and the eigenvector is $\vec{A}_m(j, \vec{k}) = \vec{e}_m(j, \vec{k})/\sqrt{M_m}$. It can be seen in Eq. (4.33) that, in the long-wavelength limit $ka \ll 1$, the acoustic branch has the expected dispersion in the form $\omega_- = v_s k$ with an oscillation mode of uniform motion $A_M = A_m$ for all atoms in PUC; the optical branch reaches the highest frequency $f_l = \omega_+(0)/2\pi$

with an oscillation mode of intense opposite motion $A_M M + A_m m = 0$ toward the center of mass in PUC.

The sound speed in a crystal can be estimated by the slope of the longitudinal acoustic dispersion in the $ka \leq 1$ region:

$$v_s = \sqrt{\frac{K}{\bar{M}}} \frac{a}{2} \sim \sqrt{\frac{10^0 - 10^1}{10^{-27}}} \times 10^{-10} \sim 10^3 - 10^4 \text{ m/s} \quad (4.34)$$

where $\bar{M}$ is the average mass of atoms in a crystal and a is the lattice constant. The estimated range $10^3 - 10^4$ m/s of sound speed in crystals agrees well with experiments, as listed in Table 4.2. In the periodic table, the sound speed has similar behaviors to the Debye temperature, because both physical quantities are proportional to $\sqrt{K/M}$ at the atomic scale in a crystal.

Ionic crystals are often used as optical materials. Sodium chloride is the most common infrared transmission crystal window at $\lambda < 20\ \mu$m for gas and liquid sample cells used in infrared and FTIR spectrophotometers, as shown in Figure 4.11. Sodium chloride is also used as an infrared reflector to produce monochrome infrared lines. These optical

Table 4.2 Longitudinal velocity of sound in crystals (unit: m/s)

Na	3200	Be	13000	B	16200	C	18350	NaCl	4780
K	2000	Mg	4602	Al	5100	Si	8430	iron	5900
Cu	3570	Ca	3810	Ga	2740	Ge	5400	wood	3500
Ag	2600	Zn	3700	In	1215	Sn	2500	glass	5500
Au	1740	Cd	2310	Tl	818	Pb	1260	Teflon	1390

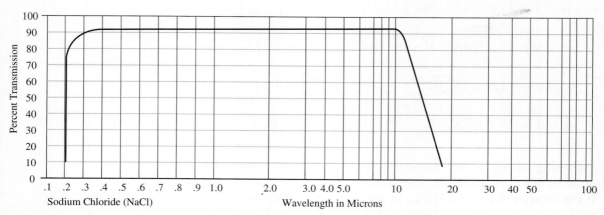

Figure 4.11 Transmission coefficient vs wavelength (unit: μm) of NaCl (from www.internationalcrystal.net)

properties are caused by photon-phonon interactions. The photon-phonon resonance must occur in the long wavelength range ($ka \ll 1$) of atomic vibrations. This is determined by both the energy conservation law $h\nu = \hbar\omega_+(0)$ and the momentum conservation law $h\nu/c = \hbar k$. The resonance frequency can be estimated by the wavelength with respect to the highest frequency f_l in the optical branch:

$$\lambda_0 = \frac{2\pi c}{\sqrt{2K(1/M + 1/m)}}$$
$$= \frac{2\pi \times 3 \times 10^8}{\sqrt{2 \times 32(1/35.45 + 1/22.99)/1.67 \times 10^{-27}}} = 36 \ \mu\text{m} \quad (4.35)$$

In the infrared transmission window $\lambda < 20 \ \mu$m of sodium chloride, the frequency of photons is higher than f_l, so the photon-phonon resonance cannot take place, and sodium chloride is transparent. In the infrared reflector of sodium chloride, the monochrome infrared line ($ka \ll 1$ and $f_l = \omega_+/2\pi$) can be produced by multi-reflections on single sodium chloride crystals, because the strong photon-phonon resonance results in the reflection coefficient $R = 1$, which will be discussed in detail in Chapter Eight.

4.3.3 Measurement of Phonon Spectra by Neutron Diffraction

The measurement of phonon spectra requires the use of neutron diffraction. The neutron has no charge, therefore it can travel through the electron shells of an atom and interact directly with the nucleus. The elastic scattering produces neutron diffraction, which reveals the structure of light elements, such as hydrogen, and heavy elements equally well. The inelastic collisions can be used to measure the atomic vibrations (phonon spectrum) and spin waves (magnon spectrum) in solids quite accurately, because the energy of the "hot" neutron in diffraction is quite small:

$$E = \left(\frac{0.286\text{Å}}{\lambda}\right)^2 \text{eV} \simeq \left(\frac{0.286\text{Å}}{1\text{Å}}\right)^2 \text{eV}$$
$$= 0.082 \text{ eV} = k_B \times 593 \text{ K} \quad (4.36)$$

The neutral nature of the neutron enables nondestructive and deeply penetrating studies of many materials which would otherwise be damaged by the harmful radiation doses of x-rays and electron beams. These are the contributions made by Chadwick, Shull, and Brockhouse, as discussed in Chapter Three.

Without a high energy accelerator, or at least a medium-power nuclear reactor, it is impossible to produce a beam with the intensity

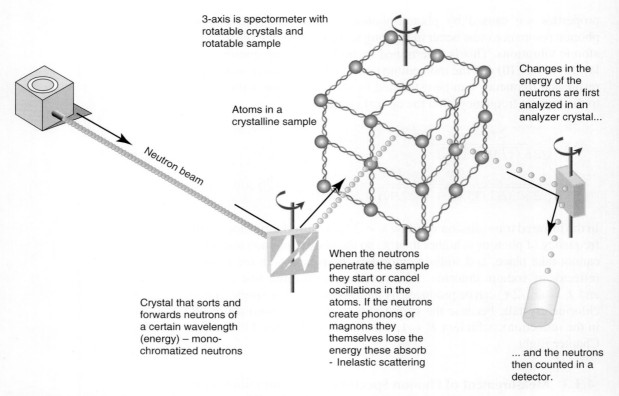

Figure 4.12 Inelastic scattering neutron diffraction (www.personal.engin.umich.edu)

necessary for useful neutron scattering. The first such neutron sources became available in 1945, and were offshoots of the nuclear advancements of the Manhatten Project. In the slow neutron nuclear reactor, the velocity of the neutron obeys the classical Maxwell-Boltzmann statistics: $f(v) \propto \exp(-\beta m v^2/2)$, which is a pure continuum spectrum. To obtain the monochrome neutron diffraction beam, a neutron selector has to be used. A Fermi-type selector consists of two disks with the distance L, rotating frequency ω, and a phase difference $\Delta \phi$. Each disk has a multilayer structure of alternating cadmium layers (high neutron absorbtion) and aluminum layers (low neutron absorbtion). The rotational axis is in the disk plane and perpendicular to the incident beam, therefore the monochrome neutron beam with a wavelength λ can be selected:

$$\lambda = \frac{h}{m_n v} = \frac{\Delta \phi}{\omega} \frac{h}{m_n L} \qquad (4.37)$$

The use of new pulsed source high-intensity neutron sources began in the 1980s, as linear accelerators and synchrotrons at the Oak Ridge National Laboratory provided the radiation necessary for the study of neutron scattering.

The inelastic neutron diffraction theory is similar to the diffraction theory introduced in Chapter Three, except that the outgoing neutron energy $\varepsilon = \hbar\Omega$ can be different from the incoming neutron energy $\varepsilon_0 = \hbar\Omega_0$. Therefore the diffraction intensity should include the time-term:

$$I = |f_{cr}e^{-i\Omega_0 t}|^2 = \left|\sum_n f_a e^{-i\vec{s}\cdot\vec{r}_n - i\Omega_0 t}\right|^2 = f_a^2 |S|^2 \tag{4.38}$$

The lattice structure factor S can be calculated by assuming that the neutron interacts with a phonon with frequency ω, wave vector $\vec{q}$, and wave function $\vec{r}_n = \vec{R}_n + \vec{A}\cos(\vec{q}\cdot\vec{R}_n - \omega t)$:

$$\begin{aligned}S &= \sum_n e^{-i\vec{s}\cdot\vec{r}_n - i\Omega_0 t} \\ &\simeq \sum_n e^{-i\vec{s}\cdot\vec{R}_n - i\Omega_0 t}\left[1 - i\vec{s}\cdot\vec{A}\cos(\vec{q}\cdot\vec{R}_n - \omega t)\right] \\ &= S_0 + \left(-\frac{i}{2}\vec{s}\cdot\vec{A}\right)\sum_n \left[e^{-i(\vec{s}-\vec{q})\cdot\vec{R}_n - i(\Omega_0+\omega)t} + e^{-i(\vec{s}+\vec{q})\cdot\vec{R}_n - i(\Omega_0-\omega)t}\right]\end{aligned}$$

(4.39)

where S_0 is the elastic scattering lattice structure factor discussed in Chapter Three. Based on Eq. (4.39), it can be seen that the inelastic scattering diffraction "points" will appear at the new positions:

$$\vec{s} = \vec{k} - \vec{k}_0 = \vec{G}_{hkl} \pm \vec{q}; \quad \Omega - \Omega_0 = \pm\omega_j(\vec{q}). \tag{4.40}$$

In the measurement, the intensity of neutrons at different energies should be counted near an elastic scattering diffraction point which satisfies the conditions $\vec{s} = \vec{G}_{hkl}$ and $F \neq 0$. In the direction $\vec{k} = \vec{k}_0 + \vec{G}_{hkl} + \vec{q}$, the phonon dispersion can be found by the intensity peaks at the outgoing neutron energies $\varepsilon = \varepsilon_0 + \hbar\omega_j(\vec{q})$.

If the crystals with Bravais lattices are measured, at a certain $\vec{q}$, two to three neutron intensity peaks are found in the energy region $\varepsilon > \varepsilon_0$ with respect to three acoustic branches, as shown in the aluminum phonon dispersion in Figure 4.13(a). If the crystals with non-Bravais lattices (two sites in basis) are tested, at a certain $\vec{q}$, four to six neutron intensity peaks are found in the energy region $\varepsilon > \varepsilon_0$, as shown in the silicon phonon dispersion in Figure 4.9 or the sodium chloride phonon dispersion in Figure 4.13(b).

The phonon density of states, or the frequency distribution $g(\omega)$, can also be measured using the neutron diffraction method. Based on the phonon dispersion of sodium chloride in Figure 4.13(b), $\omega_j(k)$ may maximize or minimize ($d\omega/dk = 0$) at the frequencies $\omega \simeq 2.4$ THz

134 *Chapter 4* Thermal Properties of Solids and Lattice Dynamics

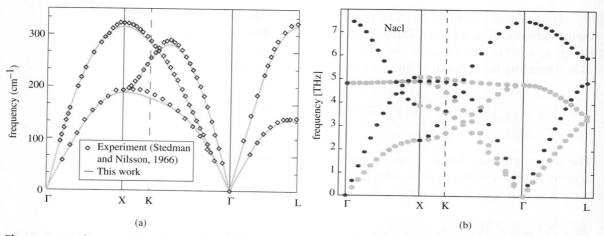

Figure 4.13 Phonon dispersions of FCC-based crystals: (a) Al (based on Stedman and Nilsson, 1966); (b) NaCl (from people.web.psi.ch/delley/nacl.html)

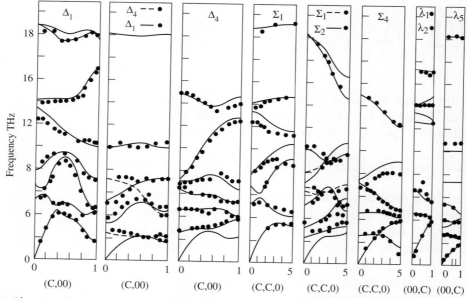

Figure 4.14 Phonon dispersion curves in tetragonal La_2CuO_4, experiments (dots and crosses) and calculations (solid and dashed lines) (from www.kiae.ru)

and 4.9 THz; therefore, by Eq. (4.10), $g(\omega)$ will have sharp maxima at $\omega \simeq 2.4$ THz and 4.9 THz too. This is proved by the measured frequency distribution of sodium chloride in Figure 4.6(b).

The phonon dispersion can be calculated using *ab initio* molecular dynamics, where proper potentials have to be chosen for a certain crystal. In Figure 4.14, the measured phonon dispersion (by neutron diffraction) and the calculation of a lanthanum-copper oxide (La_2CuO_4) single crystal are shown respectively. We can see that the phonon dispersion of the crystal is presented along the symmetry directions of tetragonal FBZ. Lanthanum-copper oxide is one of the mother compounds of high-Tc superconductors, where the electron-phonon interaction is a crucial factor; therefore it is necessary to study its phonon spectrum carefully.

Summary

This chapter introduced the quantum mechanical theory of the thermal, acoustic, and optical properties of solids. The atomic vibration wave and the phonon represent wave-particle duality, and they are the key to understanding the Einstein model, the Debye model, and neutron diffraction. To summarize in more detail:

1. Einstein phonon model: Einstein introduced the concept of phonons, i.e., the energy quanta of thermal atomic motions. The statistical analysis of phonons naturally leads to the Bose-Einstein statistics, and the tend-to-zero specific heat at very low temperatures can be explained.
2. Debye phonon model: Debye modified the Einstein model by introducing a linear spectrum for the phonon, as well as a cut-off frequency, to give a more accurate description of the low-temperature specific heat of various solids.
3. Lattice dynamics: Born and von Karman established a classical small vibration model to analyze the huge number of atoms in solids, therefore the dispersion of atomic vibrations could be calculated. Thus the phonon dispersion could also be found using the wave-particle duality principle.
4. Phonon spectrum and neutron diffraction: the phonon spectrum consists of two parts: the acoustic branch is more important for the thermal and acoustic properties, while the optical branch explains the transmission characteristics in the infrared frequencies. The phonon spectrum can be measured by the neutron spectrum in the inelastic scattering mode, which was first accomplished by Brockhouse.

References

Hall, H. E. (1983), translated by Liu, Z. *Solid State Physics* (in Chinese), Higher Education Press, Beijing.
Huang, K., and Han, R. (1988). *Solid State Physics* (in Chinese), Higher Education Press, Beijing.

von Laue, M. (1978), translated by Fan, D., and Dai, N. *History of Physics* (in Chinese), Commercial Press, Beijing.

Lewis, G. N. and Randall, M. (1961), *Thermodynamics*, 2nd Edition. McGraw-Hill Book Co, New York.

Kellermann, E. W. (1941). "On the Specific Heat of the Sodium Chloride Crystal." Proceedings of the Royal Society of London. Series A, Mathematical and Physical Sciences, Vol. 178, No. 972, 17–24.

Seitz, F. (1940). *The Modern Theory of Solids*, McGraw-Hill, New York.

Stedman, R., and Nilsson, G. (1966). "Dispersion Relations for Phonons in Aluminum at 80 and 300 K." *Phys. Rev.*, Vol. 145, 492–500.

Vocadlo, L., and Alfe, D. (2002). "Ab initio melting curve of the fcc phase of aluminum". *Phys. Rev. B*, Vol. 65, 214105.

"Debye's Contribution to Specific Heat Theory," Nave, C. R. Department of Physics and Astronomy, Georgia State University, City of Atlanta, Georgia, USA. January 1, 2006. <http://hyperphysics.phy-astr.gsu.edu/hbase/thermo/debye.html>.

"Phonon Dispersion Relations in NaCl," Delley, B. Paul Scherrer Institute, City of Villigen, Switzerland. December 15, 2005. <http://people.web.psi.ch/delley/nacl.html>.

"Surface Optical Phonons in III-V, II-VI Semiconducting Nanowires," Eklund, P. C. Physics Department, Pennsylvania State University, City of University Park, USA. December 5, 2005. <http://www.personal.psu.edu/pce3/Research%20Topics.htm>.

"Sodium Chloride (NaCl) Optical Crystal," International Crystal Laboratories, City of Garfield, New Jersey, USA. December 20, 2005. <http://www.internationalcrystal.net/iclsite3/optics_16.htm>.

"Interview with a Neutron: Using the Neutron for Scattering," Hendricks, J. L. College of Engineering, University of Michigan, City of Ann Arbor, USA. January 14, 2006. <http://www-personal.engin.umich.edu/~jlhendri/page4.html>.

"Neutron Scattering Condensed Matter Investigations," Rumiantsev, A. Yu. Russian Research Centre Kurchatov Institute, City of Moscow, Russia. Feburary 11, 2006. <http://www.kiae.ru/rus/inf/new/new9.htm>.

Exercises

4.1 Calculate the area difference between the average specific heat per phonon $c(T)$ in the Einstein model with the classic curve $\tfrac{3}{5}k_B$ by the Dulong-Petit law.

4.2 Calculate the molar specific heat of a circular 1D chain with N atoms (atomic mass m, elastic constant K): (a) by the Debye model; (b) by accurate frequency distribution $g(\omega) = \frac{2N}{\pi}\left[\frac{4K}{m} - \omega^2\right]^{-1/2}$; (c) When K and m are fixed, which specific heat, the result in (a) or (b), is larger? Use the low temperature expansion to prove your answer.

4.3 In a liquid with density ρ and surface tension coefficient σ, the dispersion relation is $\nu^2 = 2\pi\sigma/(\rho\lambda^3)$.

  a. Build a "Debye model" to calculate the total surface tension energy;

b. Find the low temperature expansion of the specific heat contributed by the surface tension wave (in liquid He).

4.4 In a diamond crystal (atomic mass 12), the Young's module is 7×10^{11} N/m and its density is 3.5 g/cm^3. Find $C_{\text{mole}}(T)$ of diamond, and discuss its characteristics.

4.5 In a 1D chain such as a polyacetylene chain $(-CH=CH-)_n$, the lattice constant is a, the mass of an atomic cluster $-CH-$ is M, and the elastic coefficients are K_1 and K_2 for single and double bonds respectively.

a. Prove that the dispersions of the polyacetylene chain are

$$\omega^2 = \frac{K_1 + K_2}{M} \left\{ 1 \pm \left[1 - \frac{2K_1 K_2 \sin^2(ka/2)}{(K_1 + K_2)^2} \right]^{1/2} \right\}.$$

(4.41)

b. Find the low-frequency and high-frequency expansions.
c. Draw the dispersions of optical and acoustic branches.

4.6 Calculate the dynamic matrix of BCC or FCC crystals.

4.7 What is the difference between the longitudinal and transverse velocity of sound in the [111] direction of the aluminum crystal?

4.8 In neutron diffraction, if the incoming neutron has an energy of 0.02 eV, the outgoing neutron emitted at an angle $2\theta = 10°$ and created a phonon with a speed 300 m/s. Find the energy loss of the neutron in the diffraction process.

Solid State Electronics Theory

5

ABSTRACT

- Ohm's law and Drude Model
- Fermi-Dirac statistics and Sommerfeld Model
- Band theory in solids

The early study of the electrical properties of solids could be said to date back to the formulation of George Simon Ohm's law in 1826, which related the resistance of a wire to its geometry. In 1864, James Clerk Maxwell's "Dynamical Theory of the Electromagnetic Field" unified the fundamental laws of electronics, magnetics, and optics. However, Maxwell's theory did not clarify how an electromagnetic field interacts with matter. In the late 19th century, Hendrik Lorentz made fundamental contributions to the study of electromagnetic phenomena of moving bodies. Based on the concept of charged particles, he formulated the Lorentz force law to describe electromagnetic phenomena in matter. The following five equations form the Maxwell-Lorentz Theory of Electromagnetic Phenomena, which was named in A. Einstein's paper on special relativity in 1905.

$$\vec{\nabla} \cdot \vec{E} = 4\pi\rho, \tag{5.1}$$

$$\vec{\nabla} \cdot \vec{B} = 0, \tag{5.2}$$

$$\vec{\nabla} \times \vec{E} = -\frac{1}{c}\frac{\partial \vec{B}}{\partial t}, \tag{5.3}$$

$$\vec{\nabla} \times \vec{B} = \frac{4\pi}{c}\vec{j} + \frac{1}{c}\frac{\partial \vec{E}}{\partial t}; \tag{5.4}$$

$$\vec{F} = q\vec{E} + \frac{q}{c}\vec{v} \times \vec{B}. \tag{5.5}$$

In 1897, the discovery of electrons by Joseph John Thomson made it clear that the negative carrier in matter is the electron. In 1900, Paul Drude proposed a unified model based on the classical theory of electrons to explain the electrical, thermal, and optical properties of solids. Drude assumed that only the negative electrons were mobile, and that these electrons formed the electron gas in metals. Lorentz "borrowed" Drude's assumption and applied the Maxwell-Boltzmann statistics on electrons. This was more accurate but still could not explain the following anomalies in the Drude model: (1) the excessive specific heat of electrons in metals, and (2) the contradiction between the infinitely large mean free path and the finite ion-ion distance.

In 1925, Enrico Fermi wrote a paper in which he applied Wolfgang Pauli's exclusion principle to electrons in a gas, which was similar to a proposal made by Dirac a few months earlier, resulting in the Fermi-Dirac statistics. In 1928, Arnold Sommerfeld applied the Fermi-Dirac statistics to the free electron gas in the Drude-Lorentz model, which introduced some important concepts, such as *Fermi gas* and *Fermi spheres*, into solid state electronics theory.

In 1926, Erwin Schrodinger's quantum wave mechanics gave rise to a revolution in the basic physics and chemistry relating to the theory of matter. In 1928, a young Swiss student at the University of Leipzig, Felix Bloch, submitted his doctoral thesis detailing the theory of electrons in solids, arguing that the interference between the de Broglie electron wave and the periodic lattice potential leads to energy bands for electrons. The band theory is the basis of quantum solid state electronics theory, therefore Bloch is often referred to as the father of solid state physics. In 1963, Walter Kohn formulated the density functional theory (DFT), which is the basic theory used to accurately compute electron bands in elements and compounds, and has thus become part of the foundation of quantum chemistry and computational materials science.

In free space, an electron in an atom may have discretized energy, but in solids the behavior of an electron is related to the behavior of all the other particles around it. When N atoms are brought together to form a solid, these discrete energy levels form energy bands due to interactions among particles, as shown in Figure 5.1. In a solid, the NZ electrons (Z being the average atomic number) in N atoms occupy a series of bands. Most of the bands are fully filled by electron quantum states, but one or several bands are partially occupied, while all the higher bands are empty. As in quantum transitions in the Bohr model, electrons in a solid may transfer from one energy level in a given band to another in the same or another band, often crossing an intervening gap of forbidden energies in the process. Studies of such quantum transitions in

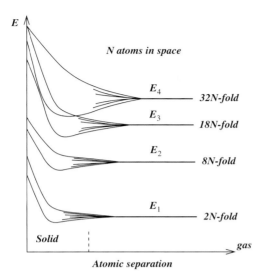

Figure 5.1 Schematics of the electronic structure of N atoms: from gas to solid

solids interacting with photons (visual light or X-rays), energetic electrons and the like confirm the general validity of the band theory, and provide detailed information about allowed and forbidden energy bands of electrons in solids.

In this chapter, the Drude model of classical free electrons, the Sommerfeld model for quantum free electrons, and the Bloch theory and other early band theories are discussed; and density functional theory is introduced. The related transport properties in conductors, semiconductors, and superconductors are mainly discussed in the next chapter.

5.1 Drude Model: Free Electron Gas

Paul Karl Ludwig Drude (1863–1906) was a German physicist. In 1888, he began studying the Maxwell electrodynamic theory. In the early 1900s, many physicists, inspired by the discovery of electrons, became interested in the practical applications of electron theory. Drude had a deep understanding of the optical, thermal, and electrical properties of metals, from which he developed his own electron theory.

In Drude's theory, every metal contains a large number of free electrons and, based on the mature kinetic theory of gases, these free electrons are treated as a gas. The essential difference between the conductors and nonconductors is that the nonconductors contain relatively few free electrons. In the beginning, Drude assumed that both positive

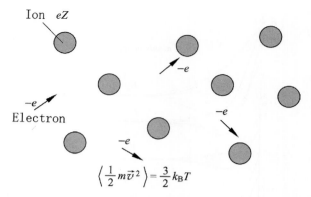

Figure 5.2 Drude model: an electron gas among the plums of positive ions

and negative particles in a metal were part of the gas, but in a later version he proposed a new assumption that only negative electrons were mobile. Electrons travel among the randomly distributed ions; therefore it is also called a plum model.

At temperature T, the equilibrium energy of N electrons can be calculated using the Maxwell-Boltzmann statistics:

$$\bar{\varepsilon} = \frac{\iiint d^3\vec{p}\,\varepsilon e^{-\varepsilon/k_B T}}{\iiint d^3\vec{p}\,e^{-\varepsilon/k_B T}} = \frac{\int d\varepsilon\,\varepsilon^{3/2} e^{-\varepsilon/k_B T}}{\int d\varepsilon\,\varepsilon^{1/2} e^{-\varepsilon/k_B T}} = \frac{3}{2}k_B T \quad (5.6)$$

where the energy $\varepsilon = \vec{p}^2/2m$ and the EDOS $g(\varepsilon) \propto \sqrt{\varepsilon}$. Therefore, according to the Maxwell-Boltzmann statistics, the specific heat per electron is $c_V = d\bar{\varepsilon}/dT = \frac{3}{2}k_B$, and the root-mean-square speed of a free electron should be $\bar{v} = \sqrt{2\bar{\varepsilon}/m} = \sqrt{3k_B T/m}$.

In a crystal with N_L unit cells, n_a atoms inside the CUC, and averagely Z free electrons per atom, the electron gas density is higher than the atomic density:

$$n = \frac{N}{V} = \frac{n_a Z}{a^3}\,\text{(FCC, BCC)}; \quad n = \frac{n_a Z}{3\sqrt{3}a^2 c/2}\,\text{(HCP)} \quad (5.7)$$

where a is the lattice constant in cubic crystals, and a, c are the lattice constants in hexagonal crystals. It can be seen that the electron density in a metal is very high, thus the pressure of the electron gas $P = (n/N_A)RT$ is very high (and can be up to the order of 10^3 atm). This high pressure is balanced by the surface work function.

Another important parameter of the electron gas—the radius r_s of a sphere with a volume equal to V/N—is proportional to the lattice constant:

$$r_s = \left(\frac{3}{4\pi n}\right)^{1/3} = \left(\frac{3}{4\pi n_a Z}\right)^{1/3} a \quad \text{(FCC, BCC);}$$

$$r_s = \left(\frac{9\sqrt{3}}{8\pi n_a Z} a^2 c\right)^{1/3} \quad \text{(HCP).} \qquad (5.8)$$

r_s in the alkali metals are relatively larger (3-6a_B) because they have BCC lattices; r_s in the metals with FCC or HCP structures are about 2–3a_B. The values of r_s and other parameters of free electron gas are listed in Table 5.1.

The basic assumptions of the Drude model are very important in solid state electronics theory. If electrons are treated as Fermions, the Drude model then becomes the Sommerfeld model. Furthermore, when the free electron assumption is replaced by the lattice periodic potential, the Sommerfeld model evolves into the band theory. The basic assumptions of the Drude model are:

1. Independent electron approximation: the electron-electron Coulomb repulsions $+\sum e^2/r_{ij}$ are neglected.
2. Free electron approximation: the electron-ion Coulumb attractions $-\sum Ze^2/|\vec{r}_i - \vec{R}_l|$ are neglected except at the collision.
3. Collision approximation: the electron-ion collisions are instantaneous. The electron velocity changes suddenly in a collision. The outgoing velocity is only related to the temperature, and is independent of the incoming velocity.
4. Relaxation time approximation: the average time between two electron-ion collisions is called the relaxation time τ, and the corresponding displacement is the mean free path $l = \bar{v}\tau$. In the nonequilibrium state with zero external force, the average velocity $\langle \vec{v} \rangle$ would tend to zero in the form $\exp(-t/\tau)$.
5. Classical physics is used to describe the motion of electrons. The equilibrium state of the electron gas in particular obeys the Maxwell-Boltzmann statistics.

Based on Drude's assumptions, both the DC conductivity and AC conductivity can be calculated (see Exercises). In a metal, the electron velocities are basically isotropic, and there is a small average velocity–the drift velocity of of electrons $\vec{v}_d$–along the external field. In a differential volume, the current density vector $\vec{j}$ is along the direction of the local drift velocity:

$$\vec{j} = (-e)\frac{n(\vec{v}_d \Delta t) A}{\Delta t A} = -ne\vec{v}_d \qquad (5.9)$$

where Δt is the time interval and A is the local cross-section area. The drift velocity is the average velocity $\langle \vec{v} \rangle$ of all local electrons, which is totally different from the root-mean-square speed $\bar{v} = \sqrt{3k_B T/m}$ (about 10^7 cm/s at room temperature) of the electron gas. Following Drude's assumption, the drift velocity is

$$\vec{v}_d = \langle \vec{v}_0 + \vec{a}t \rangle = \vec{a}\tau = -\frac{e\vec{E}}{m}\tau \qquad (5.10)$$

where the initial velocity $\vec{v}_0$ after collision is isotropically distributed. Based on the $\vec{v}_d - \vec{E}$ relation, the DC conductivity σ can be easily found:

$$\vec{j} = -ne\vec{v}_d = \sigma \vec{E}; \quad \sigma = \frac{ne^2}{m}\tau \qquad (5.11)$$

In the Drude conductivity $\sigma = ne^2\tau/m$, the parameters n, e, m, and τ are the density, charge, mass, and relaxation time of electrons in a metal respectively. In the band theory, the DC conductivity can be expressed in the same form as Eq. (5.11), but the explanations of n, m, and τ might differ in a conductor or a semiconductor.

The relaxation time can be estimated using the Drude formula (cgs unit):

$$\tau = \frac{mn^{-1}}{e^2\rho} = \frac{2.2}{(\rho/\mu\Omega \cdot \text{cm})}\left(\frac{r_s}{a_B}\right)^3 \times 10^{-15} \text{ s} \qquad (5.12)$$

where the MKS-cgs resistance transformation is $1\Omega = \frac{1}{9} \times 10^{-11}$ s/cm. The resistivity of metals is in the order of $10^0 - 10^1 \mu\Omega \cdot$ cm and the average radius r_s is about $2 - 5a_B$. Therefore, the relaxation time is in the order of $10^{-14} - 10^{-15}$s, and the mean free path in the Drude model is about $l = \bar{v}\tau \simeq 1 - 10$Å. This seems to be consistent with the Drude assumptions: l is the ion-ion distance in a metal.

The free electron density n, the electron charge, and the mass are all fixed, therefore, in metals, the increase in resistivity ρ versus higher temperature is due to the decrease in relaxation time. In Figure 5.3, the resistivity $\rho(T)$ and the conductivity $\sigma(T)$ are plotted for selected metals. For most of the metals (except iron), the resistivity is proportional to the temperature T; following the Drude conductivity in Eq. (5.11), the relaxation time τ should be proportional to T^{-1}. But it is contrary to the Drude model in Figure 5.2, because the ion-ion distance would not vary as T^{-1} versus temperature.

The greatest achievement of the Drude model was the successful explanation of the Wiedemann-Franz law for all metals:

$$\frac{\kappa}{\sigma T} = L \simeq 2 - 3 \times 10^{-8} \text{W} \cdot \Omega/\text{K}^2 \qquad (5.13)$$

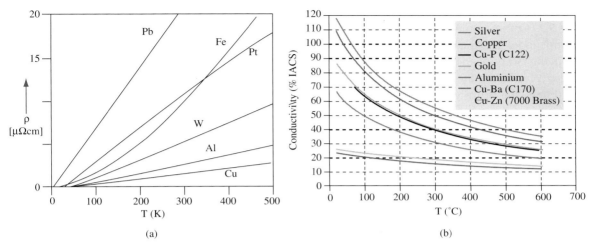

Figure 5.3 Electrical properties of selected metals: (a) resistivity versus temperature (from www.tf.uni-kiel.de); (b) conductivity versus temperature

where κ is the thermal conductivity, σ is the conductivity (the values of κ and σ are listed in Table 5.1), and L is a constant called the Lorentz number.

Thermal conductivity in a metal, like electrical conductivity, is due to the movement of free electrons, therefore κ can also be estimated using the Drude model. The energy current density at x_0 is:

$$j_x = j^q(x_0^- \to x_0^+) - j^q(x_0^+ \to x_0^-) = \frac{1}{2} n v_x [\varepsilon(x_0^-) - \varepsilon(x_0^+)]$$
$$= \frac{1}{2} n v_x \frac{d\varepsilon}{dT} \frac{dT}{dx} [(x_0 - v_x \tau) - (x_0 + v_x \tau)] = n c_V v_x^2 \tau \left(-\frac{dT}{dx}\right)$$

(5.14)

where the specific heat per electron $c_V = d\varepsilon/dT = \frac{3}{2} k_B$, and $v_x^2 = \frac{1}{3}\bar{v}^2 = k_B T/m$.

Using the expressions of κ and σ from the Drude model, the Lorentz number can be expressed by the universal constants k_B and e (cgs units):

$$L = \frac{\kappa}{\sigma T} = \frac{n c_V v_x^2 \tau}{(ne^2\tau/m)T} = \frac{\frac{3}{2} n k_B^2 T \tau/m}{(ne^2\tau/m)T} = \frac{3 k_B^2}{2e^2}$$
$$= 0.124 \times 10^{-12} \mathrm{erg}^2/\mathrm{esu}^2/\mathrm{K}^2 \quad (5.15)$$

In Drude's original work, there was a mistake by a factor of two; therefore his calculation $L = 0.248 \times 10^{-12} \mathrm{erg}^2/\mathrm{esu}^2/\mathrm{K}^2 = 2.23 \times 10^{-8} \mathrm{W} \cdot \Omega/\mathrm{K}^2$ still agreed very well with the measured Lorentz number,

Table 5.1 Common metal properties (mostly at room temperature) by the Drude model and the Sommerfeld model: Z is the valence electron number, $n/(10^{22}\,\mathrm{cm}^{-3})$ the electron density, $r_s/(\text{Å})$ the average occupying radius of an electron, $\rho/(\mu\Omega\cdot\mathrm{cm})$ the resistivity at 77 K and 273 K, $\tau/(10^{-14}\mathrm{s})$ the relaxation time, $\kappa/(\mathrm{W/m/K})$ the thermal conductivity; $\varepsilon_F/(\mathrm{eV})$ the Fermi energy, $T_F/(10^4\,\mathrm{K})$ the Fermi temperature, $v_F/(10^6\,\mathrm{m/s})$ the Fermi velocity, and $k_F/(\text{Å}^{-1})$ the Fermi wave vector (based on Ashcroft and Mermin, 1976, and www.webelements.com)

	Z	n	r_s	ρ(77 K)	ρ	τ	κ	ε_F	T_F	v_F	k_F
Li	1	4.70	1.72	1.04	8.55	0.88	85	4.74	5.51	1.29	1.12
Na	1	2.65	2.08	0.80	4.20	3.17	140	3.24	3.77	1.07	0.92
K	1	1.40	2.57	1.38	6.10	4.11	100	2.12	2.46	0.86	0.75
Rb	1	1.15	2.75	2.20	11.0	2.79	58	1.85	2.15	0.81	0.70
Cs	1	0.91	2.98	4.50	18.8	2.08	36	1.59	1.84	0.75	0.65
Cu	1	8.37	1.41	0.20	1.56	2.66	400	7.00	8.16	1.57	1.36
Ag	1	5.86	1.60	0.30	1.51	4.01	430	5.49	6.38	1.39	1.20
Au	1	5.90	1.59	0.50	2.04	2.91	320	5.53	6.42	1.40	1.21
Be	2	24.7	0.99	–	2.80	0.51	190	14.3	16.6	2.25	1.94
Mg	2	8.61	1.41	0.62	3.90	1.06	160	7.08	8.23	1.58	1.36
Ca	2	4.61	1.73	–	3.43	2.23	200	4.69	5.44	1.28	1.11
Sr	2	3.55	1.80	7.0	23.0	0.37	35	3.93	4.57	1.18	1.02
Ba	2	3.15	1.96	17.0	60.0	0.19	18	3.64	4.23	1.13	0.98
Zn	2	13.2	1.22	1.10	5.50	0.49	120	9.47	11.0	1.83	1.58
Cd	2	9.27	1.37	1.60	6.80	0.56	97	7.47	8.68	1.62	1.40
Hg	2	8.65	1.40	5.80	melt	–	8.3	7.13	8.29	1.58	1.37
Fe	2+	17.0	1.12	0.66	8.90	0.23	80	11.1	13.0	1.98	1.71
Al	3	18.1	1.10	0.30	2.45	0.80	235	11.7	13.6	2.03	1.75
Ga	3	15.4	1.16	2.75	13.6	0.17	29	10.4	12.1	1.92	1.66
In	3	11.5	1.27	1.80	8.0	0.38	82	8.63	10.0	1.74	1.51
Tl	3	10.5	1.31	3.70	15.0	0.22	46	8.15	9.46	1.69	1.46
Sn	4	14.8	1.17	2.10	10.6	0.22	67	10.2	11.8	1.90	1.64
Pb	4	13.2	1.22	4.70	19.0	0.14	35	9.47	11.0	1.83	1.58
Sb	5	16.5	1.13	8.0	39.0	0.05	24	10.9	12.7	1.96	1.70
Bi	5	14.1	1.19	35.0	107.	0.02	8	9.90	11.5	1.87	1.61

as shown in Figure 5.4 (MKS-cgs unit transformations: $\mathrm{W} = 10^7\,\mathrm{erg/s}$, $\mathrm{erg} = \mathrm{esu}^2\cdot\mathrm{cm}$, $\Omega = \frac{1}{9}\times 10^{11}\,\mathrm{s/cm}$).

Later developments in solid state physics proved that the Drude model's correct estimation of the Lorentz number was based on the cancellation of two mistakes: the room temperature specific heat of electrons $c_V = \frac{3}{2}k_B$ was about 100 times higher than the correct value, and the kinetic energy $\frac{1}{2}m\bar{v}_x^2 = k_B T/2$ was about 100 times lower.

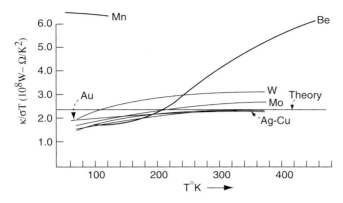

Figure 5.4 Wiedemann-Franz law: the theoretical Lorentz number 2.45×10^{-8} W · Ω/K^2 and the measured $\kappa/\sigma T$. The deviations of $\kappa/\sigma T$ of Mn and Be from the Lorentz number are due to the failure of the free electron gas assumption (based on Seitz, 1940)

5.2 Sommerfeld Model: Free Electron Fermi Gas

Arnold Sommerfeld was one of the greatest theoretical physicists and educators of the 20th century. As Max von Laue commented: "Theoretical physics is a subject which attracts youngsters with a philosophical mind who speculate about the highest principles without sufficient foundations. It was just this type of beginner that Sommerfeld knew how to handle, leading them step by step to a realization of their lack of actual knowledge and providing them with the skill necessary for fertile research." His most famous students included Peter Debye (who was awarded his doctorate in 1908, and won the Nobel Prize in Chemistry in 1936), Wolfgang Pauli (doctorate in 1921, Nobel Prize in Physics in 1945), Werner Heisenberg (doctorate in 1923, Nobel Prize in Physics in 1932), and Hans Bethe (doctorate in 1928, Nobel Prize in Physics in 1967). The young students and scholars who worked in his group included Léon Brillouin (who visited in 1913), Linus Pauling (who visited in 1926, and won the Nobel Prize in Chemistry in 1954), and Max von Laue (who attended in 1909, and won the Nobel Prize in Physics in 1914).

Sommerfeld's first great contribution to physics was made in 1916. He replaced the circular orbits of the Niels Bohr atom with elliptical orbits (several orbits in for a quantum number n). He also introduced the magnetic quantum number, and, four years later, the inner quantum number. It was his theoretical work on the inner quantum number that led to the discovery of electron spin. In 1928, he replaced the Drude-Lorentz model with the Sommerfeld model, in which he treated electrons in a metal as a degenerate electron gas. The basic assumptions of the Sommerfeld model are:

1. Independent electron approximation.
2. Free electron approximation: the electron-ion Coulumb attractions are neglected except at the collision.
3. Collision approximation: it is necessary to include collision to calculate resistivity and other physical measurable quantities of electron gas.
4. Relaxation time approximation: the relaxation time is needed to calculate the conductivity, thermal conductivity, and Hall coefficient of metals.
5. Fermi-Dirac statistics are used in the analysis of the degenerate electron gas, which is formally called the free electron Fermi gas.

The energy density of states of the free electron Fermi gas is similar to that of the classical free electron gas, which can be found by the DOS $\rho(k) = \frac{V}{2\pi^2}k^2$ discussed in Chapter Four (free electron energy $\varepsilon = \hbar^2 k^2/2m$, spin degeneracy 2):

$$g(\varepsilon) = \frac{2}{V}\rho(k)\frac{dk}{d\varepsilon} = \frac{1}{\pi^2}k^2\frac{m}{\hbar^2 k} = \frac{1}{2\pi^2}\left(\frac{2m}{\hbar^2}\right)^{3/2}\sqrt{\varepsilon} \qquad (5.16)$$

The EDOS can be measured by the inelastic scattering of high energy particles: an inner shell electron is knocked out by a photon or a high-energy electron, and a valance electron in the electron Fermi gas jumps into the inner shell and a phonon is emitted (usually within the soft X-ray spectrum). In Figure 5.5, the measured EDOS of lithium is shown. In metallic lithium, an electron in the 1s shell stays at an energy level $E_1 < 0$, but a valence electron in the electron gas has the energy $E(\vec{k}) > 0$; therefore the energy of the photo-emitted soft X-ray photon can be found, referring to the Moseley formula of X-ray line spectrum introduced in Chapter Three:

$$E = \varepsilon(\vec{k}) - \varepsilon_1^0 = \frac{\hbar^2 \vec{k}^2}{2m} + \frac{Ry}{1^2}(Z-1)^2 \qquad (5.17)$$

By Eq. (5.17), the photon emitted from lithium should have a minimum energy at $13.6\,\text{eV} \times 2^2 = 54.4\,\text{eV}$ (lithium $Z = 3$), slightly larger than the measurement in Figure 5.5. The intrinsic soft X-ray emission width is 4–5 eV for lithium, which is consistent with the Fermi energy of lithium discussed later on. The tail in the range of 54–55 eV is mainly caused by the band width of 1s electrons. The measured EDOS is larger than the free electron $g(\varepsilon) = c\sqrt{\varepsilon}$ in Eq. (5.16) by a constant factor, due to the heavier effective mass of electrons in crystals. All deviations from the Sommerfeld model have to be explained by the band theories of electrons.

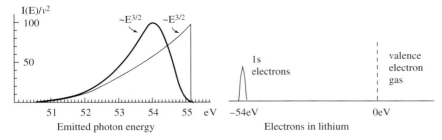

Figure 5.5 Measured and free electron EDOS of lithium: $I(\varepsilon)/\nu^2 \propto \varepsilon g(\varepsilon) \propto \varepsilon^{3/2}$

At temperature T, in the w-fold degeneracy energy level ε, the average occupation number $\bar{n}$ can be calculated by minimizing the difference between the Helmholtz free energy and the Gibbs free energy $F - G = E - TS - \mu n$ of n identical electrons (where T, V, μ are fixed and μ is not zero in the free electron Fermi gas):

$$0 = \frac{d}{dn}\left\{n\varepsilon - k_B T \ln \frac{w!}{n!(w-n)!} - \mu n\right\}$$
$$= \varepsilon - k_B T \ln \frac{w - \bar{n}}{\bar{n}} - \mu \qquad (5.18)$$

$$\bar{n} = w \frac{1}{e^{\beta(\varepsilon - \mu)} + 1} = w f_{F-D}(\varepsilon) \qquad (5.19)$$

$f_{F-D}(\varepsilon)$ is the Fermi-Dirac distribution, and was successively formulated by Fermi and Dirac in 1925, based on Pauli's exclusion principle for all fundamental particles with half-integer spins.

At 0 K, the Fermi-Dirac distribution is a pure step function: $f_{F-D}(\varepsilon)$ equals one when $\varepsilon \leq \mu$, and zero elsewhere. This means that the energy of any electron state must be less than the chemical potential $\mu(0)$ at 0 K, which is called the Fermi energy ε_F. In the Sommerfeld model, a free electron has the energy $\varepsilon = \hbar^2 k^2/2m$. At 0 K, $\varepsilon \leq \varepsilon_F$ for any electron implies that the de Broglie wavevector $\vec{k}$ in the reciprocal space should stay inside a "sphere," which is called the Fermi sphere. The radius of the Fermi sphere is called the Fermi wavevector k_F, which is directly related to the density n of the free electron Fermi gas:

$$n = \frac{N}{V} = \frac{2}{V}\int_0^{k_F} dk\, \rho(k) = \frac{1}{\pi^2}\int_0^{k_F} dk\, k^2 = \frac{1}{3\pi^2}k_F^3 \qquad (5.20)$$

Using Eq. (5.8) and Eq. (5.20), the Fermi wavevector and the Fermi energy can be directly related to the electron's average occupying radius

r_s in the free electron gas:

$$k_F = (3\pi^2 n)^{1/3} = \frac{3.63}{r_s/a_B}(\text{Å}^{-1});$$

$$\varepsilon_F = \frac{\hbar^2}{2m}(3\pi^2 n)^{2/3} = \frac{50.1}{(r_s/a_B)^2}(\text{eV}). \quad (5.21)$$

In metals, r_s is in the range 2–6a_B, therefore in the Sommerfeld model, the Fermi energy should stay in the range 1.5–15 eV, as listed in Table 5.1. The measured ε_F is smaller than the ε_F of the free electron in Table 5.1, due to the band structures of electrons in a lattice. However, the free electron gas model is still quite successful in roughly estimating the trends.

It is also interesting to check the Fermi temperature T_F:

$$T_F = \frac{\varepsilon_F}{k_B} = \frac{\hbar^2}{2mk_B}(3\pi^2 n)^{2/3} = \frac{58.2 \times 10^4}{(r_s/a_B)^2}(\text{K}) \quad (5.22)$$

The Fermi temperature is in the range of $1.8\text{--}17 \times 10^4$ K. At room temperature, the ratio $k_B T/\varepsilon_F = T/T_F \simeq 1/100$, therefore following Eq. (5.19), the Fermi statistics equate closely to the step function at 0 K shown in Figure 5.6:

$$\begin{aligned}
(\varepsilon - \mu)/k_B T \ll 1, \quad & f_{F-D}(\varepsilon) = 1 \\
(\varepsilon - \mu)/k_B T \gg 1, \quad & f_{F-D}(\varepsilon) = 0 \\
(\varepsilon - \mu)/k_B T \sim 1, \quad & \frac{df_{F-D}}{d\varepsilon} \simeq -\delta(\varepsilon - \mu)
\end{aligned} \quad (5.23)$$

At room temperature, $f_{F-D}(\varepsilon)$ is nearly a step function, therefore $df_{F-D}/d\varepsilon$ is non-zero in a very thin shell $\Delta\varepsilon \sim k_B T \ll \varepsilon_F$ near the Fermi sphere and it can be expressed by the former δ-function.

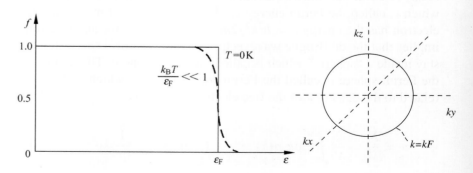

Figure 5.6 Fermi-Dirac statistics at different temperature: '$f_{F-D}(\varepsilon)$ is close to a step function around $\varepsilon = \mu$ even at room-T

The macroscopic physical quantities of free electron Fermi gas, such as the equilibrium energy and specific heat, can be found using the Sommerfeld expansion method, which takes advantage of the δ-function characteristics of $df/d\varepsilon$. The macroscopic value X of the microscopic quantity $x(\varepsilon)$ is calculated as:

$$X = \int_0^\infty d\varepsilon\, g(\varepsilon) f(\varepsilon) x(\varepsilon); \quad \frac{dy(\varepsilon)}{d\varepsilon} = g(\varepsilon) x(\varepsilon)$$

$$= \int_0^\infty d\varepsilon \left[-\frac{df}{d\varepsilon}\right] y(\varepsilon);$$

$$= \int_0^\infty d\varepsilon \left[-\frac{df}{d\varepsilon}\right] \left[y(\mu) + \sum_{n=1}^\infty \frac{1}{n!} \left.\frac{d^n y}{d\varepsilon^n}\right|_\mu (\varepsilon - \mu)^n\right]$$

$$= y(\mu) + \sum_{m=1}^\infty a_m \left.\frac{d^{2m} y}{d\varepsilon^{2m}}\right|_\mu (k_B T)^{2m} \tag{5.24}$$

$(-df/d\varepsilon)$ is an even function of $(\varepsilon - \mu)$, therefore $\int d\varepsilon [-df/d\varepsilon] (\varepsilon - \mu)^n$ vanishes when n is an odd number. If $y = 0$ is assumed in the range of $\varepsilon < 0$, the Sommerfeld expansion coefficients are:

$$a_n = \int_{-\infty}^\infty du \left[-\frac{d}{du}\frac{1}{e^u + 1}\right] \frac{u^{2n}}{(2n)!}$$

$$= (2 - 2^{2-2n}) \zeta(2n) = (2^{2n} - 2) \frac{\pi^{2n}}{(2n)!} B_n \tag{5.25}$$

where $\zeta(2n)$ is the Euler number and B_n is the Bernoulli number ($B_1 = \frac{1}{6}$, $B_2 = \frac{1}{30}$). Therefore the first order Sommerfeld expansion coefficient is $a_1 = \pi^2/6$.

The most important physical quantity of free electron Fermi gas is the chemical potential at temperature T, which can be found by calculating the density n:

$$n = \int_0^\infty d\varepsilon\, g(\varepsilon) f(\varepsilon) = \frac{2}{3} g(\mu) \mu + \frac{\pi^2}{6} g'(\mu) (k_B T)^2$$

$$\mu(T) = \varepsilon_F - \frac{\pi^2}{6} \frac{g'(\varepsilon_F)}{g(\varepsilon_F)} (k_B T)^2 = \varepsilon_F \left(1 - \frac{\pi^2}{12} \left(\frac{k_B T}{\varepsilon_F}\right)^2\right) \tag{5.26}$$

where the function $y(\varepsilon) = \frac{2}{3} g(\varepsilon) \varepsilon$ or $y'(\varepsilon) = g(\varepsilon) = c \sqrt{\varepsilon}$ is defined by the EDOS in Eq. (5.16). The chemical potential at 0K $\mu(0)$ is just ε_F. At room temperature, the difference $\mu(T) - \varepsilon_F$ is only 1/10,000 of the Fermi energy, which is tiny.

5.2.1 Specific Heat of Electrons

The heat capacity of free electrons in metals can be found using the formula $C_V = d\bar{E}/dT$. The equilibrium energy $\bar{E}$ of free electron Fermi gas can also be found using the Sommerfeld expansion and the chemical potential in Eq. (5.26):

$$\bar{E}/V = \int_0^\infty d\varepsilon\, g(\varepsilon)\, f(\varepsilon)\, \varepsilon; \quad \frac{dy(\varepsilon)}{d\varepsilon} = g(\varepsilon)\varepsilon$$

$$= \frac{2}{5}g(\mu)\mu^2 + \frac{\pi^2}{6}\frac{3}{2}g(\mu)(k_B T)^2$$

$$= \frac{2}{5}\left[g(\varepsilon_F)\varepsilon_F^2 + \left(\frac{5}{2}g(\varepsilon_F)\varepsilon_F\right)\left(-\frac{\pi^2}{12}\frac{(k_B T)^2}{\varepsilon_F}\right)\right]$$

$$+ \frac{\pi^2}{6}\frac{3}{2}g(\varepsilon_F)(k_B T)^2$$

$$= \frac{3}{5}n\varepsilon_F + \frac{\pi^2}{6}g(\varepsilon_F)(k_B T)^2 \tag{5.27}$$

Then the specific heat per volume c_V and the molar specific heat C_{mole}^e (with respect to 1 mole of atoms and $N_A Z$ valence electrons in a solid) of the free electron Fermi gas is proportional to the temperature T:

$$c_V = \frac{\pi^2}{3}g(\varepsilon_F)k_B^2 T = \frac{\pi^2}{3}\frac{3n}{2\varepsilon_F}k_B^2 T = \frac{\pi^2}{2}nk_B\frac{T}{T_F}$$

$$C_{\text{mole}}^e = \frac{N_A}{N/V}c_V = \frac{\pi^2}{2}ZR\frac{T}{T_F} \tag{5.28}$$

where the free electron energy density of states at the Fermi energy $g(\varepsilon_F)$ has the form

$$g(\varepsilon_F) = \frac{1}{2\pi^2}\left(\frac{2m}{\hbar^2}\right)^{3/2}\sqrt{\frac{\hbar^2}{2m}(3\pi^2 n)^{2/3}} = \frac{1}{2\pi^2}\frac{3\pi^2 n}{\frac{\hbar^2}{2m}(3\pi^2 n)^{2/3}} = \frac{3n}{2\varepsilon_F}$$

$$\tag{5.29}$$

At room temperature, the specific heat of electrons C_{mole}^e in the Sommerfeld model is much less than the specific heat of atomic vibrations C_{mole} in the Debye model. In metals, the temperature ratio T_F/T is about 100 at room temperature, which explains why the specific heat for electrons estimated in the Drude model was about 100 times higher than the real value. The intrinsic reason is that most electrons inside the Fermi sphere ($\varepsilon < \varepsilon_F - k_B T$) cannot be thermally excited from one energy level to another, because all the neighboring quantum states are full, and any transition $\varepsilon \to \varepsilon + k_B T$ is forbidden by Pauli's exclusion principle. That is a conclusion following the Fermi-Dirac statistics.

At low temperatures, C^e_{mole} of electrons in metals is comparable to or even higher than C_{mole} in the Debye model:

$$\frac{C_{mole}}{C^e_{mole}} = \frac{\frac{12\pi^4}{5}R\left(\frac{T}{\Theta_D}\right)^3}{\frac{\pi^2}{2}ZR\frac{T}{T_F}} = \frac{12\pi^2}{5Z}\left(\frac{T}{\Theta_D}\right)^3\frac{T_F}{T} \propto \left(\frac{T}{T_e}\right)^2 \to 0 \quad T \to 0 \tag{5.30}$$

The characteristic temperature is $T_e = \sqrt{\Theta_D^3/T_F}$. When $T \ll T_e$ the specific heat of electrons is dominant in a metal.

Measured specific heat has the same trend as the theoretical results in Eq. (5.28); however the measured specific heat is always higher. In non-magnetic metals, the measured low-temperature molar specific heat has the form $C_P \simeq C_V = \gamma T + A_2 T^3 + A_3 T^5$ (based on Martin, 1965)

$$C^e_{mole} = \frac{\pi^2}{2}ZR\frac{T}{T_F'} + 3Rf_D\left(\frac{T}{\Theta_D}\right); \quad f_D(u) = 3u^3 \int_0^{1/u} dx \frac{x^4 e^x}{(e^x-1)^2} \tag{5.31}$$

where $f_D(u)$ is the Debye function with respect to the specific heat of atomic vibrations; T_F' is the Fermi temperature fitted by the measurement of specific heat in metals. For alkali and noble metals, the measured Debye temperature Θ_D agrees well with the data listed in Table 4.1; the measured Fermi temperature T_F' in Table 5.2 has the same trends as the free electron Fermi temperature T_F in Table 5.1, but T_F' is lower by a factor of 1–2.

The differences between T_F' and T_F can only be explained by the band theories for electrons, which are discussed later in this chapter. In Eq. (5.22) of T_F, the only "tunable" parameter is the electron mass m; therefore the effective mass m^* of electrons can be measured

Table 5.2 Measured specific coefficient γ, Fermi temperature T_F', and the theoretical T_F in the Sommerfeld model for alkali metals and noble metals (based on Martin, 1965 and 1973)

	Li	Na	K	Rb	Cs	Cu	Ag	Au
γ/(mJ/mole/K^2)	1.63	1.38	2.08	2.41	3.19	0.69	0.64	0.69
$T_F'/(10^4 K)$		2.52	2.97	1.97	1.70	1.28	5.94	6.12	5.96
$T_F/(10^4 K)$	5.51	3.77	2.46	2.15	1.84	8.16	6.38	6.42

by comparing the experimental and theoretical low-temperature specific heats:

$$\frac{m^*}{m_e} = \frac{\gamma}{\frac{\pi^2}{2}ZR/T_F} = \frac{T_F}{T'_F} \quad (5.32)$$

where $m_e = 9.110 \times 10^{-31}$ kg is the bare mass of electrons. The physical meaning of the effective mass m^* will be discussed later.

The measured specific heat of transition metals is much larger than would be expected from the Sommerfeld model. In Figure 5.7, the specific heat coefficients γ are plotted for transition metals in the fifth and sixth periods.

The measured specific heat coefficients of transition metals in the fourth period are listed in Table 5.3, and they show similar trends to the data of the other two periods in Figure 5.7. The specific heat of transition metals is about ten times that of noble metals. The specific heat coefficient γ has the expression:

$$\gamma = \frac{\pi^2}{2}ZR\frac{1}{T_F}\frac{m^*}{m_e} = Z\frac{10^4 \text{K}}{T_F}\frac{m^*}{m_e} \times 4.10\,\text{mJ/mole/K}^2 \quad (5.33)$$

In the Sommerfeld model, if $Z = 2$ and $T_F = 13.0 \times 10^4$ K are chosen for α−Fe, the effective mass is $m^* = 7.91\,m_e$, which is consistent with the measured m^* in other experiments. The effective masses of transition

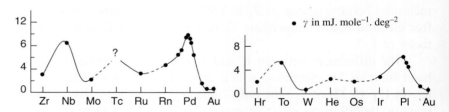

Figure 5.7 Measured specific heat coefficient $\gamma/(\text{mJ/mole/K}^2)$ in transition metals (based on Coles, 1964)

Table 5.3 Measured specific heat coefficient γ of transition metals from Ti to Cu (based on Heer and Erickson, 1957, and Friedberg, Estermann and Goldman, 1952)

	Ti	V	Cr	Mn	Fe	Co	Ni	Cu
$\gamma/(\text{mJ/mole/K}^2)$	3.47	5.85	1.56	1.36	4.99	4.74	7.23	0.69

metals are very large because the complex combinations of d-bands and s-bands result in an elaborate non-spherical Fermi sphere, which is very different from the simple Fermi sphere for free electron gas given by the Sommerfeld model.

5.2.2 Electrical and Thermal Conductivity

In the measurement of electrical or thermal conductivity in metals, external fields or temperature differences have to be applied. When an external field is applied, the electrons are no longer in equilibrium; and the non-equilibrium statistics, equivalent to Boltzmann equations, have to be analyzed first.

When a small electric field is applied, the energy of an electron contains the electrical potential energy: $\varepsilon' = \varepsilon - eV$, and the non-equilibrium statistics $g(\vec{r}, \vec{k}, t)$ can be expanded to the first order:

$$g(\vec{r}, \vec{k}, t) = f_{\text{F-D}}(\varepsilon) + \frac{df}{d\varepsilon}\Delta\varepsilon; \quad \Delta\varepsilon = -eV = e\vec{E}\cdot\Delta\vec{r}$$

$$= f_{\text{F-D}}(\varepsilon) - \left(-\frac{df}{d\varepsilon}\right)\left(e\vec{E}(\vec{r}, t)\cdot\vec{v}(\vec{k})\tau\right) \quad (5.34)$$

where the velocity $\vec{v}(\vec{k}) = \hbar\vec{k}/m$ in the Sommerfeld model. The former non-equilibrium statistics can be visualized as a tiny "drift" of the Fermi sphere toward the direction of $-\vec{E}$, the drift distance being proportional to the relaxation time τ.

The electrical conductivity can be found using the non-equilibrium statistics in Eq. (5.34). In an external electric field $\vec{E}$, the density of states of free electrons depends on the directions of $\vec{k}$, thus the number of states in the reciprocal space is $dn = 2d^3\vec{k}/(2\pi/L)^3 = V d^3\vec{k}/(4\pi^3)$ instead of the isotropic form $dn = 2\rho(k)dk$. The current density in a constant external field $\vec{E}$ is:

$$\vec{j} = \iiint \frac{d^3\vec{k}}{4\pi^3} g(\vec{r}, \vec{k}, t)(-e\vec{v}); \quad \iiint \frac{d^3\vec{k}}{4\pi^3} f_{\text{F-D}}(\varepsilon)(-e\vec{v}) = 0$$

$$= -\iiint \frac{d^3\vec{k}}{4\pi^3}\left(-\frac{df}{d\varepsilon}\right)\left(e\vec{E}\cdot\vec{v}\tau\right)(-e\vec{v})$$

$$= \left(\frac{e^2\tau_F}{m}\right)\iiint \frac{d^3\vec{k}}{4\pi^3}\left(-\frac{df}{d\varepsilon}\right)m\vec{v}\vec{v}\cdot\vec{E}$$

$$(5.35)$$

where τ_F is the relaxation time at the Fermi sphere. In a zero field $\vec{E} = 0$, $\vec{j}$ must be zero, because $f_{\text{F-D}}(\varepsilon)$ is an even function of $\vec{k}$ and $\vec{v} = \hbar\vec{k}/m$ is an odd function of $\vec{k}$. The non-equilibrium integral in Eq. (5.35) is only non-zero for diagonal elements, and it can be transformed back into the integral of the kinetic energy $mv_x^2 = mv_y^2 = mv_z^2 = \frac{2}{3}\varepsilon$. Then the conductivity in the Sommerfeld model is

$$\sigma = \frac{e^2 \tau_F}{m} \int d\varepsilon\, g(\varepsilon) \left(-\frac{df}{d\varepsilon}\right) \frac{2}{3}\varepsilon = \frac{e^2 \tau_F}{m} g(\varepsilon_F) \frac{2}{3}\varepsilon_F = \frac{ne^2 \tau_F}{m} \quad (5.36)$$

which has the same form as the Drude conductivity, except for the explanation of τ.

When a temperature difference is applied in the x-direction, the force acting on an electron can be derived from the first law of thermodynamics and the flow of entropy (at equilibrium the entropy $S/N = (\varepsilon - \mu)/T$ by Eq. 5.18):

$$\vec{F} \cdot d\vec{r} = \frac{1}{N}(dU - T\,dS) = -eV + \frac{1}{N} S\,dT$$
$$= e\vec{E}_T \cdot \vec{r} + \frac{\varepsilon - \mu}{T}(\vec{\nabla}T) \cdot \vec{r} \quad (5.37)$$

where $\vec{E}_T$ is the electric field caused by the temperature difference, which is quite small and can be neglected. Then the non-equilibrium statistics $g(\vec{r}, \vec{k}, t)$ due to temperature difference can be expanded to the first order:

$$g(\vec{r}, \vec{k}, t) = f_{F-D}(\varepsilon) - \left(-\frac{df}{d\varepsilon}\right)\left(\frac{\varepsilon - \mu}{T} \vec{\nabla}T \cdot \vec{v}\tau\right) \quad (5.38)$$

These non-equilibrium statistics can be visualized as the "drift" of the Fermi sphere towards the direction of $-\vec{\nabla}T$, which also has a very small value.

The thermal conductivity can be found using the non-equilibrium statistics in Eq. (5.38). The energy current density of electrons in a metal is:

$$\vec{j}_q = \iiint \frac{d^3\vec{k}}{4\pi^3} g(\vec{r}, \vec{k}, t)[(\varepsilon - \mu)\vec{v}];$$
$$= -\iiint \frac{d^3\vec{k}}{4\pi^3} \left(-\frac{df}{d\varepsilon}\right)\left(\frac{\varepsilon - \mu}{T}\vec{\nabla}T \cdot \vec{v}\tau\right)[(\varepsilon - \mu)\vec{v}]$$
$$= \left(\frac{\tau_F}{mT}\iiint \frac{d^3\vec{k}}{4\pi^3}\left(-\frac{df}{d\varepsilon}\right)(\varepsilon - \mu)^2 m\vec{v}\vec{v}\right) \cdot (-\vec{\nabla}T) \quad (5.39)$$

In equilibrium ($\vec{\nabla} T = 0$), $\vec{j}_q$ must be zero, because $f_{F-D}(\varepsilon)(\varepsilon - \mu)$ is an even function of $\vec{k}$ and $\vec{v} = \hbar\vec{k}/m$ is an odd function of $\vec{k}$. Similarly, the integral of the matrix $m\vec{v}\vec{v}$ could be transformed back into the integral of the kinetic energy $mv_x^2 = mv_y^2 = mv_z^2 = \frac{2}{3}\varepsilon$. Then the thermal conductivity in the Sommerfeld model would be

$$\kappa = \frac{\tau_F}{mT} \int d\varepsilon \, g(\varepsilon) \left(-\frac{df}{d\varepsilon}\right) (\varepsilon - \mu)^2 \frac{2}{3}\varepsilon$$

$$= \frac{n\tau_F}{mT} 2a_1 (k_B T)^2 = \frac{\pi^2}{3} \frac{n\tau_F}{m} k_B^2 T \qquad (5.40)$$

where $a_1 = \frac{\pi^2}{6}$ is the first Sommerfeld coefficient. The thermal conductivity κ in the Sommerfeld model is also similar to that in the Drude model.

Using the expressions κ and σ from the Sommerfeld model, the Lorentz number can also be expressed by the universal constants k_B and e (cgs unit):

$$L = \frac{\kappa}{\sigma T} = \frac{\frac{\pi^2}{3} \frac{n\tau_F}{m} k_B^2 T}{\frac{ne^2 \tau_F}{m} T} = \frac{\pi^2 k_B^2}{3e^2} = 0.272 \times 10^{-12} \, \text{erg}^2/\text{esu}^2/\text{K}^2$$

$$(5.41)$$

Therefore the Lorentz number in the Sommerfeld model is $L = 2.45 \times 10^{-8}$ W·Ω/K^2, which explains the Wiedemann-Franz law quantitatively (MKS-cgs unit transformations: W·Ω/K$^2 = \frac{1}{9} \times 10^{-4}$erg^2/esu^2/K^2), as shown in Figure 5.4.

5.2.3 Thermionic Emission of Electrons from Metal Surfaces

The thermionic emission of electrons from metal surfaces was first discovered by Thomas Edison in 1883. It has important practical applications in photoemission, X-ray generation, electron beam generation, and vacuum diodes and triodes, therefore numerous studies have been carried out in this field.

Based on the Sommerfeld model of free electron Fermi gas, O. W. Richardson and Max von Laue summarized an equation of emission:

$$I = AT^2 e^{-b_0/T} = AT^2 e^{-W/k_B T} \qquad (5.42)$$

where A is a universal constant (based on Dushman, 1930) and W is the work function characterizing a metal. W is the energy barrier at a flat clean surface of a metal due to the asymmetry of chemical bonds at the surface, which can be changed by external fields.

The thermionic emission current in Eq. (5.42) is the total flow of electrons at a surface $x > x_0$, the energy of which is higher than the barrier $\varepsilon = \mu + W + \varepsilon_\perp > \mu + W$:

$$\begin{aligned} I &= \frac{1}{2} \iiint \frac{\mathrm{d}k_x \mathrm{d}k_y \mathrm{d}k_z}{4\pi^3} f_{\text{F-D}}(\varepsilon)[-ev_x]; \quad v_x = \frac{\partial \varepsilon}{\partial p_x} = \frac{1}{\hbar} \frac{\partial \varepsilon}{\partial k_x} > 0 \\ &= -\frac{em}{4\pi^2 \hbar^3} \int_0^\infty \mathrm{d}\varepsilon_\perp \int_{\mu+W+\varepsilon_\perp}^\infty \mathrm{d}\varepsilon \frac{1}{e^{(\varepsilon-\mu)/k_B T} + 1} \\ &= \frac{em k_B T}{2\pi^2 \hbar^3} \int_0^\infty \mathrm{d}\varepsilon_\perp \ln\left[1 + e^{-(W+\varepsilon_\perp)/k_B T}\right] \\ &= \frac{em}{4\pi^2 \hbar^3} (k_B T)^2 e^{-W/k_B T} \end{aligned}$$

(5.43)

where $\varepsilon_\perp = (p_y^2 + p_z^2)/2m$; the Fermi-Dirac statistics at the tail $\varepsilon \ll \mu$ are the Maxwell-Boltzmann statistics. The following universal thermionic emission constant was first derived by S. Dushman:

$$\begin{aligned} A &= \frac{em k_B^2}{4\pi^2 \hbar^3} = \frac{1.6 \times 10^{-19} \times 9.1 \times 10^{-31} \times (1.38 \times 10^{-23})^2}{4\pi^2 (1.055 \times 10^{-34})^3} \\ &= 60 \text{ A/cm}^2/\text{K}^2 \end{aligned}$$

(5.44)

In Figure 5.8, the thermionic emission current of tungsten in cesium gas is plotted at different temperatures of the bulb. Electrons have a probability $r(T)$ of being reflected, thus the $\log_{10} I - (1/T)$ curve should have the form by Eq. (5.42):

$$\log_{10} I = (\log_{10} e)[\ln[A(1-r)] - 2\ln(1/T) - W/k_B T] \quad (5.45)$$

When the temperature of tungsten is low (large $1000/T$), the emission current $I \propto \exp(-W/k_B T)$, thus the work function W can be found by the slope between the points "A" and "B" in Figure 5.8. When the temperature of tungsten is very high (small $1000/T$), the reflection coefficient $r(T)$ tends to one, and the term $\ln[A(1-r)]$ tends to $-\infty$, therefore the emission current I must decrease at higher temperatures of the filament ($T > 1000$ K).

The measured constants A and the work functions W of different materials are listed in Table 5.4. Sometimes the measured constant A agrees well with the universal constant in Eq. (5.44); however, sometimes it shows large deviations, which might be caused by the anisotropic property of crystals or the surface effect.

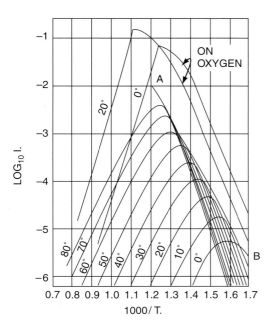

Figure 5.8 Thermionic emission of tungsten in cesium gas. The temperature marked at different curves is the bulb temperature (based on Dushman, 1930)

Table 5.4 Measured thermionic emission constants A and W of materials (based on Dushman, 1930)

	C	Ca	Zr	Mo	Cs	Hf	Ni	Ta	W	Pt
$A/(A/cm^2/K^2)$	5.93	60.2	330	60.2	162	14.5	26.8	60.2	60.2	1.7×10^4
$W/(eV)$	3.93	2.24	4.13	4.44	1.81	3.53	2.27	4.07	4.52	6.27

5.2.4 Hall Effect

The Hall effect was discovered by an American physicist, Edwin Hall, in 1879. In a perpendicular magnetic field, the current in a thin film will flow to the +x direction due to the balance of the electric force $F_e^y > 0$ and the Lorentz magnetic force $F_m^y < 0$, as shown in Figure 5.9. The Hall coefficient can be defined as

$$R_\mathrm{H} = \frac{E_y}{j_x B_z} = \frac{F_e^y/q}{(nqv_x)(-F_m^y/(qv_x))} = \frac{1}{nq} \quad (5.46)$$

In the cgs unit $R_\mathrm{H} = 1/nqc$. Therefore the Hall coefficient is a function of the carrier charge q and the carrier density n in matter. The former

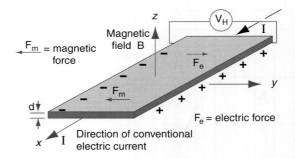

Figure 5.9 Schematics of Hall effect of negative carriers such as electrons

derivation is true for both the negative ($q < 0$) and positive ($q > 0$) carriers.

The Hall coefficient changes sign in accordance with the charge carrier, and therefore provides an important means of investigating the electronic structure of the solid state. In particular, the positive Hall coefficients exhibited by metals, such as zinc and chromium, are a clear indication that a naive picture of a Fermi sea of free conduction electrons is inappropriate, because the majority of the carriers are clearly positively charged. The Hall coefficient of semiconductors can easily change sign by doping, which will be discussed in the next chapter.

The Hall coefficient can be calculated using the Sommerfeld model. In a driving external field E_x, a transverse electric field E_y, and a vertical magnetic field B_z, the non-equilibrium statistics $g(\vec{r}, \vec{k}, t) = f(\varepsilon) + v_x \eta_1 + v_y \eta_2$ are quite complex (based on Seitz, 1940):

$$g(\vec{r}, \vec{k}, t) = f_{F-D}(\varepsilon) - \left(-\frac{df}{d\varepsilon}\right)(e\vec{E} \cdot \vec{v}\tau)$$

$$+ \frac{df}{dv_x}\frac{ev_y H_z}{mc}\tau + \frac{df}{dv_y}\frac{-ev_x H_z}{mc}\tau$$

$$\eta_1 = \left(\frac{df}{d\varepsilon}\right)(eE_x\tau) - \eta_2 \frac{eH_z}{mc}\tau;$$

$$\eta_2 = \left(\frac{df}{d\varepsilon}\right)(eE_y\tau) + \eta_1 \frac{eH_z}{mc}\tau; \qquad (5.47)$$

$$\eta_1 = \frac{E_x - \omega_c \tau E_y}{1 + (\omega_c \tau)^2}\left(\frac{df}{d\varepsilon}\right)(e\tau)$$

$$\eta_2 = \frac{E_y + \omega_c \tau E_x}{1 + (\omega_c \tau)^2}\left(\frac{df}{d\varepsilon}\right)(e\tau) \qquad (5.48)$$

where the Lorentz force $\vec{F} = -e\vec{E} - \frac{e}{c}\vec{v} \times \vec{B}$ and the synchrotron frequency $\omega_c = eH_z/mc$ are used. Therefore $g(\vec{r}, \vec{k}, t)$ depends on both the electric and magnetic field.

The current density in the Hall system can be found using the non-equilibrium statistics in Eqs. (5.47) and (5.48):

$$\vec{j} = e\tau_F \iiint \frac{d^3\vec{k}}{4\pi^3} \left(-\frac{df}{d\varepsilon}\right)$$
$$\times \left(\frac{E_x - \omega_c \tau E_y}{1 + (\omega_c \tau)^2} v_x + \frac{E_y + \omega_c \tau E_x}{1 + (\omega_c \tau)^2} v_y\right)(e\vec{v}) \quad (5.49)$$

$$j_x = E_x \sigma_{11} - E_y \sigma_{12}; \quad \sigma_{11} = \frac{e^2}{m} \iiint \frac{d^3\vec{k}}{4\pi^3} \left(-\frac{df}{d\varepsilon}\right) \frac{m v_x v_x \tau}{1 + (\omega_c \tau)^2}$$
(5.50)

$$0 = E_x \sigma_{12} + E_y \sigma_{11}; \quad \sigma_{12} = \frac{e^2}{m} \iiint \frac{d^3\vec{k}}{4\pi^3} \left(-\frac{df}{d\varepsilon}\right) \frac{m v_x v_x \omega_c \tau^2}{1 + (\omega_c \tau)^2}$$
(5.51)

where the transverse current j_y must be zero in the Hall effect.

It is easy to solve the pair of linear equations Eqs. (5.50) and (5.51). The Hall coefficient can be found by comparing the longitudinal conductivity σ_{11} and transverse conductivity σ_{12} (cgs units):

$$E_x = \frac{\sigma_{11}}{\sigma_{11}^2 + \sigma_{12}^2} j_x, \quad \frac{1}{\sigma} = \frac{E_x}{j_x} = \frac{\sigma_{11}}{\sigma_{11}^2 + \sigma_{12}^2} = \frac{m}{ne^2 \tau_F} \quad (5.52)$$

$$E_y = \frac{-\sigma_{12}}{\sigma_{11}^2 + \sigma_{12}^2} j_x = \frac{1}{\sigma}\left(-\frac{\sigma_{12}}{\sigma_{11}}\right) j_x = \frac{1}{\sigma}(-\omega_c \tau_F) j_x \quad (5.53)$$

$$R_H = \frac{E_y}{j_x H_z} = \frac{1}{\sigma}\left(-\frac{e}{mc}\tau_F\right) = -\frac{e\tau_F}{\sigma m}\frac{1}{c} = -\frac{1}{nec}$$
$$= -\frac{6.94 \times 10^{-24}}{n/(10^{22}\text{cm}^{-3})}(\text{s/G})$$
$$= -\frac{1}{n/(10^{22}\text{cm}^{-3})} \times 6.25 \times 10^{-10}(\Omega \cdot \text{m/G}) \quad (5.54)$$

where in the former equation of R_H, the Drude conductivity $\sigma = ne^2 \tau_F / m$ is utilized.

The measured Hall coefficient is compared with the estimation from the Sommerfeld model using the free electron data in Table 5.1, as listed in Table 5.5. The Sommerfeld model explains R_H for alkali metals quite well, and the estimations of R_H for noble metals are basically correct, only deviating by a factor. The estimations of R_H using the Sommerfeld model for bivalent metals and transition metals are wrong in many cases, and can be better explained using band theories.

Table 5.5 Measured $R'_H/(10^{-6}\mu\Omega\cdot\text{cm/G})$ and theoretical R_H by Eq. (5.54).

	Li	Na	K	Rb	Cs	Cu	Ag	Au
R'_H	−1.7	−2.5	−4.2	−5.0	−7.3	−0.55	−0.88	−0.72
R_H	−1.33	−2.36	−4.46	−5.43	−6.87	−0.74	−1.07	−1.06
	Be	Mg	Ca	Sr	Ba	Zn	Cd	Hg
R'_H	2.44	−0.83	−1.78	–	–	0.33	0.55	−0.8
R_H	−0.25	−0.73	−1.36	−1.76	−1.98	−0.47	−0.67	−0.72
	Fe	Al	Ga	In	Sn	Pb	Sb	Bi
R'_H	0.25	−0.34	–	−0.07	−0.04	0.09	−20	-10^3
R_H	−0.37	−0.35	−0.40	−0.54	−0.42	−0.47	−0.38	−0.44

5.3 Band Theory

The Drude model and the Sommerfeld model are both free electron models, and they provided very good explanations of the electrical and thermal properties of metals. However, as discussed in the last section, the predictions by the free electron Fermi gas model have quantitative, and sometimes even qualitative, deviations from the results of experiments in the specific heat, conductivity, thermal conductivity, thermionic emission, and Hall effects in metals.

A more accurate solid state electronics theory is the band theory, which explains the electrical properties of conducting and non-conducting solids in the same frame. A series of revolutions in physics and chemistry took place after Schrodinger's work in quantum wave mechanics in 1926. The band theory was formulated at almost the same time as the Sommerfeld model. In 1927, the concept of the band theory was first mentioned by a German physicist, M. J. O. Strutt, and reliably established by Felix Bloch's theorem of electron energy bands in crystals in 1928. In the period from 1928 to 1929, Rudolf Peierls proposed the concepts of carriers in near-filled or half-filled bands. In 1930, Léon Brillouin discussed the band gaps and introduced the concept of the Brillouin zones. In 1946, Walter Kohn, who was born in Austria, first learned about the band theory in Harvard. Kohn worked at Bell Laboratories in 1953 together with the pioneers in semiconductor transistors, and developed the density functional theory (DFT) in 1963 when he visited the École Normale Supérieure in France. The electronic structure of crystals could be computed using the DFT, and it became a subject of intense interest in physics, chemistry, and materials science. Walter Kohn was awarded the Nobel Prize in Chemistry in 1998.

Figure 5.10 Schematics of the density functional theory (from http://troubadix.physik.tu-berlin.de/~petz0433/dft/eindex.html)

The visualized representation of electron bands is similar to that of the phonon dispersions discussed in Chapter Four, and they should be plotted along a series of symmetry directions in the first Brillouin zone (FBZ); and the BCC, FCC, or HCP crystals would have the respective characteristic band views. The differences between conductors, insulators, and semiconductors can be distinguished by the zero, large, or small energy band gap respectively near the Fermi surface.

Two basic models of energy bands, as the two tractable limits of Bloch's theorem, are the weak potential approximation and the tight-binding model. The tight-binding model is important for understanding the inner bands with respect to the discrete energy levels in free atoms. The band of electrons with a very weak potential is crucial to the understanding of the valence bands in metals.

In density functional theory, the energy bands of an alloy are completely characterized by its electronic density distribution $n(\vec{r}) = |\Psi(\vec{r})|^2$. Based on the variational principle, it can be proved that there is an explicit elementary relationship between the single particle periodic potential $v(\vec{r})$ and the multi-electron density $n(\vec{r})$ of the ground state; i.e., the density $n(\vec{r})$ completely determines the potential $v(\vec{r})$, the total Hamiltonian $\mathcal{H}$, and hence all properties derived from $\mathcal{H}$, including the properties of excited states. The computational methods based on DFT have been successfully used to calculate the energy bands of matter. The greatest achievements arising from band theory are in the field of semiconductors and solid electronic devices, which are discussed in the next chapter.

5.3.1 Bloch's Theorem

In 1928, Felix Bloch established a theorem about the quantum energy distributions of electrons in crystals. In the early stages, the band theory was not as "useful" as the Sommerfeld model, which could calculate various physical properties of metals; however Bloch's theorem provided a solid basis for the future applications of the band theory in semiconductors and other materials. The basic tenets of Bloch's band theory are:

1. Independent electron approximation: A single electron state is considered.
2. An electron moves in a periodic electric potential in a crystal, which is caused by other electrons and ions. The free electron approximation is no longer used.
3. Quantum mechanics and Fermi-Dirac statistics are used.

Electrons have wave-particle duality in nature, and Schrodinger's equation can be used to calculate the eigen-energies of the de Broglie wavefunction ψ:

$$\left[-\frac{\hbar^2}{2m} \vec{\nabla}^2 + V(\vec{r}) \right] \psi = \varepsilon \psi \tag{5.55}$$

Electrons moving in the periodic potential $V(\vec{r})$ are called the Bloch electrons.

Based on Max Born's statistical explanation and the periodic nature of crystals, the density of electrons $n(\vec{r}) = |\psi(\vec{r})|^2$ should not vary under

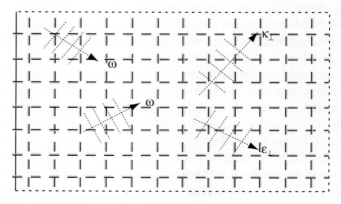

Figure 5.11 Bloch theory: electron's de Broglie wave in a periodic lattice

the translational symmetry operation $\mathcal{T}_{\vec{R}}$ of a lattice:

$$\mathcal{T}_{\vec{R}} n(\vec{r}) = n(\vec{r} + \vec{R}) = n(\vec{r}) \quad (\forall \vec{R} = \sum_{\alpha} n_{\alpha} \vec{a}_{\alpha}) \tag{5.56}$$

Therefore, the values of the wavefunctions at the same positions of neighboring unit cells should only differ by a simple phase factor:

$$\psi(\vec{r} + \vec{a}_{\alpha}) = e^{i\theta_{\alpha}} \psi(\vec{r}) \quad (\forall \alpha = 1, 2, 3) \tag{5.57}$$

If the Born-Karman condition is used in a crystal with $L_1 \times L_2 \times L_3$ primitive unit cells, the unknown phases θ_{α} in the former equation can be found:

$$\psi(\vec{r} + L_{\alpha} \vec{a}_{\alpha}) = \psi(\vec{r}) \to e^{iL_{\alpha}\theta_{\alpha}} = 1 \to \theta_{\alpha} = \frac{2\pi}{L_{\alpha}} l_{\alpha} \tag{5.58}$$

where l_{α} are integers. Then the first form of Bloch's theorem can be obtained:

$$\psi(\vec{r} + \vec{R}) = \psi\left(\vec{r} + \sum_{\alpha} n_{\alpha} \vec{a}_{\alpha}\right) = e^{i \sum_{\alpha} \frac{2\pi}{L_{\alpha}} l_{\alpha} n_{\alpha}} \psi(\vec{r}) = e^{i\vec{k}\cdot\vec{R}} \psi(\vec{r})$$
$$\tag{5.59}$$

where the form of the wavevector $\vec{k} = \sum_{\alpha} \frac{l_{\alpha}}{L_{\alpha}} \vec{a}_{\alpha}^{*}$ is for the Born-Karman condition, as discussed in Chapter Three.

The second form of Bloch's theorem, namely that the electron wavefunction in a lattice is a modulated plane wave $\psi_{\vec{k}}(\vec{r}) = u_{\vec{k}}(\vec{r}) e^{i\vec{k}\cdot\vec{R}}$, can easily be derived from the first form of Bloch's theorem:

$$u_{\vec{k}}(\vec{r} + \vec{R}) = e^{-i\vec{k}\cdot(\vec{r}+\vec{R})} \psi_{\vec{k}}(\vec{r} + \vec{R}) = e^{-i\vec{k}\cdot\vec{r}} \psi_{\vec{k}}(\vec{r}) = u_{\vec{k}}(\vec{r}) \tag{5.60}$$

Therefore the modulated amplitude $u_{\vec{k}}(\vec{r})$ is a periodic function in the lattice, which is consistent with the periodic distribution of the electron density in Eq. (5.56).

The energy spectrum of a single electron in a lattice is a function of the wavevector $\vec{k}$, and can be solved by a static eigen-equation of the modulated amplitude $u_{\vec{k}}(\vec{r})$ of the wavefunction of electrons:

$$\left[\frac{(\vec{p} + \hbar\vec{k})^2}{2m} + V(\vec{r})\right] u_{n\vec{k}}(\vec{r}) = \varepsilon_n(\vec{k}) u_{n\vec{k}}(\vec{r}) \tag{5.61}$$

Note that the quantum commutator $[\vec{p}, f(\vec{r})] = -i\hbar \vec{\nabla} f(\vec{r})$ is used in the derivation. The Hilbert space of the Hamiltonian operator

$\mathcal{H}_{\vec{k}} = (\vec{p} + \hbar\vec{k})^2/2m + V(\vec{r})$ spans a series of eigenfunctions $u_{n\vec{k}}(\vec{r})$ with respect to eigenvalues $\varepsilon_n(\vec{k})$ from $n = 0, 1, \ldots, \infty$:

$$u_{n\vec{k}}(\vec{r}) = \langle \vec{r} \mid n, \vec{k} \rangle; \quad \mathcal{H}_{\vec{k}} \mid n, \vec{k} \rangle = \varepsilon_n(\vec{k}) \mid n, \vec{k} \rangle. \tag{5.62}$$

From the view of the energy spectrum $\varepsilon_n(\vec{k})$, all electron eigen-energies with a fixed quantum number n should form a three-dimensional "surface" in four-dimensional $(\varepsilon, \vec{k})$ space, which is impossible to visualize in the real world 3D space.

The energy spectrum $\varepsilon_n(\vec{k})$ of electrons has the periodicity in the reciprocal space

$$\varepsilon_n(\vec{k}) = \varepsilon_n(\vec{k} + \vec{G}); \quad \forall \vec{G} \tag{5.63}$$

where $\vec{G} = \sum_\alpha h_\alpha \vec{a}_\alpha^*$ is a reciprocal position vector in the reciprocal space. The wavefunction with respect to the $|n, \vec{k} + \vec{G}\rangle$ quantum state is

$$\psi_{\vec{k}+\vec{G}}(\vec{r}) = e^{i(\vec{k}+\vec{G})\cdot\vec{r}} u_{\vec{k}+\vec{G}}(\vec{r}) = e^{i\vec{k}\cdot\vec{r}} \left(e^{i\vec{G}\cdot\vec{r}} u_{\vec{k}+\vec{G}}(\vec{r}) \right) = e^{i\vec{k}\cdot\vec{r}} \psi'_{\vec{k}}(\vec{r})$$

$$\tag{5.64}$$

The wavefunction $\psi_{\vec{k}+\vec{G}}(\vec{r})$ satisfies Schrodinger's equation in Eq. (5.55), therefore the unknown wavefunction $\psi'_{\vec{k}}(\vec{r})$ must satisfy the static quantum wave equation Eq. (5.61). In a lattice, $\exp(i\vec{G}\cdot\vec{R}) = 1$ is always true, so $\psi'_{\vec{k}}(\vec{r})$ and $u_{\vec{k}}(\vec{r})$ satisfy the same periodicity and the same static eigen-equation. In results based on the uniqueness of the solution to the differential equation, the equivalence

$$\psi'_{\vec{k}}(\vec{r}) = e^{i\vec{G}\cdot\vec{r}} u_{\vec{k}+\vec{G}}(\vec{r}) = u_{\vec{k}}(\vec{r}) \tag{5.65}$$

is a natural conclusion. Furthermore, by the second form of Bloch's theorem, the wavefunction $\psi_{\vec{k}+\vec{G}}(\vec{r})$ is also equivalent to $\psi_{\vec{k}}(\vec{r})$, and the corresponding equivalence of the eigen-energies $\varepsilon(\vec{k} + \vec{G})$ and $\varepsilon(\vec{k})$ is proved.

The energy band $\varepsilon_n(\vec{k})$ should also be invariant under the rotational or reflectional symmetry operation $\mathcal{R}$ using the relations $\vec{k} \cdot (\mathcal{R}\vec{r}) = (\mathcal{R}^{-1}\vec{k}) \cdot \vec{r}$ and $u_{\vec{k}}(\mathcal{R}\vec{r}) = u_{\vec{k}}(\vec{r})$:

$$\mathcal{R}^{-1}\mathcal{H}\mathcal{R} = \mathcal{H}$$
$$\rightarrow \mathcal{H}[\mathcal{R}\Psi_{n\vec{k}}(\vec{r})] = \varepsilon_n(\vec{k})[\mathcal{R}\Psi_{n\vec{k}}(\vec{r})] \rightarrow \varepsilon_n(\mathcal{R}^{-1}\vec{k}) = \varepsilon_n(\vec{k}) \tag{5.66}$$

The symmetry of 2D energy bands was first recognized by Eugene Wigner in 1931. Wigner was subsequently rewarded with a Nobel Prize in 1963. The lattice translation, rotation, and reflection symmetries of energy bands imply that characteristics of $\varepsilon_n(\vec{k})$ can be expressed by plotting a diagram along the symmetry directions in the FBZ, or the Wigner-Seitz cell of the reciprocal lattice.

5.3.2 Tight-Binding Model

The tight-binding approximation or the tight-binding model was the first band-calculation method, devised by Felix Bloch in 1928 (based on Bloch, 1928). Bloch's theorem is true for both conducting and bound electrons in a crystal. In a solid, most electrons are tightly bound to the inner shells of atoms, but a few valence electrons are shared among atoms in chemical bonds. The tight-binding model can explain the energy bands of the inner shell electrons, and some bands of the valence electrons.

It is very hard to analytically solve Schodinger's Eq. (5.55) or Eq. (5.61) with a periodic potential. Therefore the perturbation theory has to be used in one way or another to simplify the problem. In the tight-binding model, the zeroth order Hamiltonian $\mathcal{H}^0$ is a single electron Hamiltonian $\mathcal{H}_{\vec{R}}^a(\vec{r},\vec{p}) = \vec{p}^2/2m + V(\vec{r}-\vec{R})$ in an isolated atom located at $\vec{R}$; but in the total Hamiltonian, the electron is influenced by the periodic potential $\sum_{\vec{R}} V(\vec{r}-\vec{R})$ contributed by all atoms in a crystal:

$$\mathcal{H} = \frac{\vec{p}^2}{2m} + \sum_{\vec{R}} V(\vec{r}-\vec{R}) = \mathcal{H}^0 + \mathcal{H}^1 = \mathcal{H}_{\vec{R}}^a + \sum_{\vec{R}'\neq\vec{R}} V(\vec{r}-\vec{R}');$$

(5.67)

$$\Phi_n(\vec{r}-\vec{R}) = \langle \vec{r}-\vec{R}|n'lm\sigma\rangle; \quad \mathcal{H}_{\vec{R}}^a \Phi_n(\vec{r}-\vec{R}) = \varepsilon_n^0 \Phi_n(\vec{r}-\vec{R}).$$

(5.68)

The zeroth order eigenfunction $\Phi_n(\vec{r}-\vec{R})$ is just the single electron atomic orbits $|n'lm\sigma\rangle$, and the eigen-energy ε_n^0 relates to the 1s, 2s, 2p, 3s, 3p ... energy levels of an isolated atom located at $\vec{R}$. In this model, the element crystal is assumed, thus the energy levels ε_n^0 are independent of $\vec{R}$.

Felix Bloch proposed the Bloch wavefunction of the single electron satisfying the basic requirements of the Bloch's theorem. The Bloch wavefunction is a linear combination of atomic orbits (LCAO) of the n'th electron orbits in all atoms:

$$\psi_{n\vec{k}}(\vec{r}) = \frac{1}{\sqrt{N_L}} \sum_{\vec{R}} \Phi_n(\vec{r}-\vec{R}) e^{i\vec{k}\cdot\vec{R}}; \quad \mathcal{H}\psi_{n\vec{k}}(\vec{r}) = \varepsilon_n(\vec{k})\psi_{n\vec{k}}(\vec{r});$$

(5.69)

$$\langle \Phi_n(\vec{r} - \vec{R}) \mid \Phi_{n'}(\vec{r} - \vec{R}')\rangle \simeq \delta_{\vec{R}\vec{R}'}\delta_{nn'}; \quad \langle \psi_{n\vec{k}}(\vec{r}) \mid \psi_{n'\vec{k}}(\vec{r})\rangle \simeq \delta_{nn'}.$$

(5.70)

where N_L is the number of primitive unit cells in an element crystal, and the approximate orthogonal condition of zeroth order eigenstates is used. Bloch's theorem is satisfied for $\psi_{n\vec{k}}(\vec{r})$ due to the Born-Karman condition in a crystal:

$$\psi_{n\vec{k}}(\vec{r} + \vec{R}_0) = \frac{1}{\sqrt{N_L}} \sum_{\vec{R}} \Phi_n(\vec{r} - (\vec{R} - \vec{R}_0))e^{i\vec{k}\cdot(\vec{R}-\vec{R}_0)} e^{i\vec{k}\cdot\vec{R}_0}$$
$$= \psi_{n\vec{k}}(\vec{r})e^{i\vec{k}\cdot\vec{R}_0}$$

(5.71)

The energy band can be found by calculating the average energy $\langle \Psi_{n\vec{k}} | \mathcal{H} | \Psi_{n\vec{k}} \rangle$:

$$\varepsilon_n(\vec{k}) = \frac{1}{N_L} \sum_{\vec{R}_1} \sum_{\vec{R}_2} \left\langle \Phi_n(\vec{r} - \vec{R}_1) \right.$$
$$\left. \times \left| \frac{\vec{p}^2}{2m} + \sum_{\vec{R}} V(\vec{r} - \vec{R}) \right| \Phi_n(\vec{r} - \vec{R}_2) \right\rangle e^{-i\vec{k}\cdot(\vec{R}_1 - \vec{R}_2)} \quad (5.72)$$

Both the potential $V(\vec{r} - \vec{R})$ and the zeroth order wave function $\Phi_n(\vec{r} - \vec{R})$ are localized around the atom at $\vec{R}$; therefore it is reasonable to assume that:

$$\langle \Phi_n(\vec{r} - \vec{R}_1) | V(\vec{r} - \vec{R}) | \Phi_n(\vec{r} - \vec{R}_2) \rangle = 0; \quad (\vec{R}_1 \neq \vec{R}_2 \neq \vec{R})$$

(5.73)

In other words, the matrix element of the potential V is assumed to be non-zero when any one of the three conditions $\vec{R}_1 = \vec{R}_2$, $\vec{R}_2 = \vec{R}$ or $\vec{R}_1 = \vec{R}$ is satisfied. The n'th energy band can then be found by Eqs. (5.72) and (5.73):

$$\varepsilon_n^0 | \Phi_n(\vec{r} - \vec{R}) \rangle = \left(\frac{\vec{p}^2}{2m} + V(\vec{r} - \vec{R}) \right) | \Phi_n(\vec{r} - \vec{R}) \rangle \quad (5.74)$$

$$\varepsilon_n(\vec{k}) = \varepsilon_n^0 + \bar{V}_n$$
$$+ \frac{1}{N_L} \sum_{\vec{R}_1 \neq \vec{R}_2} \langle \Phi_n(\vec{r} - \vec{R}_1) | V(\vec{r} - \vec{R}_2) | \Phi_n(\vec{r} - \vec{R}_2) \rangle e^{-i\vec{k}\cdot(\vec{R}_1 - \vec{R}_2)}$$

$$
\begin{aligned}
&= \varepsilon_n^0 + \bar{V}_n \\
&+ \frac{1}{N_L} \sum_{\vec{R}_1 \neq \vec{R}_2} \langle \Phi_n(\vec{r} - \vec{R}_1) | V(\vec{r} - \vec{R}_1) | \Phi_n(\vec{r} - \vec{R}_2) \rangle e^{-i\vec{k}\cdot(\vec{R}_1-\vec{R}_2)} \\
&= \varepsilon_n^0 + \bar{V}_n - \sum_{\vec{d}} J_{n\vec{d}}^{(1)} e^{i\vec{k}\cdot\vec{d}}
\end{aligned}
$$

(5.75)

where $\vec{d}$ is a displacement vector between neighboring atoms, and there are 6, 8, 12, and 12 different $\vec{d}$ in a SC, BCC, FCC, and HCP lattice respectively. The energy band of the tight binding model in Eq. (5.75) has the parameters:

$$\bar{V}_n = \langle \Phi_n(\vec{r} - \vec{R}_1) | \sum_{\vec{R}} V(\vec{r} - \vec{R}) | \Phi_n(\vec{r} - \vec{R}_1) \rangle; \tag{5.76}$$

$$J_{n\vec{d}}^{(1)} = -\langle \Phi_n(\vec{r} - \vec{R} - \vec{d}) | V(\vec{r} - \vec{R}) | \Phi_n(\vec{r} - \vec{R}) \rangle; \tag{5.77}$$

The minus sign in the definition of $J_{n\vec{d}}^{(1)}$ is used because the potential $V(r - \vec{R})$ felt by a single electron at the $\vec{R}$ atom is usually negative. The physical meaning of $\bar{V}_n$ is the modification of the n'th energy level of an isolated atom by the total potential; while $J_{n\vec{d}}^{(1)}$ describes the overlapping of the wavefunction for neighboring atoms. The band width is proportional to $J_{n\vec{d}}^{(1)}$, and it must be wider for higher n-shells.

In a Bravais lattice, the tight-binding band in Eq. (5.75) has periodicity in the reciprocal lattice, because the displacement between neighboring atoms $\vec{d}$ is a position vector:

$$\varepsilon_n(\vec{k} + \vec{G}) = \varepsilon_n^0 + \bar{V}_n - \sum_{\vec{d}} J_{n\vec{d}}^{(1)} e^{i\vec{k}\cdot\vec{d}} e^{i\vec{G}\cdot\vec{d}} = \varepsilon_n(\vec{k}) \tag{5.78}$$

The periodicity in the reciprocal lattice implies that the energy band $\varepsilon_n(\vec{k})$ can simply be expressed in the FBZ. So, how many electron states are there in an energy band? As discussed in Chapter Three, there are N_L different values of $\vec{k}$ in the FBZ, therefore totally $2N_L$ quantum states $\psi_{n\vec{k}}(\vec{r}) = \langle \vec{r} | n, \vec{k}, \sigma \rangle$ can be filled in the n'th band.

The expression of tight-binding energy bands $\varepsilon_n(\vec{k})$ depends on the lattice structure. First, we shall look at the tight-binding s-band and p-band in an element crystal with a SC lattice. The six neighboring displacements in an SC lattice are

$$\vec{d} = \pm a\hat{e}_x; \quad \vec{d} = \pm a\hat{e}_y; \quad \vec{d} = \pm a\hat{e}_z \tag{5.79}$$

By Eq. (5.75), the isotropic s-band with constants $\bar{V}_s$ and $J_s^{(1)} > 0$ has the form:

$$\varepsilon_s(\vec{k}) = \varepsilon_s^0 + \bar{V}_s - 2J_s^{(1)}(\cos(k_xa) + \cos(k_ya) + \cos(k_za)) \quad (5.80)$$

The p-bands are a little more complicated, because the probability of electrons in the p_x, p_y, p_z orbits has the dumbbells shape along the $\pm x$, $\pm y$, $\pm z$ directions respectively. By Eq. (5.75), the anisotropic p-bands has the form:

$$\varepsilon_{p_x}(\vec{k}) = \varepsilon_p^0 + \bar{V}_p - 2\left(J_{p\sigma}^{(1)}\cos(k_xa) + J_{p\pi}^{(1)}\cos(k_ya) + J_{p\pi}^{(1)}\cos(k_za)\right);$$

(5.81)

$$\varepsilon_{p_y}(\vec{k}) = \varepsilon_p^0 + \bar{V}_p - 2\left(J_{p\pi}^{(1)}\cos(k_xa) + J_{p\sigma}^{(1)}\cos(k_ya) + J_{p\pi}^{(1)}\cos(k_za)\right);$$

(5.82)

$$\varepsilon_{p_x}(\vec{k}) = \varepsilon_p^0 + \bar{V}_p - 2\left(J_{p\pi}^{(1)}\cos(k_xa) + J_{p\pi}^{(1)}\cos(k_ya) + J_{p\sigma}^{(1)}\cos(k_za)\right).$$

(5.83)

Where $J_{p\sigma}^{(1)} < 0$ describes the head-on-head overlapping of dumbbells from the neighboring atoms, i.e., the σ-bond, the $J_{p\pi}^{(1)} > 0$ stands for the side-by-side overlapping of dumbbells from the neighboring atoms, i.e., the π-bond. The σ-bond is stronger than the π-bond; therefore $|J_{p\sigma}^{(1)}| > J_{p\pi}^{(1)}$ is always true.

The band width of the s-band is $12J_s^{(1)}$, which is the difference between the energy at the center Γ and the corner C of the FBZ: $\varepsilon_s(\frac{\pi}{a}, \frac{\pi}{a}, \frac{\pi}{a}) - \varepsilon_s(0, 0, 0)$. The p_x, p_y, and p_z bands twist and intersect in the FBZ. Along the [100] direction, p_x band maximizes, but p_y, p_z bands maximize at Γ, as shown in Figure 5.12. The total p-band width

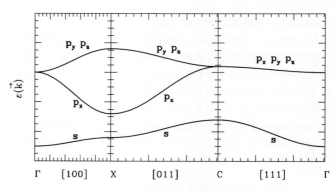

Figure 5.12 The s-band and p-bands in the FBZ of SC lattices

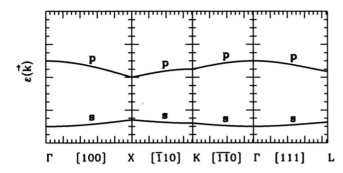

Figure 5.13 The inner s-band and p-bands in the FBZ of DIA structures

is $4|J_{p\sigma}^{(1)}| + 4J_{p\pi}^{(1)}$ in a crystal with SC structure, which is the energy difference $\varepsilon_{p_y}(\frac{\pi}{a}, 0, 0) - \varepsilon_{p_x}(\frac{\pi}{a}, 0, 0)$.

Secondly, it is interesting to look at the tight-binding bands in a silicon crystal. The diamond structure of silicon is not a Bravais lattice. The two different sets of the four neighboring displacements are

$$\vec{d}^A = \frac{a}{4}(+\hat{e}_x + \hat{e}_y + \hat{e}_z); \quad \vec{d}^B = \frac{a}{4}(-\hat{e}_x - \hat{e}_y - \hat{e}_z);$$
$$\vec{d}^A = \frac{a}{4}(-\hat{e}_x + \hat{e}_y - \hat{e}_z); \quad \vec{d}^B = \frac{a}{4}(+\hat{e}_x + \hat{e}_y - \hat{e}_z);$$
$$\vec{d}^A = \frac{a}{4}(+\hat{e}_x - \hat{e}_y - \hat{e}_z); \quad \vec{d}^B = \frac{a}{4}(-\hat{e}_x + \hat{e}_y + \hat{e}_z);$$
$$\vec{d}^A = \frac{a}{4}(-\hat{e}_x - \hat{e}_y + \hat{e}_z); \quad \vec{d}^B = \frac{a}{4}(+\hat{e}_x - \hat{e}_y + \hat{e}_z). \quad (5.84)$$

In a silicon crystal, the 3s-band and 3p-band are not distinct because of the sp^3-hybridization of valence electrons, as discussed in Chapter Two. However, the tight-binding model cannot deal with the energy bands with respect to covalent bonds correctly, because the perturbation theory no longer holds to describe the sp^3-hybrid orbits.

In the tight-binding model, the inner s-band of silicon must be an average of the energy bands in Eq. (5.75) contributed by displacement vectors around the A and B sites respectively:

$$\varepsilon_s(\vec{k}) = \frac{1}{2}\left(\varepsilon_s^A(\vec{k}) + \varepsilon_s^B(\vec{k})\right)$$
$$= \varepsilon_s^0 + \bar{V}_s - 4J_s^{(1)} \cos\left(\frac{k_x a}{4}\right) \cos\left(\frac{k_y a}{4}\right) \cos\left(\frac{k_y a}{4}\right) \quad (5.85)$$

The inner shell p_x, p_y, p_z bands of silicon are degenerate, because all of the "dumbbells" span the same angle along the bond directions.

According to Eq. (5.75), the anisotropic p-bands of silicon have the form:

$$\varepsilon_p(\vec{k}) = \frac{1}{2}\left(\varepsilon_p^A(\vec{k}) + \varepsilon_p^B(\vec{k})\right)$$
$$= \varepsilon_p^0 + \bar{V}_p - 4J_p^{(1)} \cos\left(\frac{k_x a}{4}\right) \cos\left(\frac{k_y a}{4}\right) \cos\left(\frac{k_y a}{4}\right). \tag{5.86}$$

where $J_p^{(1)} < 0$, because the dumbbells of the p-orbits from neighboring atoms have a parallel displacement. The band widths of the s-band and p-band of a silicon crystal are $4J_s^{(1)}$ and $4|J_p^{(1)}|$ respectively, and both are the difference between the energy at the center Γ and the face center X of the truncated octahedral FBZ: $|\varepsilon(\frac{2\pi}{a}, 0, 0) - \varepsilon(0, 0, 0)|$.

5.3.3 Weak Potential Approximation

In the tight-binding model, $\mathcal{H}_{\vec{R}}^a$ of an isolated atom is chosen as the zeroth order Hamiltonian in the perturbation theory; however the valence bands in ionic, covalent, or metallic crystals cannot be explained correctly by the model, because the valence electron probability in these solids is totally different from that in an atom.

In this section, another method of solving the Schodinger Eq. (5.55) or Eq. (5.61) is introduced. The free electron Hamiltonian $\mathcal{H}_0 = \vec{p}^2/2m$ is chosen as the zeroth order Hamiltonian in the perturbation theory, corresponding to the situation of the valence band in metals. The total Hamiltonian includes a very weak periodic potential $U(\vec{r})$ contributed by all ions and other electrons in a metal:

$$\mathcal{H} = \frac{\vec{p}^2}{2m} + U(\vec{r}) = \mathcal{H}^0 + \mathcal{H}^1; \tag{5.87}$$

$$\phi_{\vec{k}}(\vec{r}) = \frac{1}{\sqrt{V}} e^{i\vec{k}\cdot\vec{r}} = \langle \vec{r}|\vec{k}\rangle; \quad -\frac{\hbar^2}{2m}\vec{\nabla}^2 \phi_{\vec{k}}(\vec{r}) = \varepsilon_0(\vec{k}) \phi_{\vec{k}}(\vec{r}). \tag{5.88}$$

The zeroth order quantum state $|\vec{k}\rangle$ with an eigen-energy $\varepsilon_0(\vec{k}) = \hbar^2 \vec{k}^2/2m$ is simply the de Broglie plane wave with respect to the free electron model.

The total energy and the total quantum state $|\psi_{\vec{k}}\rangle$ of an electron in the weak potential can be found using the first and second order perturbation theory:

$$\varepsilon(\vec{k}) = \varepsilon_0(\vec{k}) + \langle \vec{k}|U(\vec{r})|\vec{k}\rangle + \sum_{\vec{k}' \neq \vec{k}} \frac{|\langle \vec{k}|U(\vec{r})|\vec{k}'\rangle|^2}{\varepsilon_0(\vec{k}) - \varepsilon_0(\vec{k}')}; \tag{5.89}$$

$$|\psi_{\vec{k}}\rangle = |\vec{k}\rangle + \sum_{\vec{k}' \neq \vec{k}} |\vec{k}'\rangle \frac{\langle \vec{k}'|U(\vec{r})|\vec{k}\rangle}{\varepsilon_0(\vec{k}) - \varepsilon_0(\vec{k}')} \tag{5.90}$$

The matrix element of the periodic potential is the key quantity. The periodic potential can be expanded into the Fourier series:

$$U(\vec{r}) = \sum_{\vec{G}} U_{\vec{G}} e^{i\vec{G}\cdot\vec{r}}; \iff U(\vec{r}+\vec{R}) = U(\vec{r}) \tag{5.91}$$

It is then easy to prove that there are selection rules for the elements $\langle \vec{k}|U(\vec{r})|\vec{k}'\rangle$:

$$\langle \vec{k}'|U(\vec{r})|\vec{k}\rangle = \frac{1}{V}\sum_{\vec{G}} U_{\vec{G}} \iiint d^3\vec{r} e^{i(\vec{G}-\vec{k}'+\vec{k})\cdot\vec{r}} = \begin{cases} 0 & \vec{k}'-\vec{k}\neq \forall \vec{G} \\ U_{\vec{G}} & \vec{k}'-\vec{k} = \vec{G} \end{cases}$$

$$\tag{5.92}$$

Meanwhile, a series of energy bands of electrons in a weak potential can be found:

$$\varepsilon_n(\vec{q}) = \frac{\hbar^2}{2m}(\vec{q}+\vec{G}^n_{hkl})^2 + \bar{U}$$
$$+ \sum_{\vec{G}'\neq 0} \frac{|U_{\vec{G}'}|^2}{\frac{\hbar^2}{2m}[(\vec{q}+\vec{G}^n_{hkl})^2 - (\vec{q}+\vec{G}^n_{hkl}+\vec{G}')^2]} \tag{5.93}$$

$$|n,\vec{q}\rangle = |\vec{q}+\vec{G}^n_{hkl}\rangle + \sum_{\vec{G}}|\vec{q}+\vec{G}^n_{hkl}+\vec{G}'\rangle\frac{U_{\vec{G}'}}{\varepsilon_0(\vec{k})-\varepsilon_0(\vec{k}')} \tag{5.94}$$

where $\varepsilon_n(\vec{q})$ is the band in the $(n+1)$'th Brillouin Zone ($n = 0, 1, 2, \ldots$) and $|n,\vec{q}\rangle$ is the respective eigenstate of the Bloch electron. The wave vector of the Bloch electron $\vec{q}$ is defined in the FBZ, and can be found by a reciprocal lattice displacement $\vec{G}^n_{hkl} = \vec{k} - \vec{q}$. The band with the lowest energy is $\varepsilon_0(\vec{q})$, where $\vec{k} = \vec{q}$.

The 1D energy bands of electrons in a weak potential are quite clear, as illustrated in Figure 5.14. In Eq. (5.93), the last term is the second order perturbation energy:

$$\varepsilon_n^{(2)}(q) = \sum_{n'\neq 0} \frac{|U_{n'}|^2}{\frac{\hbar^2}{2m}[(q\pm na^*)^2 - (q\pm na^* + n'a^*)^2]} \tag{5.95}$$

The perturbation $\varepsilon_0^{(2)}(\vec{q})$ for the first band in the FBZ must be negative, because it has the lowest energy and $q^2 < (q+n'a^*)^2$ is always true; therefore the first band $\varepsilon_0(q)$ must "bend" down. Similarly, the n'th band will bend up ($d^2\varepsilon_n(q)/dq^2$ increases) or down (with respect to $d^2\varepsilon_n(q)/dq^2 < 0$), where the perturbation $\varepsilon_n^{(2)}(\vec{q})$ is mainly contributed by a positive or negative term $[(q\pm na^*)^2 - (q\pm na^* + n'a^*)^2]^{-1}$ with respect to the minimum energy difference.

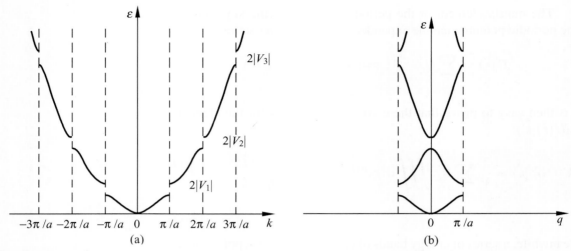

Figure 5.14 1D bands of electrons in a weak potential: (a) $\varepsilon(k)$, where k is the free electron wave number; (b) $\varepsilon_n(q)$, where q is the wave number of Bloch electrons in the FBZ

In a 3D lattice, the perturbation theory is a little more complicated. For example, for the alkali metals, the lattice structure is BCC. In the energy bands of BCC crystals, $\vec{q}$ is tracing along the symmetry points $\Gamma - H - P - \Gamma - N - H$ in the FBZ, where all $\vec{q}$ locate in the first quadrant of FBZ, whose nearest perturbed state must locate in the second BZ with a displacement $\vec{a}_3^* = (\hat{e}_x + \hat{e}_y)a^*/2$. The second order perturbation energy can still be found by Eq. (5.93):

$$\varepsilon_0(\vec{q}) \simeq \frac{\hbar^2}{2m}\vec{q}^2 + \bar{U} + \frac{|U_{\vec{a}_3^*}|^2}{\frac{\hbar^2}{2m}[\vec{q}^2 - (\vec{q} - \vec{a}_3^*)^2]};$$

$$\varepsilon_1(\vec{q}) \simeq \frac{\hbar^2}{2m}(\vec{q} - \vec{a}_3^*)^2 + \bar{U} + \frac{|U_{\vec{a}_3^*}|^2}{\frac{\hbar^2}{2m}[(\vec{q} - \vec{a}_3^*)^2 - \vec{q}^2]}, \quad (5.96)$$

In Figure 5.15, the solid line and the dotted line are the first valence band $\varepsilon_0(\vec{q})$ and second valence band $\varepsilon_1(\vec{q})$ respectively. The horizontal axis $|\vec{q}|$ has the form:

$$P - \Gamma : \vec{q} = \left(\frac{1}{2}, \frac{1}{2}, \frac{1}{2}\right)\frac{a^*}{2} + \eta\left(-\frac{1}{2}, -\frac{1}{2}, -\frac{1}{2}\right)\frac{a^*}{2}; \quad (0 < \eta < 1)$$

$$\Gamma - H : \vec{q} = \eta(0, 1, 0)\frac{a^*}{2}; \quad \Gamma - N : \vec{q} = \eta\left(\frac{1}{2}, \frac{1}{2}, 0\right)\frac{a^*}{2};$$

(5.97)

The N_L valence electrons of the ns-shell in the alkali metals will fill half of the first valence band $\varepsilon_0(\vec{q})$, because there are $2N_L$ quantum states $|n, \vec{q}\rangle$ in a band.

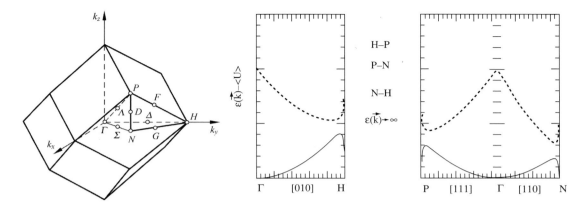

Figure 5.15 First and second 3D electron bands in a weak potential in alkali metals

It should be emphasized that the perturbation energy in Eq. (5.96) and Figure 5.15 is infinity from $N - H$ at the surface of the FBZ, because the free electron energies in quantum states $|\vec{q}\rangle$ and $|\vec{q} - \vec{a}_3^*\rangle$ are identical:

$$\vec{q}^2 - (\vec{q} - \vec{a}_3^*)^2 = \left[\frac{a^*}{2}(\hat{e}_y + \eta\hat{e}_x - \eta\hat{e}_y)\right]^2$$
$$- \left[\frac{a^*}{2}(-\hat{e}_x + \eta\hat{e}_x - \eta\hat{e}_y)\right]^2 = 0; \quad (5.98)$$

where the coefficient $\eta \in (0, 1/2)$. The perturbation theory in Eq. (5.96) gives $\varepsilon^{(2)}(\vec{q}) \to \infty$ but it cannot be used. The degenerate perturbation theory is needed to find the band gap:

$$|\psi\rangle = \alpha|\vec{q}\rangle + \beta|\vec{q} - \vec{a}_3^*\rangle, \quad \mathcal{H}|\psi\rangle = \varepsilon|\psi\rangle; \quad (5.99)$$
$$\alpha\varepsilon = \langle\vec{q}|\mathcal{H}|\psi\rangle = \alpha(\varepsilon_0 + \bar{U}) + \beta U_{\vec{a}_3^*}$$
$$\beta\varepsilon = \langle\vec{q} - \vec{a}_3^*|\mathcal{H}|\psi\rangle = \alpha U_{\vec{a}_3^*}^* + \beta(\varepsilon_0 + \bar{U}). \quad (5.100)$$

The perturbed energies from $N - H$ at the surface of the FBZ are

$$\varepsilon_\pm(\vec{q}) = \varepsilon_0(\vec{q}) + \bar{U} \pm |U_{\vec{a}_3^*}| \quad (5.101)$$

The band gap is simply $2|U_{\vec{a}_3^*}|$, where $U_{\vec{a}_3^*}$ is the Fourier coefficient of the periodic potential $U(\vec{r})$ at the wavelength $\lambda = 2\pi/|\vec{a}_3^*| = a/\sqrt{2}$. Generally, the energy gap is proportional to ΔU, the amplitude difference between the rise and fall of the potential $U(\vec{r})$.

5.3.4 Introduction to DFT and Computational Methods

In 1963, Walter Kohn visited his friend, Philippe Noziéres, at the École Normale Supérieure in Paris. There he read some of the metallurgical literature regarding the charge transfer between unit cells, which presented a different view from the de Brolie wave distributed in the whole space. Kohn started to construct the density functional theory, based on the Thomas-Fermi approximation published in 1927, which was the first theory emphasizing the electron density n. If the total energy of an electron on the Fermi surface $E_{tot} = \varepsilon_F - e\phi(\vec{r})$ is zero ($\phi(\vec{r})$ is the electric potential felt by the electron), the Thomas-Fermi self-consistent differential equation can be found by the electrostatic Poisson equation and the relationship between ε_F and $n(\vec{r})$ (based on Thomas, 1927 and Fermi, 1928):

$$\vec{\nabla}^2 \phi = 4\pi en; \; \varepsilon_F = \frac{\hbar^2}{2m}(3\pi^2 n)^{2/3} \Rightarrow \vec{\nabla}^2[n^{2/3}] = \frac{8\pi}{(3\pi^2)^{2/3} a_B} n, \tag{5.102}$$

where the multi-electron density $n(x)$ in real space is chosen as the basic variable, but not the many-body wavefunction $\Psi(x_1, x_2, \ldots, x_N)$ of multi-electrons.

In chemistry, the basic shapes of the chemical bonds in molecules, clusters, or solids are known, which means the probability density $n(\vec{r})$ of all electrons in a solid is roughly known. The basic question concerning the DFT is: "Is the single electron periodic potential uniquely determined by the multi-electron density?" In Paris, Walter Kohn met Pierre Hohenberg. Hohenberg had received very good mathematical training in the Soviet Union and France. The two of them proved the Hohenberg-Kohn theorem based on the following variational principle:

The single electron potential $v(\vec{r}_i)$ is uniquely determined by the multi-electron ground-state density $n_0(\vec{r}) = |\Psi[n_0(\vec{r})]|^2$, within an arbitrary constant. The total ground-state energy $E[n_0(\vec{r})]$ and density $n_0(\vec{r}) = |\Psi[n_0(\vec{r})]|^2$ in a solid is found via the variational principle. The total Hamiltonian and energy of the many-body electron system is (based on Kohn, Becke, and Parr, 1996):

$$\mathcal{H}(\{\vec{r}_i, \vec{p}_i\}) = K + V + U = \sum_i \frac{\vec{p}_i^2}{2m} + \sum_i v(\vec{r}_i) + \frac{1}{2} \sum_{i \neq j} \frac{e^2}{|\vec{r}_i - \vec{r}_j|}; \tag{5.103}$$

$$E[n(\vec{r})] = \iiint d^3\vec{r}\, n(\vec{r}) v(\vec{r}) + \langle \Psi[n(\vec{r})] | K + U | \Psi[n(\vec{r})] \rangle \tag{5.104}$$

The proof for the Hohenberg-Kohn theorem is not covered here as it is similar to the derivation of the Hartree-Fock theory from the variational principle.

An important practical method for band computation is the Kohn-Sham equation, derived with a constraint $\iiint d^3\vec{r}\, n(\vec{r}) = N$, which was developed by W. Kohn and his post-doctoral researcher, Lu J. Sham, when Kohn returned to University of California San Diego in 1963:

$$\left(-\frac{\hbar^2}{2m}\vec{\nabla}^2 + v(\vec{r}) + \iiint d^3\vec{r}\,'\frac{n(\vec{r}')e^2}{|\vec{r}-\vec{r}'|} + v_{xc}(\vec{r})\right)\psi_j(\vec{r}) = \varepsilon_j \psi_j(\vec{r}) \quad (5.105)$$

$$n(\vec{r}) = \sum_j |\psi_j(\vec{r})|^2; \quad v_{xc}(\vec{r})\delta n(\vec{r}) = \delta E_{xc}[n(\vec{r})] = \delta\langle\Psi|U|\Psi\rangle; \quad (5.106)$$

which has almost the same form as the Hartree-Fock-Slater theory of Eqs. (2.4) and (2.5) in Chapter Two, except that the density $n(\vec{r})$ plays the fundamental role in the DFT instead of the wavefunctions. As in the Hartree-Fock equations, the single electron local equations in Eqs. (5.105) and (5.106) have to be solved self-consistently, because there are many electrons in a PUC. This Kohn-Sham equation is used extensively by physicists, chemists, and material scientists.

The only approximation in the DFT theory is in the calculation of the exchange-correlation energy $E_{xc}[n(\vec{r})]$. The local density approximation (LDA) was formulated by Kohn and Sham based on J. C. Slater's earlier work (based on Kohn and Sham, 1965, and Slater, 1953):

$$E_{xc}[n(\vec{r})] = \iiint d^3\vec{r}\, \varepsilon_{xc}[n(\vec{r})];$$
$$\varepsilon_{xc}[n(\vec{r})] \simeq -\frac{3e^2}{2\pi}(3\pi^2 n(\vec{r}))^{1/3} n(\vec{r}) \quad (5.107)$$

where the quantity $(3\pi^2 n(\vec{r}))^{1/3}$ is in a similar form to the Fermi wavevector in Eq. (5.21) of the Sommerfeld model. The exchange-correlation energy per electron $\varepsilon_{xc}[n(\vec{r})]$ can be rewritten into a simple form $n(\vec{r})(-e^2/l_{xc})$, where the exchange-correlation length $l_{xc} \propto k_F^{-1}$ is obviously related to the average occupation radius of the electron r_s in the free electron gas model.

The real application of the Kohn-Sham theory in chemistry and physics was not realized until the early 1980s, following the development of reliable computational technology. The *ab initio* computational methods of energy bands based on the DFT and LDA have been very

Table 5.6 Practical *ab initio* computational methods of energy bands in solids (based on Feng and Jin, 2003, and Slater, 1953)

Abbr.	Name	Core concepts: wave function choices
OPW	orthogonal plane wave method	$\psi_{\vec{k}}(\vec{r}) = \sum_{\vec{G}} C_{\vec{G}} \phi_{\vec{k}+\vec{G}}(\vec{r})$; $\phi_{\vec{k}}(\vec{r}) = a_0 e^{i\vec{k}\cdot\vec{r}} - \Phi_{\vec{k}}(\vec{r})$ the correction from all inner shells $\Phi_{\vec{k}}(\vec{r}) = \sum_n c_{n\vec{k}} \psi_{n\vec{k}}(\vec{r})$ the Bloch wave function $\psi_{n\vec{k}}(\vec{r}) = \frac{1}{\sqrt{N_L}} \sum_{\vec{R}} \Phi_n(\vec{r} - \vec{R}) e^{i\vec{k}\cdot\vec{R}}$
PSP	pseudo potential method	$\psi_{\vec{k}}(\vec{r}) = \psi^{\text{ps}}(\vec{r}) - \sum_n c_{n\vec{k}} \psi_{n\vec{k}}(\vec{r})$; n is for all inner shells the Shroedinger's equation $(\frac{\vec{p}^2}{2m} + V^{\text{ps}}) \psi^{\text{ps}}(\vec{r}) = \varepsilon(\vec{k}) \psi^{\text{ps}}(\vec{r})$ the pseudo potential $V^{\text{ps}} = V_{tot} + \sum_n (\varepsilon(\vec{k}) - \varepsilon_n) |n\rangle\langle n|$ is weak
APW	augmented plane wave method	the plane wave is expanded around the j'th atom in solid: $a_0 e^{i\vec{k}\cdot\vec{r}} = e^{i\vec{k}\cdot\vec{R}_j} \sum_l (2l+1) i^l P_l(\cos\theta) j_l(kr_j)$; the inner shell wave functions are $\sum_n a_{nl}^j u_{nl}^j(r_j) = a_0 e^{i\vec{k}\cdot\vec{R}_j} (2l+1) i^l j_l(kr_j)$

successful. In Table 5.6, the practical LDA-DFT computational methods are summarized. The tight-binding model and the weak potential approximation were used in the construction of the OPW and PSP methods respectively. The APW method was introduced by J. C. Slater in the 1950s to calculate the "muffin tin orbitals" (MTOs) in a potential $V(\vec{r} - \vec{R}) = 0$ ($|\vec{r} - \vec{R}| > r_0$) around each atom in a crystal.

5.3.5 Real Bands and Fermi Surfaces

The band structure $\varepsilon_n(\vec{k})$ in crystals can be measured by inelastic scattering of electrons or photons using various types of equipment. The theory behind these experiments is similar to the inelastic-scattering neutron diffraction theory used in the measurement of phonon spectra, as introduced in the last section of Chapter Four.

The measured and calculated energy bands of electrons $\varepsilon_n(\vec{k})$ are comparable in simple crystals. In this section, typical energy bands for alkali metals, noble metals, alkali-earth metals, and transition metals are introduced.

All the alkali metals have the BCC structure, where the potential $U(\vec{r})$ felt by a valence electron is very weak. The energy bands of BCC metals calculated using the weak potential approximation are illustrated in Figure 5.15 (note that the calculation is wrong near the surface of FBZ). The more accurate valence electron bands of sodium are plotted in Figure 5.16. It can be seen that a small band gap appears at the points N and P, which is different from the prediction in Figure 5.15.

The energy band of sodium is quite close to the free electron band. The Λ_1 band in Figure 5.16 corresponds to the valence band

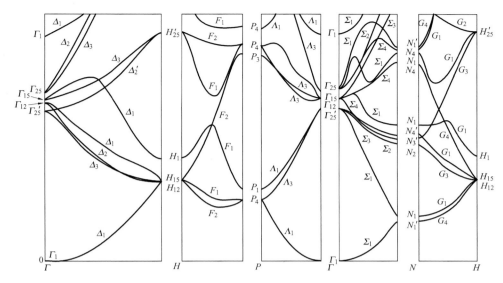

Figure 5.16 Valence electron bands of the BCC crystal Na (based on Burns, 1985)

for 3s-electrons in sodium. In a sodium crystal with N_L atoms, there are N_L valence electrons, therefore the Λ_1 band is half-filled. The Fermi wavevector k_F can be labeled with respect to k_N between the points N and Γ in FBZ. It can be proved that (exercise 5.14):

$$k_F = \frac{(6\pi^2)^{1/3}}{\sqrt{2\pi}} k_N \simeq 0.8773 k_N \qquad (5.108)$$

The Fermi surface in sodium and other alkali metals is simply a sphere with a radius k_F.

The noble metals have the FCC crystal structure, but the valence electrons can still be described quite well in terms of free electron Fermi gas, as discussed in the Sommerfeld model. The calculated electron bands of copper are plotted in Figure 5.17(a). Below the Fermi surface ε_F, there are six bands plotted, which correspond to five 3d-bands for $3d^{10}$ and one 4s-band for $4s^1$ in a copper atom.

In Figure 5.17(a), it can be seen that there are only two intersections on the Fermi surface of the 4s-band: one between $\Gamma - X$ and the other between $\Gamma - K$. The Fermi surface does not have any intersection from $\Gamma - L$, because the $\varepsilon_{4s}(\vec{k})$ band bends down as discussed in the weak potential approximation; therefore there is a "hole" near each L point in the FBZ on the near-spherical Fermi surface in noble metals, which is illustrated in Figure 5.17(b). The eight holes near the eight L points can be measured using the de Haas–van Alphen effect, which is a magneto-electric effect which will be discussed further in Chapter Seven.

In a copper crystal with N_L atoms, there are also N_L valence electrons, therefore the 4s-band is also half-filled. The Fermi wavevector k_F

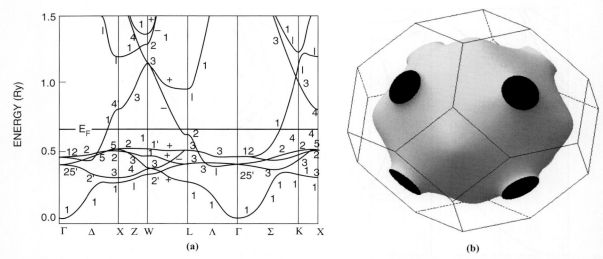

Figure 5.17 Calculated electronic structure of Cu (from www.phys.nara–wu.ac.jp): (a) energy bands; (b) Fermi surface (based on Choy, Naset, Chen, Hershfield, and Stanton, 2000)

can be labeled with respect to k_L between the distance L and Γ in FBZ. It can be proved that (exercise 5.15):

$$k_F = \frac{(12\pi^2)^{1/3}}{\sqrt{3\pi}} k_L \simeq 0.9025 k_L \tag{5.109}$$

The Fermi surface of copper and other noble metals is a sphere with eight holes.

The structures of alkali-earth metals (beryllium and magnesium: HEX; calcium and strontium: FCC; barium: BCC) have diversities caused by the complex electronic structures for 2A elements. The calculated electron bands of the FCC calcium crystal by the LDA-DFT methods are plotted in Figure 5.18(a). Below the Fermi surface ε_F, there are two valence bands plotted: one is the nearly-filled 4s-band, and the other is the nearly-empty 3d-band, which is only filled near the L point in FBZ.

If there is no overlapping among bands, the $2N_L$ valence electrons in a crystal with N_L calcium atoms should just fill the 4s-band; however, in reality, the lowest 3d-band in calcium has lower energy than the 4s-band near the L points, and the electrons start to fill the 3d-band before fully filling the 4s-band. Therefore the Fermi surfaces in calcium and other alkali-earth metals are "broken" into pieces. In Figure 5.18(b) and (c), the Fermi surface measured using the de Haas-van Alphen effect is shown for the higher-energy 3d-bands and the lower-energy 4s band in the calcium crystal respectively. In the 4s band, the Fermi surface is a sphere with 14 big holes in the eight $\langle 111 \rangle$ and six $\langle 100 \rangle$ directions;

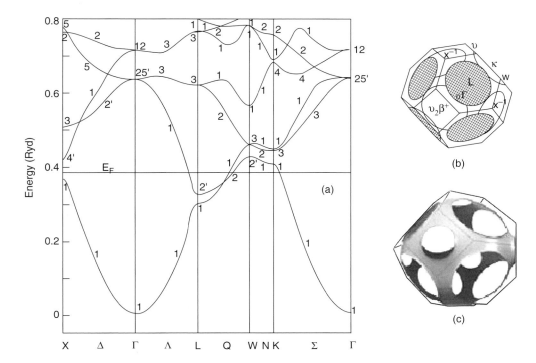

Figure 5.18 (a) Calculated electronic structure of the FCC crystal Ca (based on Jan and Skriver, 1981); (b) measured Fermi surface in the 3d-band (based on Condon and Marcus, 1964); (c) the Fermi surfaces in the 4s-band (based on Choy, Naset, Chen, Hershfield, and Stanton, 2000)

while in the lowest 3d-band, the Fermi surfaces are like eight round "cookies" whose flat surfaces are in the positions bisecting $\vec{G}_{111}$ near the eight L points in the FBZ.

The structures of trivalent metals also have diversities (aluminum: FCC; gallium: ORC; indium: TET; thallium: HEX), because the atoms of IIIA families contain both ns and np bands. The calculated electron bands of the FCC aluminum crystal are plotted in Figure 5.19(a), where the 3s band is fully-filled, and both 3p bands intersect with the Fermi surface ε_F. The 3p-band with lower energy has a minimum at the L point; while another 3p-band has a minimum at the X point in the FBZ. The Fermi surfaces in the aluminum crystal are also "broken" into pieces. Among these Fermi surfaces, the largest piece is continuous, and in the corresponding 3p band, the electron states fill from the surface of the truncated octahedral inward to a place between $L - \Gamma$ points in the FBZ. In Figure 5.19(b), the two sets of Fermi surfaces with respect to the two 3p bands are shown in the same plot.

The electronic structures of transition metals are the most complex, because the five d-bands and the s-band are mixed together, and all might have intersections with the Fermi surface. The measured (circles)

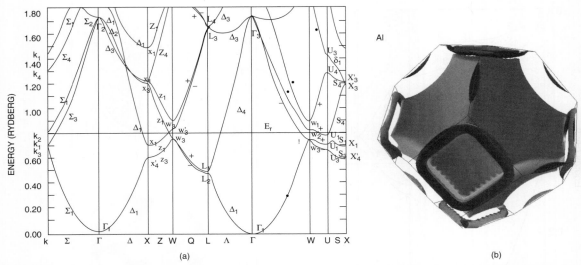

Figure 5.19 Calculated electronic structure of the FCC crystal Al: (a) energy bands (based on Singhal and Callaway, 1977); (b) Fermi surfaces (based on Choy, Naset, Chen, Hershfield, and Stanton, 2000)

and calculated (lines) electron bands of BCC-Fe, FCC-Co, and FCC-Ni crystals are plotted in Figure 5.20(a), (c), (d) respectively. In iron, cobalt, and nickel atoms, there are eight, nine, and ten electrons in the $3d$-$4s$-shells respectively; therefore in iron, cobalt, and nickel crystals, about 67%, 75%, and 83% of the electron states are filled in the six $3d$-$4s$-bands in the corresponding crystals respectively.

In the calculated bands of iron there are six $3d$-$4s$-bands for spin-up majority electrons; but in Figure 5.20(a), along $\Gamma - P - H$ symmetry directions in the FBZ, only five bands appear for spin-up electrons, and another band in the $H - \Gamma$ direction is not shown. In Figure 5.20(a), the Fermi surfaces appear in three bands for spin-up and two bands for spin-down electrons. In fact, the Fermi surface should appear in four bands for spin-up and four bands for spin-down electrons if the missed bands in Figure 5.20(a) in the $H - \Gamma$ direction are included. Therefore the Fermi surfaces in iron are extremely complex.

The computation of energy bands in ferro-magnetic materials should use a special exchange-correlation energy in the LDA-DFT methods. This ferromagnetic exchange-correlation energy is derived by J. Callaway from the Hartree-Fock exchange potential (Callaway, 1955), which is similar to the LDA potential in Eq. (5.106) brought up by J. C. Slater (Slater, 1953):

$$\Delta E_{xc} = V_+([n_+(\vec{r})]) - V_-([n_-(\vec{r})]);$$
$$V_\sigma([n_\sigma(\vec{r})]) = -3e^2\sigma \left(\frac{3}{4\pi}n_\sigma(\vec{r})\right)^{1/3} \qquad (5.110)$$

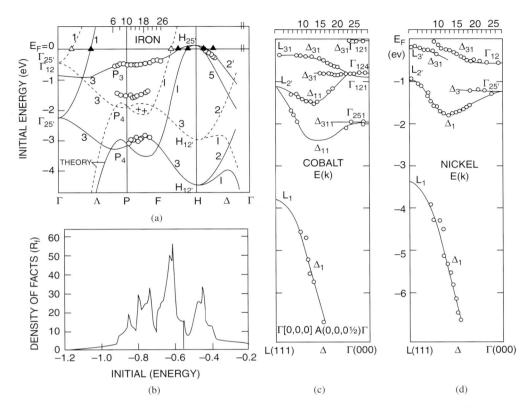

Figure 5.20 Electronic structure of Fe, Co, Ni: (a) calculated (solid/dashed lines stand for spin up/down or majority/minority spins respectively) and measured bands of Fe (based on Eastman, Himpsel, and Knapp, 1980, and Moruzzi, Janak, and Williams, 1978); (b) calculated total density of states of Fe (based on Callaway and Wang, 1977); (c)/(d) measured (based on Eastman, Himpsel, and Knapp, 1980) and calculated (based on Callaway, 1955) energy bands of Co and Ni

where $n_\sigma(\vec{r})$ is the density of electrons with $\sigma = \pm\frac{1}{2}\hbar$ spins. Obviously the spin-up electron has lower energy than the spin-down electron in the above potential.

The structures of insulators are numerous and complex, typified by the diamond structure and ABO_3 structures. In Figure 5.21(a), the calculated energy bands of diamond are shown; in Figure 5.21(b), the calculated energy bands of strontium titanium oxide ($SrTiO_3$), crystal with the ABO_3 structure, are plotted. It can be seen that the most significant characteristic of the bands in insulators is the wide band gap. The E_g of diamond is 5.48 eV and the E_g of strontium titanium oxide is 3.55 eV. In insulators, the Fermi energy ε_F locates within the energy gap, therefore there is no Fermi surface in the insulating crystals. The energy bands of semiconductors are similar to those of insulators, but the band gaps are smaller. The real bands of semiconductor crystals will be discussed in the next chapter.

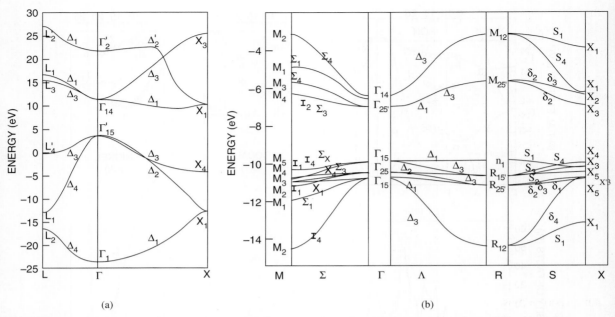

Figure 5.21 Energy bands of insulators: (a) diamond (based on Brener, 1975); (b) SrTiO$_3$ (based on Kahn and Leyendecker, 1964)

5.3.6 Semiclassical Model and Effective Mass

In the previous sections of this chapter, free electron models and band theory were introduced. In the Sommerfeld model of free electron Fermi gas, various physical properties could be calculated using the equilibrium Fermi-Dirac statistics or non-equilibrium statistics, and the energy density of states $g(\varepsilon)$ could be found based on the free electron energy spectrum $\varepsilon = \hbar^2 k/2m_e$.

If macroscopic physical quantities are calculated using band theory, a new approximation is needed to find the non-equilibrium statistics for electrons in bands. In 1928, W. A. Houston and Felix Bloch developed a semiclassical model based on the quantum theory of metallic conduction (based on Seitz, 1940). Houston correctly pointed out that the mean free path of electrons in a lattice would be infinitely long if the atomic vibrations were neglected. The electron with a Bloch wavefunction in the n'th band would have a constant velocity:

$$\vec{v}_n(\vec{k}) = \frac{1}{\hbar} \vec{\nabla}_{\vec{k}} \varepsilon_n(\vec{k}) \tag{5.111}$$

The velocity would last forever without the collision of phonons or other particles in a perfect crystal. Of course, in ordinary metals, even if there are no external fields, there will always be phonons and imperfections, therefore the electrons will always be scattered, and the mean free path will be finite.

The band electron can be seen as a new quasi-particle in solids, because the energy-momentum dispersion of the electron $\varepsilon_n(\vec{k})$ in a crystal no longer satisfies the non-relativistic dispersion of fundamental particles $\varepsilon = \vec{p}^2/2m$ at all de Broglie wavelengths. In the Δk neighbor of wavevector $\vec{k}$, a superposition of the eigenstates for the Bloch electron would be a wave packet with a size $\Delta x = \pi/\Delta k$, which is a wave-particle duality of quasi-particles. The motion equation analogy to Newton's second law can be derived by the work-energy principle with an external force $\vec{F}$:

$$\begin{aligned} \mathrm{d}\varepsilon_n(\vec{k}) &= \vec{F} \cdot \mathrm{d}\vec{r} \\ \mathrm{d}\vec{k} \cdot (\vec{\nabla}_{\vec{k}}\varepsilon_n(\vec{k})) &= \vec{F} \cdot \vec{v}_n(\vec{k})\mathrm{d}t \\ \frac{\mathrm{d}(\hbar\vec{k})}{\mathrm{d}t} &= \vec{F} \end{aligned} \qquad (5.112)$$

An effective mass matrix of band electron can be defined thereafter:

$$\frac{\mathrm{d}v_\alpha}{\mathrm{d}t} = \frac{\mathrm{d}}{\mathrm{d}t}\left(\frac{1}{\hbar}\frac{\partial \varepsilon_n}{\partial k_\alpha}\right) = \left(\frac{1}{\hbar^2}\frac{\partial^2 \varepsilon_n}{\partial k_\alpha \partial k_\beta}\right)\frac{\mathrm{d}(\hbar k_\beta)}{\mathrm{d}t} = [(m^*)^{-1}]_{\alpha\beta} F_\beta$$

(5.113)

where the Einstein notation is used for the index $\beta = 1, 2, 3$. As a quasi-particle, the effective mass matrix of a band electron is different from the fundamental particle, an electron in a vacuum, as it depends on the wavevector $\vec{k}$ and the specific band n:

$$m_n^*(\vec{k}) = \hbar^2 \left(\vec{\nabla}_{\vec{k}}\vec{\nabla}_{\vec{k}}\varepsilon_n(\vec{k})\right)^{-1} \qquad (5.114)$$

If the main shafts are chosen in a crystal along the high symmetry directions, the effective mass matrix can be diagonalized into the effective mass $m_{n\alpha}^*(\vec{k})$ ($\alpha = 1, 2, 3$) of electrons in the n'th band of a crystal.

In Figure 5.22, the schematics of the band $\varepsilon(k)$, the effective velocity $v(k) \propto \mathrm{d}\varepsilon/\mathrm{d}k$, and the inverse effective mass $(m^*)^{-1} \propto \mathrm{d}^2\varepsilon/\mathrm{d}k^2$ are plotted for a valence band and an inner band respectively. For most of $\vec{k}$ in the FBZ except at the edge, the valence band is similar to the energy dispersion of the free electron $\varepsilon_v(k) \propto k^2$, as discussed in the weak potential approximation; therefore the effective mass m^* is simply a constant within this range. Near the edge of the FBZ, $\varepsilon(k)$ bends down and tends to be a constant. The effective mass could then be negative.

If the phrase "effective mass of metal A" is used, it is usually m_F^*, but in the formula of conductivity, it is the average m^* in valence bands. In the Sommerfeld model, the measured specific heat deviates from the

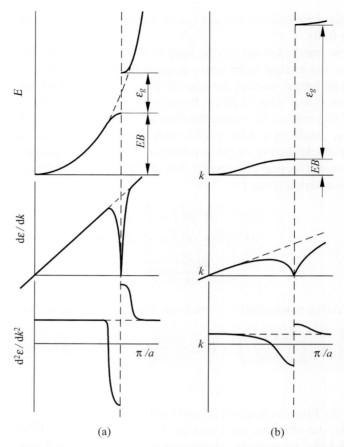

Figure 5.22 Schematics of energy bands, effective velocity, and inverse effective mass: (a) near-free electron valence band; (b) tight-binding inner band (based on Burns, 1985)

free electron estimation, as discussed in Section 5.2.1. The experimental low-temperature specific heat coefficient γ is a direct, and even the best, measurement of the effective mass:

$$\frac{m_F^*}{m_e} = \frac{\sum_n \hbar^2 \langle (d^2\varepsilon_n/dk^2)_{\varepsilon_F}^{-1} \rangle}{m_e} = \frac{\gamma/(\text{mJ/mol/K}^2)}{0.0705Z} \left(\frac{a_B}{r_s}\right)^2 \quad (5.115)$$

where Z and r_s are the values of free electron Fermi gas in Table 5.1.

The measured effective mass in metals by low-temperature specific heat are listed in Table 5.7. It can be seen that the effective mass m_F^* at Fermi energy is between 1 and $1.5\,m_e$ for most metals. A large m_F^* of the transition metals, like manganese or iron, is a sum of the contributions from multi-bands which have intersections with the Fermi surface

Table 5.7 Measured low temperature specific heat coefficient $\gamma/(\text{mJ/mol/K}^2)$ (based on Ashcroft and Mermin, 1976, and Martin, 1965 and 1973) and the calculated effective mass at Fermi surface m_F^*/m_e of common metals

Elements	Li	Na	K	Rb	Cs	Cu	Ag	Au
γ	1.63	1.38	1.96	2.41	3.19	0.67	0.67	0.69
r_s/a_B	3.24	3.92	4.85	5.19	5.62	2.66	3.02	3.00
m_F^*/m_e	2.20	1.27	1.18	1.27	1.43	1.34	1.04	1.09
Elements	Be	Mg	Ca	Sr	Ba	Zn	Cd	Hg
γ	0.21	1.33	2.72	3.64	2.72	0.59	0.71	2.09
r_s/a_B	1.87	2.66	3.26	3.40	3.70	2.30	2.58	2.64
m_F^*/m_e	0.42	1.34	1.81	2.23	1.41	0.79	0.75	2.12
Elements	Mn	Fe	Al	Ga	Sn	Pb	Sb	Bi
γ	16.7	5.00	1.25	0.63	1.84	2.83	0.63	0.084
r_s/a_B	2.14	2.12	2.07	2.19	2.21	2.30	2.13	2.25
m_F^*/m_e	25.90	7.89	1.38	0.62	1.33	1.90	0.39	0.047

$\varepsilon = \varepsilon_F$, just like that in energy bands of iron in Figure 5.20. In some metals, such as zinc, cadmium, antimony, and bismuth, the m_F^* is smaller than one. This is also an effect of the cancelation in multi-bands, where some of the bands contribute a positive curvature $(d^2\varepsilon_n/dk^2)_{\varepsilon_F}^{-1} > 0$, but others have negative curvatures at the Fermi surface.

In an external electromagnetic field, the semiclassical equation for motion of electrons is:

$$\hbar\frac{dk_\alpha}{dt} = m^*_{\alpha\beta}\frac{dv_\beta}{dt} = -e(\vec{E} + \vec{v}\times\vec{B})_\alpha \tag{5.116}$$

where the Lorentz force is used. In an external electric field at the $+x$ direction, the motion of band electrons can be visualized as the "drift" of the wavevector to the $-x$ direction in the FBZ. In the band theories, the non-equilibrium statistics and the relaxation time approximation are needed to calculate the electrical conductivity, which will be discussed further in Chapter Six.

This brings us to the end of this chapter, in which the classical free electron model of metals, the quantum free electron Fermi gas in metals, and the band theories have been covered. In the next chapter, the transport properties of solids will be discussed, which is the area in which band theory has been most successfully implemented.

Summary

This chapter has introduced the basic theories for the analysis of electronic structures in solids, and has shown how the evolution from the classical Drude model of electron gas to the quantum mechanical Sommerfeld model and the band theories has played an important part in the development of solid state physics. It has also covered the accurate calculation of energy bands and its dependance on the density functional theory developed by Kohn, and how this correlates to the Hartree-Fock theory introduced in Chapter Two. To summarize, the key areas covered in this chapter are:

1. Drude model: the concept of electron gas confined in metals as the basic physical picture of the electronic structure. Based on classical physics, the Drude model successfully explained the relationship between electrical and thermal conductivity in metals. The concepts of the Drude model, such as electron density and relaxation time, still survive in all electronic theories of solids.
2. Sommerfeld model: when the wave-particle duality and the Fermi-Dirac statistics were established as fundamental principles, Sommerfeld modified the Drude model and introduced the concept of electron Fermi gas. The energy distribution of electrons in metals was clarified. Many physical quantities could then be calculated: specific heat, electrical and thermal conductivity, thermionic emission of electrons, and the Hall effect in metals.
3. Band theory: the band theory was first proposed by Bloch in 1928. It assumes that electrons are moving in a periodic electronic potential. Initially, only the two extreme approximations, the tight-bind model, and the weak potential approximation, could be solved. Kohn's density functional theory, together with the development of computational science, made it possible to calculate the energy bands in a way comparable to measured real electronic bands.
4. Semiclassical model: an effective single-electron motion picture was derived from the collective many-body electronic states in real energy bands. For the semiclassical model, the effective mass becomes a matrix, and is dependent on the electron matter wavevector. The average effective mass at the Fermi surface is especially important for explaining the electrical properties of solids.

References

Ashcroft, N. W., and Mermin, N. D. (1976). *Solid State Physics*, Holt, Rinehart and Winston, New York.

Bloch, F. (1928). "Uber die quantenmechanik der electronen in kristallgittern," *Z. Physik.*, Vol. 52, 555.
Brener, N. E. (1975). "Correlated Hartree-Fock Energy Bands for Diamond," *Phys. Rev. B*, Vol. 11, 929–934.
Burns, G. (1985). *Solid State Physics*, Academic Press, Orlando
Callaway, J. (1955). "Electronic Energy Bands in Iron," *Phys. Rev.*, Vol. 99, 500–509.
Callaway J. and Wang C. S. (1977). *Energy Bands in Ferromagnetic Iron, Phys. Rev. B*, Vol. 16, 2095–2105.
Choy T.-S., Naset J., Chen J., Hershfield S., and Stanton C. "A database of fermi surface in virtual reality modeling language (vrml)", *Bulletin of The American Physical Society*, 45(1):L36 42, 2000.
Coles, B. R. (1964). "Electronic Structure and Superconductivity of Transition Metals and their Alloys," *Rev. Mod. Phys.*, Vol. 36, 139–145.
Condon, J. H., and Marcus, J. A. (1964). "Fermi Surface of Calcium by the de Haas-van Alphen Effect," *Phys. Rev.*, Vol. 134, A446–A452.
Dushman, S. (1930). "Thermionic Emission," *Rev. Mod. Phys.*, Vol. 2, 381–476.
Eastman, D. E., Himpsel, F. J., and Knapp, J. A. (1980). "Experimental Exchange-split Energy-band Dispersions for Fe, Co, and Ni," *Phys. Rev. Lett.*, Vol. 44, 95–98.
Fermi, E. (1928). "Sulla Deduzione Statistica di Alcune Proprietá dell atomo. Applicazione alla Teoria del Sistema Periodico Degli Elementi," *Z. Physik*, Vol. 48, 73.
Feng, D., and Jin, G. (2003). *Condensed Matter Physics* (in Chinese), Higher Education Press, Beijing.
Friedberg, S. A., Estermann, I., and Goldman, J. E. (1952). "The Electronic Specific Heat in Chromium and Magnesium," *Phys. Rev.*, Vol. 85, 375–376.
Heer, C. V., and Erickson, R. A. (1957). "Hyperfine Coupling Specific Heat in Cobalt Metal," *Phys. Rev.*, Vol. 108, 896–898.
Huang, K., and Han, R. (1988). *Solid State Physics* (in Chinese), Higher Education Press, Beijing.
Jan, J. P., and Skriver, H. L. (1981). "The Electronic Structure of Calcium," *J. Phys. F: Met. Phys.*, Vol. 11, 805–820.
Kahn, A. H., and Leyendecker, A. J. (1964). "Electronic Energy Bands in Strontium Titanate," *Phys. Rev.*, Vol. 135, A1321–A1325.
Kohn, W., Becke, A. D., and Parr, R. G. (1996). "Density Functional Theory of Electronic Structure," *J. Phys. Chem.*, Vol. 100, 12974–12980.
Kohn, W., and Sham, L. J. (1965). "Self-consistent Equations including Exchange and Correlation Effects," *Phys. Rev.*, Vol. 140, A1133–A1138.
von Laue, M. (1978), translated by Fan, D., and Dai, N. *History of Physics* (in Chinese), Commercial Press, Beijing.
Martin, D. L. (1965). "Analysis of Alkali-Metal Specific-Heat Data," *Phys. Rev.*, Vol. 139, 150–160.
Martin, D. L. (1973). "Specific Heat of Copper, Silver, and Gold below 30° K," *Phys. Rev. B*, Vol. 8, 5357–5360.
Moruzzi, V. L., Janak, J. F., and Williams, A. R. (1978). *Calculated Electronic Properties of Metals*, Pergamon, New York.
Seitz, F. (1940). *The Modern Theory of Solids*, McGraw-Hill, New York.
Singhal, S. P., and Callaway, J. (1977). "Self-consistent Energy Bands in Aluminum: an Improved Calculation," *Phys. Rev. B*, Vol. 16, 1744–1745.
Slater, J. C. (1953). "An Augmented Plane Wave Method for Periodic Potential Problem," *Phys. Rev.*, Vol. 92, 603–608.
Thomas, L. H. (1927). "The Calculation of Atomic Fields," *Proc. Camb. Phil. Soc.*, Vol. 23, 542.
Nobel Laureates for physics. Nobel Prize Organization, City of Stockholm, Sweden. 14 November 2005. <http://nobelprize.org/>

"Electronic Structure of Copper." Department of Physics, Nara Women's University. City of Kitauoyahigashi-machi, Japan. 27 February 2006. <http://www.phys.nara-wu.ac.jp/in_kamoku/suzuki/electronic_structre.html>

"WebElements Periodic Table." University of Sheffield and WebElements Ltd, City of Sheffield, UK. 1 May 2005. <http://www.webelements.com/>

Exercises

5.1 Use the Drude model to calculate the AC conductivity in a uniform external magnetic field. If a metal is put in a magnetic field $\vec{H} = H_z \hat{e}_z$, and a rotating AC electric field $\vec{E} = (E_x \hat{e}_x + E_y \hat{e}_y) \exp(i\omega t)$ ($E_y = \pm i E_x$), then:

   a. Prove that the AC current density is

   $$j_x = \frac{\sigma_0}{1 - i(\omega \mp \omega_c)\tau} E_x; \quad j_y = \pm i j_x \quad j_z = 0 \quad (5.117)$$

   where $\sigma_0 = ne^2\tau/m$ is the Drude DC conductivity and $\omega_c = eH_z/mc$ is the synchrotron frequency.

   b. By the Maxwell equations in Eqs. (5.1) to (5.4), prove that the dispersion relation in a metal is $k^2c^2 = \epsilon\omega$, where the dielectric constant is $\epsilon(\omega) = 1 + i\frac{4\pi\sigma}{\omega}$; and prove that the AC wave solution is $E_x = E_0 e^{i(kz-\omega t)}$, $E_y = \pm i E_x$, $E_z = 0$.

   c. Draw a plot of the real and imaginary part of the dielectric constant $\epsilon(\omega)$.

   d. When $\omega \ll \omega_c$, the low-frequency Helicon wave dispersion is $\omega = \omega_c(k^2c^2/\omega_p^2)$ where ω_p is the plasma frequency. If the wavelength is 1cm and the magnetic field is 1T, estimate the Helicon wave frequency ω.

5.2 Metallic lithium has a BCC structure with a lattice constant 3.5Å. Calculate the emission width in the characteristic soft X-ray spectrum of lithium.

5.3 In the Sommerfeld model, prove that the kinetic energy in a metal with N electrons is simply $\frac{3}{5}N\varepsilon_F$ and derive the expression for the pressure P and the volume elastic module $B = -V(\partial P/\partial V)$. Calculate the elastic module B in lithium (BCC, lattice constant 3.5Å), and compare it to the order of Young's module 10^{11} N/m².

5.4 In a two-dimensional free electron Fermi gas,

   a. Find the relationship between the density and the Fermi wavevector: $n - k_F$.

   b. Find the relationship between the Fermi wavevector and the average occupation radius of electrons r_s: $k_F - r_s$.

c. Calculate the EDOS $g(\varepsilon)$.
d. Prove that the "Sommerfeld expansion" of electron density in a 2D gas is independent of temperature.

5.5 The atom He^3 with a spin $\frac{1}{2}\hbar$ is the isotope of helium. He^3 is a fermion. Near the absolute zero temperature, the liquid He^3 has the density of 0.081 g/cm^3. If the Sommerfeld model is used as an approximation to analyze the liquid He^3, find its Fermi energy ε_F and the Fermi temperature T_F.

5.6 The measured low temperature specific heat coefficients γ of alkali metals are listed in Table 5.2. Compare the EDOS at the Fermi surface $g'(\varepsilon_F)$ and $g(\varepsilon_F)$ found by the measured γ and by the free electron Fermi gas model respectively.

5.7 The metal silver (Ag) has the density of 10.5 g/cm^3 and the atomic weight 107.87. If the Sommerfeld model is used to analyze the valence electrons in silver,

a. Calculate the Fermi energy ε_F, Fermi temperature T_F, Fermi wavevector k_F and Fermi velocity v_F.
b. If the resistivity of silver is $1.61\,\mu\Omega\cdot\text{cm}$ and $0.038\,\mu\Omega\cdot\text{cm}$ at temperature 295 K and 20 K respectively, calculate the mean free path for electrons at Fermi surface at the two temperatures.

5.8 At very high temperatures, only tungsten wires could be used for the thermionic emission of electrons. What is the specific physical quantity that prevents other metals from being used at high temperatures?

5.9 The Moseley formula says that the frequency of the K_α line in the X-ray spectrum is proportional to $(Z-1)^2$, and the corresponding width of K_α line of molybdenum(Mo) target has been given in Figure 3.25(b). What would be the rough estimation of the tight-binding energy parameter J_{2p}^1 of molybdenum (BCC, lattice constant $a = 3.15$Å)?

5.10 In a 1D lattice, if the lattice constant is a, and a single electron feels the 1D muffin-tin periodic potential as

$$V(x) = \begin{cases} \frac{1}{2}m\omega^2[b^2 - (x-na)^2] & |x-na| \leq b \\ 0 & \text{others} \end{cases} \quad (5.118)$$

where the non-zero potential appears in a $2b = a/2$ area around each atom in the lattice. Calculate the first and second band gaps by the weak potential approximation.

5.11 In a 2D square lattice, the lattice constant is a and the periodic potential is

$$U(x, y) = -4U\cos(2\pi x/a)\cos(2\pi y/a) \quad (5.119)$$

Calculate the first band gap at the point $(\pi/a, \pi/a)$ in the FBZ using the weak potential approximation.

5.12 A tight-binding energy band in the SC lattice along [100] direction in an FBZ has the form

$$\varepsilon(k) = \frac{\hbar^2}{ma^2}\left(\frac{7}{8} - \cos(ka) + \frac{1}{8}\cos(2ka)\right) \quad (5.120)$$

  a. Calculate and plot the group velocity of electrons along this direction.
  b. Calculate the effective mass at Γ and X in the FBZ.

5.13 Use weak potential approximation to analyze metals with two valence electrons:

  a. In a 2D system, prove that the zeroth order perturbation energy at the corner C in the FBZ is twice as large as the energy at the edge-center X in the FBZ.
  b. In a 3D system, what is the ratio of the zeroth order perturbation energy at the corner C over the energy at the face-center X in the FBZ?
  c. In the valence $4s$-band of zinc(Zn), what is the probable way to fill in the $4s$-band, if the conclusion in (b) is utilized?
  d. Based on the weak potential approximation, explain why the effective mass m_F^* of Zn is less than the bare electron mass m_e.

5.14 Prove that for alkali metals, the Fermi wavevector is $k_F = \frac{(6\pi^2)^{1/3}}{\sqrt{2}\pi}k_N \simeq 0.8773\, k_N$, where N is the face-center point in FBZ.

5.15 Prove that for noble metals, the Fermi wavevector is $k_F = \frac{(12\pi^2)^{1/3}}{\sqrt{3}\pi}k_L \simeq 0.9025\, k_L$, where L is the [111] face-center point in FBZ.

5.16 In a copper zinc (Cu_xZn_{1-x}) alloy, what is the specific atomic ratio x that the Fermi sphere $k = k_F$ of the free electron Fermi gas in the alloy just equals k_L in the copper crystal's FBZ?

5.17 In Figure 5.18(a), the valence electron energy bands of Ca are shown. Explain qualitatively why the effective mass $m_F^*/m_e = 1.8$ is larger than one.

5.18 In Figure 5.20(a), one spin-up band and two spin-down bands are missing along the $H - \Gamma$ direction of FBZ. If the effective mass m_F^*/m_e was about 8 for iron, what would the shape or curvature of the missed bands along the $H - \Gamma$ direction be?

Electrical Transport Properties of Solids

6

> **ABSTRACT**
>
> - Conductors: relaxation time in band theories
> - Semiconductors: basics of solid state devices
> - Superconductors: conventional and high-Tc materials

The electrical properties of solids are probably those which have led to the widest range of applications. In the mid-19th century, Michael Faraday, the great inventor, had an opportunity to show his first electric motor to King William IV. The king was fascinated by his invention, but asked, "Of what use is it?" One hundred and fifty years later, it is almost impossible to imagine life without electricity.

The information technology industry comprises four major areas: processing, storage, transmission, and input/output of information. Solid state integrated circuits have been used in information processing since the patents of Jack Kilby and Robert Noyce in 1959. The electronic components in integrated circuits are classified into passive devices and active devices. The passive devices—resistors, capacitors, and inductors—have been around since Faraday's time in the early 19th century. The active devices—diodes and triodes—originally had the form of vacuum tubes, but were replaced by solid state devices following the revolutionary work of William Shockley, John Bardeen, and Walter Brattain in 1947. Historically, the specific band structures of semiconductors were originally analyzed by studying the electrical properties of pure and doped silicon. These initial studies were carried out by Bardeen and his colleagues at the Bell Laboratory, and were clarified further following the development of DFT calculations.

Superconductivity is a particularly interesting electrical transport property of solids, which was first found in mercury at very low temperatures by Heike Kamerlingh-Onnes in 1911. Traditional superconductors are metals at room temperature, and have a very low critical temperature. The high-Tc superconductors discovered by Georg Bednorz and Alex

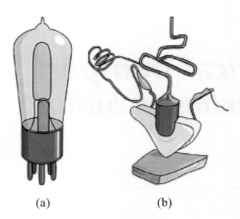

(a)　　　　　　　(b)

Figure 6.1 The transistor, a "smart" amplification device: (a) shows a vacuum tube transistor; (b) shows the first solid state transistor, which made use of the semiconductor germanium

Muller in 1986 are insulators at room temperature. Superconductors have zero resistivity below certain critical temperatures, and their most important application is to provide sustainable current, for example, creating high uniform magnetic fields using superconductor coils.

In this chapter, the basic electrical transport properties of solids will be discussed by presenting a unified view based on the energy band theory of electrons in solids. The relaxation time approximation in conductors, the basic solid state devices of semiconductors, and conventional and high-Tc superconductors will be the major focus.

6.1 Conductors

The strict classification of conductors, semiconductors, and insulators has to be implemented by analyzing the total electric current density $\vec{j}$ of all electrons in a solid. In the last chapter, it was shown that in the Somerfeld model, non-equilibrium statistics had to be used to analyze any transport property of a metal. For the band theories of electrons, the total current density $\vec{j}$ in a solid can also be found using the non-equilibrium statistics $g_n(\vec{r}, \vec{k}, t)$ over all electron bands:

$$\vec{j} = \frac{1}{V} \iiint d^3\vec{r} \sum_n \iiint \frac{d^3\vec{k}}{4\pi^3} g_n(\vec{r}, \vec{k}, t)[-e\vec{v}_n(\vec{k})] \quad (6.1)$$

$$g_n(\vec{r}, \vec{k}, t) \simeq f_{\text{F-D}}[\varepsilon_n(\vec{k})] - \left(-\frac{df}{d\varepsilon}\right)(e\vec{E}(\vec{r}, t) \cdot \vec{v}_n(\vec{k})\tau_e(\vec{k})) \quad (6.2)$$

where V is the volume of the crystal; $\vec{E}(\vec{r}, t)$ is the external electric field; $\vec{v}_n(\vec{k})$ is the velocity and $\tau_e(\vec{k})$ is the relaxation time of a Bloch electron in the n'th band.

In a crystal with N_L primitive unit cells, if there are n_a atoms in a PUC and an average of Z electrons in an atom, the total number of electrons is $Zn_a N_L$. The quantum states of all electrons would fill bands, from low to high energies, i.e., the 1s band is filled first and then all the way from inner bands to the bands of valence electrons. In a non-ferromagnetic solid (no band splitting for up/down spins), there are $2N_L$ electron states in a band, therefore the total number of bands is approximately $Zn_a/2$. In a crystal, the bands of inner-core electrons are fully-filled, but one or several bands of valence electrons are half-filled.

In a fully-filled band, the non-equilibrium distribution $g_n(\vec{r}, \vec{k}, t)$ in the uniform electric field $\vec{E}$ is the same as the Fermi statistics $f_{F-D}(\varepsilon_n)$. In the semiclassical model described in the last chapter, it was shown that the wavevector of electrons must move against the E-field: $d\vec{k} \propto -\vec{E}$; however, the periodicity of the energy band $\varepsilon_n(\vec{k}) = \varepsilon_n(\vec{k} + \vec{G})$ implies that the total statistics in a fully-filled band are invariant under the uniform motion of $\{\vec{k}\}$ of all electron states, as illustrated in Figure 6.2(a). Then the total current density contributed by electrons in a fully-filled band is

$$\vec{j}_n = \iiint \frac{d^3\vec{k}}{4\pi^3} f_{F-D}[\varepsilon_n(\vec{k})](-e\vec{v}_n(\vec{k})) \equiv 0. \quad (6.3)$$

The formal integral is zero because the Fermi-Dirac statistic $f_{F-D}(\varepsilon_n)$ is an even function of $\vec{k}$, while the group velocity $\vec{v}_n(\vec{k}) = \hbar^{-1}\vec{\nabla}_{\vec{k}}\varepsilon_n(\vec{k})$ is an odd function of $\vec{k}$ in the FBZ. In other words, a fully-filled band contributes nothing to electrical conduction.

In insulators and semiconductors, at near 0 K, all energy bands are fully-filled or empty. In a semiconductor, under the influence of thermal energy $k_B T$ or other external fields, a few electrons are excited from the highest fully-filled band, i.e., the valance band, to the lowest

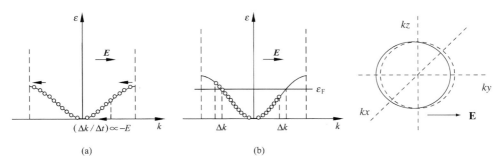

Figure 6.2 Non-equilibrium distributions in (a) fully-filled band and (b) half-filled band

empty band, i.e., the conduction band; therefore both the valence band and the conduction band become "half-filled" bands. In other words, the semiconductor has a certain ability to conduct electrically. In an insulator, the band gap between the valence band and the conduction band is so large that the external force cannot excite the electrons from the valence band to the conduction band; therefore there is no detectable electrical conduction. It can be seen that there is no clear demarcation line between semiconductors and insulators.

In a half-filled band, the non-equilibrium distribution $g_n(\vec{r}, \vec{k}, t)$ does differ from $f_{F-D}(\varepsilon)$, because there is asymmetry in g_n near the Fermi surface, as seen in Figure 6.2(b). The asymmetry is tiny and does not increase with time, because statistically $g_n(\vec{r}, \vec{k}, t)$ goes back to $f_{F-D}(\varepsilon_n)$ after the relaxation time τ_F:

$$eEl_F \ll \varepsilon_F; \Leftrightarrow \Delta k \simeq eE\tau_F/\hbar \ll k_F, \tag{6.4}$$

where $l_F = v_F \tau_F$ is the mean free path of electrons on the Fermi surface.

In Section 5.3.5, the real bands of typical metals are illustrated. The energy bands in metals have a common characteristic: there must be a half-filled band. This characteristic is the key to the electrical conductivity of metals. The total current density contributed by Bloch electrons in a half-filled band is:

$$\vec{j}_n = e^2 \tau_F \iiint \frac{d^3\vec{k}}{4\pi^3} \left(-\frac{df}{d\varepsilon}\right) \vec{v}_n(\vec{k}) \vec{v}_n(\vec{k}) \cdot \vec{E}. \tag{6.5}$$

In metals, $\vec{j}_n$ no longer disappears because both the statistical fluctuation $\Delta g_n(\vec{r}, \vec{k}, t) = g - f$ in a half-filled band and the velocity $\vec{v}_n(\vec{k})$ are odd functions of $\vec{k}$ in the FBZ. It is therefore evident that the total current density $\vec{j}$ in a solid is entirely due to the half-filled bands (labeled n') in the band theory, so the total conductivity matrix is:

$$\begin{aligned}
\sigma_{\alpha\beta} &= e^2 \tau_F \sum_{n'} \iiint \frac{d^3\vec{k}}{4\pi^3} \left(-\frac{df}{d\varepsilon}\right) v_{n'}^{\alpha}(\vec{k}) v_{n'}^{\beta}(\vec{k}) \\
&= e^2 \tau_F \sum_{n'} \iiint \frac{d^3\vec{k}}{4\pi^3} \left(-\frac{df}{d\varepsilon} \hbar^{-1} \frac{\partial \varepsilon_{n'}}{\partial k_{\alpha}}\right) v_{n'}^{\beta}(\vec{k}) \\
&= e^2 \tau_F \sum_{n'} \iiint \frac{d^3\vec{k}}{4\pi^3} f_{F-D}[\varepsilon_{n'}(\vec{k})] \hbar^{-2} \frac{\partial^2 \varepsilon_{n'}}{\partial k_{\alpha} \partial k_{\beta}} \\
&= e^2 \tau_F \sum_{n'} \iiint \frac{d^3\vec{k}}{4\pi^3} f_{F-D}[\varepsilon_{n'}(\vec{k})] (m^*)^{-1}_{\alpha\beta}(n', \vec{k}) \\
&\simeq \frac{ne^2}{m^*} \tau_F \quad m^* = \langle (m^*)^{-1}_{\alpha\beta}(n', \vec{k}) \rangle^{-1}
\end{aligned} \tag{6.6}$$

Eq. (6.6) is very similar to the expression of Drude's conductivity in Eq. (5.12), except that the physical meaning of the carrier density n and the effective mass m^* are modified in the band theories. In a metal, carriers are simply valence electrons, therefore the total density of conducting electrons is a constant:

$$n = \frac{N}{V} = \sum_{n'} \iiint \frac{d^3\vec{k}}{4\pi^3} f_{F-D}[\varepsilon_n(\vec{k})] \qquad (6.7)$$

where N is the total number of the quantum states below the Fermi level in all half-filled bands. It should be noted that the average effective mass m^* for conductivity in Eq. (6.6) is different from the effective mass at the Fermi surface m_F^*:

$$g(\varepsilon_F) = \sum_{n'} \iiint \frac{d^3\vec{k}}{4\pi^3} \delta(\varepsilon - \varepsilon_F) \simeq \frac{3n}{2\varepsilon_F}; \quad \Rightarrow \quad \varepsilon_F = \frac{\hbar^2}{2m_F^*}(3\pi^2 n)^{2/3}$$

(6.8)

where, near the Fermi surface of a metal, the energy spectrum of electrons is expanded as $\varepsilon \simeq \hbar^2 k^2 / 2m_F^*$. Therefore m_F^* depends on the detailed structures of the Fermi surface; but the effective mass m^* stands for the average properties of the half-filled band and is relatively closer to the free electron mass m_e.

The Wiedemann-Franz law in metals can also be proved using the band theory, but it is so similar to the proof using the Sommerfeld model (from Eq. (5.35) to Eq. (5.41)) that it is not worth repeating here. The characteristic physical quantity for conductors is obviously the resistivity, and the temperature dependence of this parameter in various metals is shown in Figure 6.3. Although the $\rho(T)$ curve does not have an exact linear form ηT, it can still be concluded that the average relaxation time τ_F of electrons is roughly inversely proportional to the temperature.

The physical explanation of the relaxation time in materials should include two parts: (1) the contributions of the defects in most polycrystalline materials; and (2) the contributions of the atomic vibrations in crystal regions, i.e., the electron-phonon interactions and the electron-phonon scattering. Matthiessen's rule says that the total resistivity is a direct sum of contributions (1) and (2), so the total relaxation time can be written as:

$$\rho = \frac{m^*}{ne^2\tau} = \frac{m^*}{ne^2\tau_F} + \frac{m^*}{ne^2\tau_d} = \rho_{cr} + \rho_d; \quad \Rightarrow \quad \tau^{-1} = \tau_F^{-1} + \tau_d^{-1}$$

(6.9)

In normal metals, the resistivity ρ_{cr} contributed by electron-phonon scattering is dominant at room temperature, because the crystal grain size

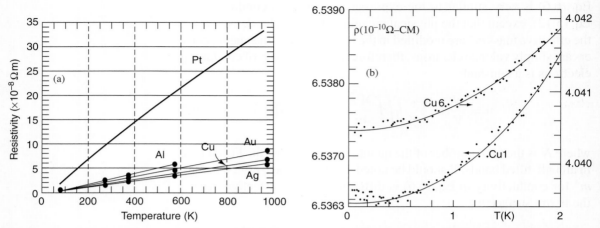

Figure 6.3 Resistivity versus temperature: (a) common metals (from www.npl.co.uk); (b) two kinds of Cu, from 0-2 K (Khoshenevisan, Pratt, Schroeder, and Steenwyk, 1979)

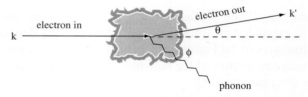

Figure 6.4 Electron-phonon scattering in solids

is usually larger than the mean free path $l_F \sim 1\text{–}10\,\text{nm}$, and $\tau_d \gg \tau_F$ is satisfied; however, at near 0 K, τ_F is very large and thus the resistivity $\rho \sim 10^{-3}\text{–}10^{-4}\,\mu\Omega\cdot\text{cm}$ is mainly contributed by ρ_d of defects, which differ with preparation conditions, as seen in Figure 6.3(b).

The relaxation time relating to defects τ_d differs from one material to another; but the relaxation time relating to the electron-phonon scattering τ_F can be calculated by analyzing the deviation of $g(\vec{k})$ from the Fermi-Dirac statistics (based on Ashcroft and Mermin, 1976):

$$\frac{g - f_{F-D}}{\tau_F} = \iiint \frac{d^3\vec{k}'}{4\pi^3} P_{\vec{k},\vec{k}'}[g(\vec{k}) - g(\vec{k}')]$$

$$\Rightarrow \tau_F^{-1} = \iiint \frac{d^3\vec{k}'}{4\pi^3} P_{\vec{k},\vec{k}'}(1 - \hat{k}\cdot\hat{k}') \qquad (6.10)$$

where the scattering probability $P_{\vec{k},\vec{k}'}$ relates to the statistics of incoming electrons, outgoing electrons, and phonons in the scattering process (phonon wave vector $\vec{q} = \vec{k}' - \vec{k}$):

$$P_{\vec{k},\vec{k}'} \propto f_{F-D}[\varepsilon_n(\vec{k})](1 - f_{F-D}[\varepsilon_n(\vec{k}')])f_{B-E}[\hbar\omega(\vec{q})] \quad (6.11)$$

The Fermi-Dirac statistics of incoming and outgoing electrons ensure that only the electrons on the Fermi surface can participate in the electron-phonon scattering, i.e., $k = k' = k_F$; the Bose-Einstein statistics of the low-energy phonons ($\hbar\omega \ll k_B T$) are simply proportional to the temperature:

$$f_{B-E}[\hbar\omega(\vec{q})] = \frac{1}{e^{\beta\hbar\omega(\vec{q})} - 1} \simeq [\beta\hbar\omega(\vec{q})]^{-1} \propto T \quad (6.12)$$

In the Debye phonon model, it has been shown that the low-energy phonons are the main contributors to the thermal properties of a solid; therefore these phonons are also dominant in the electron-phonon scattering. This is the reason why $P_{\vec{k},\vec{k}'}$ and τ_F are proportional to T^{-1}, and the resistivity in most metals is proportional to the temperature, as shown in Figure 6.3.

Before the end of this section, the order of resistivity ρ, relaxation time τ_F, and mean free path l_F should be repeated for conductors. At room temperature, for most metals, ρ is in the range of 1–100 $\mu\Omega\cdot$cm; at the Fermi surface, τ_F is in the order of 10^{-14}–10^{-15} s and l_F is about 10–1 nm. When the temperature T is higher than a very low T_{crit}, ρ_{cr} in a single crystal metal is proportional to temperature T because both τ_F and l_F are proportional to T^{-1}.

6.2 Semiconductors

Semiconductor materials have been studied in laboratories since as early as 1830, when electromagnetism was just being developed. The resistivity of semiconductor materials was then considered "extraordinary" compared to metals: when the temperature rose, $\rho(T)$ did not increase with ηT, but decreased quickly. Furthermore, the conductivity of some semiconductor materials could be largely enhanced by external stimuli, for example light shining on the sample.

By 1874, electricity was being used to carry not only power, but also information. The telegraph, telephone, and later the radio, were the earliest electronic devices. The inventor of cathode-ray tubes, German scientist Karl Ferdinand Braun, discovered the one-way conduction in lead sulfide crystals, and the first semiconductor device was born. In 1926, two Americans, L. O. Grondahl and P. H. Geiger, found rectifying properties in a copper-copper oxide(Cu-CuO) metal-insulator junction.

 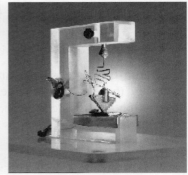

Figure 6.5 Shockley, Bardeen, and Brattain and their test board at Bell labs

This was the predecessor of the p-n junction, the theoretical explanation of which was given by Walter Schottky in 1938.

John Bardeen, Walter Brattain, and William Shockley met after World War II, when Bell Laboratories asked Shockley to build a research group for developing a solid state amplifier. In this group, Brattain did the experiments, Bardeen interpreted the results, and Shockley was the overseer. Near Christmas Eve in 1947, they completed what was called the "point-contact transistor." In 1949, Shockley described the p-n junction, while Bardeen analyzed the band structure of doped silicon by its basic electric properties. They received the Nobel prize in Physics in 1956. The three separated after this important invention: Shockley founded the Shockley Semiconductor Laboratories in California, one of the earliest successes in Silicon Valley; and Bardeen went into academia and won the second Nobel Prize in the field of superconductor theory.

Until 1959, all electronic components—resistors, capacitors, inductors, diodes, and triodes—were discrete. New technologies emerged that year, and integrated circuits were invented at Texas Instruments. In the same year, at Fairchild Semiconductor, a company founded in 1957 by eight engineers who had resigned from Shockley Semiconductor Laboratories, silicon was first used in the processing of integrated circuits instead of germanium. This made the commercial production of integrated circuits possible. By the end of the 1960s, nearly 90% of all components manufactured were integrated circuits.

The development of semiconductors was one of the most significant milestones in science and technology in the 20th century. In the pre-semiconductor era, K. F. Braun had won the Nobel Prize for his work on vacuum tubes in 1909. Nine years after the invention of the transistor, John Bardeen, Walter Brattain, and William Shockley won the Nobel Prize in Physics in 1956; and in 2000, Jack Kilby was awarded the Nobel

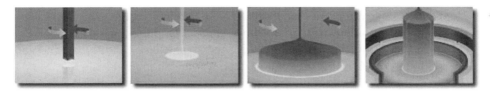

Figure 6.6 Czochralski process for single crystal silicon wafers (from www.engr.sjsu.edu)

Prize for his contributions to the invention of integrated circuits at Texas Instruments.

After Jean Hoerni and Robert Noyce of Fairchild Semiconductor developed the planar technology for fabricating silicon wafers, silicon became the most widely used semiconductor material. Silicon is not expensive and can be obtained from sand. The silicon wafers were made using the CZ process, invented in 1916 by a Polish chemist, Jan Czochralski, in which the crystalline solidification of atoms from a liquid phase was used to grow crystals. Integrated circuits are built on single-crystal silicon substrates with few defects, so the information era is also often called the silicon era.

6.2.1 Characteristics of Semiconductors

A number of materials are classified as semiconductors, which are elements or compounds of the families 2B to 6A in the periodic table, as listed in Table 6.1. Most semiconductor materials have diamond structures (group IV elements) or zinc blende structures (group III–V compounds or II–VI polarized semiconductors). In the diamond/zinc blende structure, the A and B sites are occupied by same/different atoms respectively. In semiconductors with a narrow band gap E_g, such as the infrared optic semiconductor mercury cadmium telluride (HgCdTe), the A sites are taken by either mercury or cadmium atoms at the probability x. In semiconductors with wide gaps, namely the third generation semiconductor silicon carbonate (SiC) and the blue laser material gallium nitride (GaN), there might be two to three different phases. The abbreviations "W" and "ZB" in Table 6.1 stand for the Wurtzite (HCP) and ZincBlende structures, and the atomic arrangements for the two structures are shown in Figure 6.7 respectively.

The thermal conductivity κ of semiconductors is almost of the same order as that of good conductors, but it is quite different from ceramic insulators. In metals, the thermal conductivity κ^m is due to the transport of electrons, and can be estimated by the Wiedemann-Franz law; in semiconductors or insulators, the thermal conductivity κ is due to the

Table 6.1 Semiconductor materials and their basic physical properties: κ/(W/cm/K) the room-T thermal conductivity; ϵ the dielectric constant; ρ/($\Omega \cdot$cm) the intrinsic room-T resistivity E_g/(eV) the band gap; (based on Chu, 2005, and www.semi1source.com/materials/)

Type	Name	Structure	κ	ϵ	ρ	E_g
IV	6H-SiC	HEX, $a=3.08$Å	4.90	9.7	$\sim 10^{10}$	3.03
IV	3C-SiC	CUB, $a=4.36$Å	3.60	9.7	10^5–10^{10}	2.30
IV	Si	DIA, $a=5.43$Å	1.30	11.7	2.3×10^5	1.12
IV	Ge	DIA, $a=5.65$Å	0.58	16.2	47	0.66
III–V	GaN	W, $a=3.19$Å	1.30	8.9	$> 10^{10}$	3.40
III–V	GaN	ZB, $a=5.19$Å	1.10	9.7	$> 10^{10}$	3.20
III–V	GaP	ZB, $a=5.45$Å	1.10	11.1	$\sim 10^9$	2.26
III–V	GaAs	ZB, $a=5.65$Å	0.55	12.9	3.3×10^8	1.43
III–V	InP	ZB, $a=5.86$Å	0.68	12.1	10^6–10^8	1.35
III–V	InSb	ZB, $a=6.47$Å	0.18	18.0	7×10^{-3}	0.18
II–VI	ZnSe	ZB, $a=5.66$Å	–	8.1	10^8–10^{12}	2.58
II–VI	$Hg_{1-x}Cd_xTe$	ZB, $a=6.46$Å	–	15.2–7.5	–	0–1.5

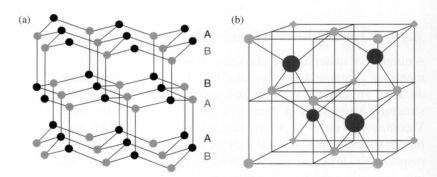

Figure 6.7 (a) Wurtzite structure; (b) Zinc Blende structure

transport of phonons:

$$\kappa^m = L\sigma T \simeq \frac{2.45 \times 10^{-8} \text{W} \cdot \Omega/\text{K}^2 \times 300\,K}{\rho/(\mu\Omega \cdot \text{cm}) \times 10^{-8}\Omega \cdot \text{cm}} \simeq 10^2 - 10^0 \text{ W/m/K}$$

$$\kappa = C_V v_x^2 \tau = \frac{1}{3} C_V v(v\tau) = \frac{1}{3} \frac{C_{\text{mole}}}{V_{\text{mole}}} vl \simeq 10^2 - 10^0 \text{ W/m/K}$$

(6.13)

In the derivation of κ, Eq. (5.15) is used, where C_{mole} is the molar specific heat, V_{mole} the molar volume, v is the sound speed, and l is the mean free path of the phonons. The thermal conductivity values κ of semiconductors listed in Table 6.1 are very high. This is because the semiconductors with DIA or ZB structures have significantly fewer defects than ceramic materials, and the mean free paths l of the phonons are large. The Joule heat generated by millions of integrated circuits can be dissipated due to the high thermal conductivity of silicon crystals.

Semiconductors are distinguished from conductors by the existence of band gap, and differentiated from insulators by the order of resistivity (low-T superconducting phase not included), as listed in Table 6.2. The borderline between semiconductors and insulators is indistinct. For example, the band gaps in silicon carbonate (SiC) gallium nitride (GaN) are larger than 3 eV, so the two materials can be regarded either as semiconductors or as semi-insulators.

In Section 5.3.5, real bands of various metals and insulators were introduced; in Figure 6.8, the energy bands of important semiconductors, namely germanium, silicon, gallium arsenide, and 3C-SiC are shown. If the position of the minimum energy in conduction band (the conduction band edge) $\vec{k}_c$ is aligned with the position of the maximum energy in valence band (the valence band edge) $\vec{k}_v$, it is a direct semiconductor (labeled "d"); otherwise it is an indirect semiconductor (labeled "i").

The band gap E_g is the energy difference between the conduction band edge $\varepsilon_c(\vec{k}_c)$ and the valence band edge $\varepsilon_v(\vec{k}_v)$. At 0 K, in insulators or semiconductors, the conduction band is empty and the valence band is fully-filled; both bands have no ability to conduct electricity. At room temperature, if the band gap is not too large, the electrons in the valence band can be thermally excited to the conduction band. Following the Boltzmann factor $\exp(-\varepsilon/k_B T)$, the resistivity ρ must be exponentially proportional to the band gap E_g in semiconductors, as shown in Table 6.1. The band gap E_g in indium antimonide (InSb) is only 0.18 eV; its room-temperature resistivity is $7 \times 10^3 \, \mu\Omega \cdot \text{cm}$, approaching the metal-semiconductor borderline.

The band gaps of direct and indirect semiconductors, as well as some insulators, are listed in Table 6.3. Only the direct semiconductors can be used as optical semiconductors, because the conservation of

Table 6.2 Classification of conductors, semiconductors, and insulators by the band gap and the resistivity ρ at 4 K and 300 K respectively

Solid type	Position of ε_F	E_g	ρ(4 K)	ρ(300 K)/($\mu\Omega \cdot$cm)
Conductors	in valence bands	$\simeq 0$	$\simeq 0$	1–100
Semiconductors	between two bands	0~3 eV	∞	10^3–10^{15}
Insulators	between two bands	> 3 eV	∞	10^{16}–10^{28}

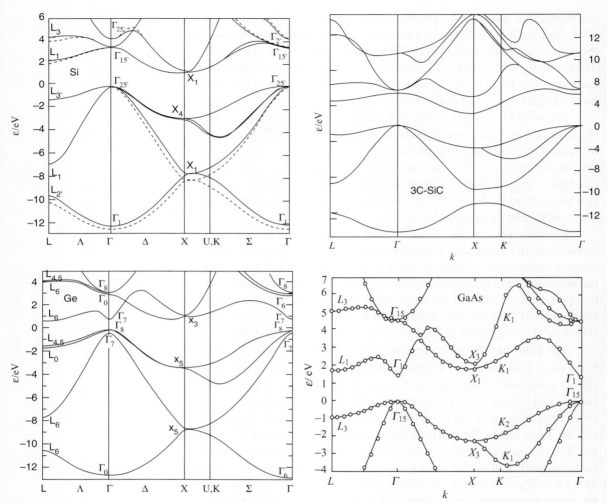

Figure 6.8 Energy bands of semiconductors: Si, Ge, GaAs, and 3C-SiC (based on Bruns, 1985, Chelikowsky and Cohen, 1976, and www.ioffe.rssi.ru)

momentum $\hbar|\vec{k}_c - \vec{k}_v| = h/\lambda$ has to be satisfied in the carrier-photon interaction. When emitting visual lights $\lambda = 300 - 700$ nm, the required wavevector difference $|\vec{k}_c - \vec{k}_v| \sim 10^4$ cm^{-1} is much less than the size $a^*/2 = 2\pi/a \sim 10^8$ cm^{-1} of the FBZ, therefore the conduction band edge and the valence band edge have to be "directly" aligned for electron excitations by photons.

The carriers in semiconductors are classified into electrons in the conduction band and holes in the valence band. At 0 K, the valence band is the highest fully-filled band, and the conduction band is the lowest empty band; therefore the Fermi energy ε_F must be located between the valence band and the conduction band. According to the characteristics

Table 6.3 Band gaps at 0 K of insulators, and indirect and direct semiconductors (based on Chu, 2005, Burns, 1985, and Dexter, Zeiger, and Lax, 1956)

		E_g(eV)			E_g(eV)			E_g(eV)
C	i	5.48	6H-SiC	i	3.03	ZB-GaN	d	3.20
ZnS	d	3.85	3C-SiC	i	2.30	GaP	i	2.26
ZnO	d	3.44	Si	i	1.17	GaAs	d	1.52
AgCl	i	3.25	Ge	i	0.74	InP	d	1.42
AgI	i	3.02	αSn	d	0.092	InAs	d	0.42
TiO_2	d	3.03	Te	d	0.33	InSb	d	0.24
Cu_2O	d	2.17	$Hg_xCd_{1-x}Te$	d	0–1.5	$Al_xGa_{1-x}As$	d	1.4–2.2

Table 6.4 Analogy of the electron-positron in a vacuum and the electron-hole in a solid

Physical properties	electron-positron	electron-hole
Annihilation process	$e + p \to 2\gamma$, γ is photon	$e + h \to$ full valence band
Energy conservation	$\varepsilon_e + \varepsilon_p = 2h\nu$	$\varepsilon_e(\vec{k}_e) + \varepsilon_h(\vec{k}_h) = 0$
Momentum conservation	$\vec{p}_e + \vec{p}_p = 0$	$\hbar\vec{k}_e + \hbar\vec{k}_h = 0$
Group velocity	$\vec{v}_e = -\vec{v}_p$	$\vec{v}_e = \vec{v}_h = \hbar^{-1}d\varepsilon_e/d\vec{k}_e$
Mass and charge	$(m_e, -e)$ (m_e, e)	$(m_e^*, -e)$ $(-m_e^*, e)$

of the Fermi-Dirac statistics, at 0 K, it is nearly a step function at $E = \varepsilon_F$. At finite temperature, the excited electrons occupy a very small percentage of the total $2N_L$ states in the conduction band.

The concept of a "hole" was devised to describe a small number of vacancies left in a valence band by the electrons excited to the conduction band. The excitation of an electron from the valence band can be viewed as the particle-antiparticle creation process, as listed in Table 6.4. In a fully-filled valence band, the total crystalline momentum $\sum \hbar \vec{k}$ is zero due to the central symmetry of the FBZ. The direction of the wavevector $\vec{k}_h$ of the valence band with one vacancy (hole) must be opposite to the wave vector $\vec{k}_e$ of the excited electron. If the valence band edge is set as $\varepsilon_e = 0$, conventionally the energy of the hole is defined as $\varepsilon_h = -\varepsilon_e$.

The effective mass of semiconductors is more complicated than that of metals. In Sections 5.2, 5.3.6, and 6.1, it was shown that, in metals, m_F^* at the Fermi surface and the average m^* in valence bands are frequently used definitions of effective mass for the expression of energy bands and various physical properties. In semiconductors, the specific elements of effective mass matrices and their averages are often used instead. Near the band edges $\varepsilon_c(\vec{k}_c)$ and $\varepsilon_v(\vec{k}_v)$, there are one conduction band and two to three valence bands respectively, therefore it is natural to define one effective mass matrix $\tilde{m}_c^*$ in the conduction band and at least two effective mass matrices $\tilde{m}_{lh}^*$ and $\tilde{m}_{hh}^*$ in the valence bands.

The valence band edge $\varepsilon_v(\vec{k}_v)$ always locates at the center Γ of the FBZ. The effective mass matrices of the light hole $\tilde{m}_{lh}^*$ and the heavy hole $\tilde{m}_{hh}^*$ are both diagonal, therefore the equipotential surfaces near the band edge in both valence bands are spherical. The light-hole and heavy-hole bands near $\vec{k}_v = 0$ can be expanded as

$$\varepsilon_e^{lh}(\vec{k}) = \varepsilon_v(\vec{k}_v) - \frac{\hbar^2 \vec{k}^2}{2m_{lh}^*}; \quad \varepsilon_e^{hh}(\vec{k}) = \varepsilon_v(\vec{k}_v) - \frac{\hbar^2 \vec{k}^2}{2m_{hh}^*}, \qquad (6.14)$$

which are plotted in Figure 6.9. It should be emphasized that the energy in the view of electron $\varepsilon_e(\vec{k})$ is equivalent to the energy in the view of hole $-\varepsilon_h(-\vec{k})$; therefore the light-hole mass m_{lh}^* and the heavy-hole mass m_{hh}^* can simply be calculated by $\hbar^2(d^2\varepsilon_{lh}/d^2k)^{-1}$ and $\hbar^2(d^2\varepsilon_{hh}/d^2k)^{-1}$ respectively.

In the $\Gamma - K$ direction, a third valence band of split-off holes can be seen, the maximum energy of which is lower by $\varepsilon_{so}(0)$ than the band edge of light and heavy holes $\varepsilon_v(\vec{k}) = 0$. The effective mass matrix of split-off holes is non-diagonal and the element along $\Gamma - K$ is labeled as m_{so}^*. Due to their high generation rates, the split-off holes in GaAs are of practical importance in infrared pico-second optical devices.

In the conduction band, the effective mass matrix $\tilde{m}_c^*$ is diagonal for direct semiconductors, but non-diagonal for indirect semiconductors such as silicon, germanium, SiC, and GaP. In direct semiconductors,

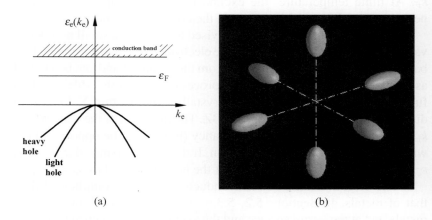

Figure 6.9 (a) light holes and heavy holes; (b) equipotential surface near $\varepsilon_c(\vec{k}_c)$ of Si

such as GaAs, the conduction band edge $\varepsilon_c(\vec{k}_c)$ locates at Γ; therefore an effective mass m_c is used to describe the equipotential surface near the conduction band edge $\vec{k}_c = 0$:

$$\varepsilon_e(\vec{k}) = \varepsilon_c(\vec{k}_c) + \frac{\hbar^2 \vec{k}^2}{2m_c}. \tag{6.15}$$

However, in indirect semiconductors, such as silicon, the energy band $\varepsilon_e(\vec{k})$ is asymmetric at any band edge point $\vec{k}_c = 0.85\vec{k}_X$ (there are six equivalent $\vec{k}_c$ in the FBZ); therefore two effective mass values, m_L^* (longitudinal to the $\Gamma - X$ direction) and m_T^* (transverse to the $\Gamma - X$ direction), should be introduced to describe the equipotential surface near a conduction band edge located at $\vec{k}_c = (0.85k_X, 0, 0)$:

$$\varepsilon_e(\vec{k}) = \varepsilon_c(\vec{k}_c) + \frac{\hbar^2[(\vec{k} - \vec{k}_c) \cdot \hat{e}_x]^2}{2m_L^*} + \frac{\hbar^2[(\vec{k} - \vec{k}_c) \times \hat{e}_x]^2}{2m_T^*}. \tag{6.16}$$

In GaAs, the $\Gamma - \Gamma$ band gap is 1.42 eV; but the $\Gamma - L$ and $\Gamma - X$ band gaps are 1.71 eV and 1.90 eV respectively. Obviously, the band structures are asymmetric, and the longitudinal and transverse effective mass m_L^* and m_T^* should also be used.

In silicon, near the six valleys $\vec{k}_c = 0.85\vec{k}_X$ of conduction band edges, the equipotential surfaces are all elliptical. In germanium, near the eight valleys $\vec{k}_c = \vec{k}_L$, the equipotential surfaces are half-elliptical, just like eight tulips, and the valley degeneracy is effectively $4 = 8/2$. In 3C-SiC and GaP, the equipotential surfaces at the six valleys $\vec{k}_c = \vec{k}_X$ are also half-elliptical and the valley degeneracy is effectively $3 = 6/2$. The band edges, the conduction valley degeneracy, and the effective masses of electrons and holes for main semiconductors are listed in Table 6.5.

Measurement of effective mass in semiconductors can be done using the Shubnikov-de Haas oscillation method or cyclotron resonance (based on Shockley, 1953, and Dexter, Zeiger, and Lax, 1956). The Shubnikov-de Haas oscillation method is used to measure the cyclotron resonance frequency of the carrier $\omega_c = eB/m^*c$ in the magnetic field $\vec{B} = H_z \hat{e}_z$ with the radio-frequency electric field $\vec{E} = E_x(\hat{e}_x \pm i\hat{e}_y)e^{i\omega t}$ perpendicular to $\vec{B}$, which was analyzed in Exercise 5.1 in Chapter Five and will be discussed further in Chapter Seven. The cyclotron resonance method can only be used in pure semiconductors at low temperatures, because the carriers have to travel for many circles within the relaxation time $\tau \sim 10^{-11}$s.

The measured mechanical, thermal, acoustic, optical, and electrical properties of the main semiconductors are listed in Table 6.6. The mechanical properties include density, bulk modulus, thermal

Table 6.5 Band edges and effective masses (unit: electron mass m_e in vacuum) in the conduction bands and valence bands of main semiconductors (based on Persson and Lindefelt, 1996, and www.ioffe.rssi.ru)

Physical Properties	3C-SiC	Si	Ge	GaP	GaAs	InSb
Conduction band edges $\vec{k}_c$	$\vec{k}_X$	$0.85\vec{k}_X$	$\vec{k}_L$	$\vec{k}_X$	Γ	Γ
Number of valleys near $\varepsilon_c(\vec{k}_c)$	6	6	8	6	1	1
Valley degeneracy M_c	3	6	4	3	1	1
Long. effective mass m_L^* or m_c	0.68	0.98	1.590	1.12	0.063	0.014
Trans. effective mass m_T^*	0.25	0.19	0.082	0.22		
Valence band edge $\vec{k}_v$	Γ	Γ	Γ	Γ	Γ	Γ
Heavy hole effective mass m_{hh}^*	1.01	0.46	0.280	0.79	0.510	0.430
Light hole effective mass m_{lh}^*	0.34	0.16	0.043	0.14	0.082	0.015
Split-off hole effective mass m_{so}^*	0.51	0.23	0.084	–	0.15	0.190
Split-off energy gap $\varepsilon_{so}(0)/(\text{eV})$	0.01	0.044	0.028	0.08	0.34	0.800

Table 6.6 Physical properties of main semiconductors (based on Burns, 1985, and www.ioffe.rssi.ru)

Physical Properties	3C-SiC	Si	Ge	GaP	GaAs	InSb
Density/(g/cm^3)	3.17	2.33	5.33	4.13	5.31	5.77
Bulk modulus/(10GPa)	25.0	10.2	7.7	8.8	7.5	4.7
Thermal Expansion/(10^{-6}/K)	2.77	2.60	5.80	4.65	6.40	5.37
Melting temperature/(°C)	2830	1412	937	1457	1238	536
Debye temperature/(K)	1200	650	370	450	340	200
Optical phonon $\hbar\omega(0)$/(meV)	102.8	64.2	37.4	45–50	33–36	22–24
Acoustic speed/(10^5 cm/s)	2.9–10	4.7–9.4	2.8–5.5	3.1–6.6	2.5–5.4	1.6–3.9
Breakdown field/(10^5 V/cm)	10	3	1	10	4	0.01
e diffusion/(cm^2/s)	20	36	100	6.5	200	2000
h diffusion/(cm^2/s)	8	12	50	4.0	10	22
e mobility/(10^3 cm^2/s/V), 300 K	0.8–0.9	1.50	3.9	0.25	8.5	77
e mobility/(10^3 cm^2/s/V), 77 K	~1	25	35	3	200	2000
h mobility/(10^3 cm^2/s/V), 300 K	0.2–0.3	0.50	1.9	0.15	0.4	0.85
h mobility/(10^3 cm^2/s/V), 77 K	~1	6	15	2	7	10
Band gap E_g/(eV), 300 K	2.360	1.124	0.663	2.267	1.424	0.180
Band gap E_g/(eV), 77 K	2.393	1.167	0.738	2.338	1.510	0.228
Band gap E_g/(eV), 0 K	2.396	1.170	0.744	2.350	1.519	0.235
Conduction edge degeneracy	3	6	4	3	1	1
DOS mass m_c^0/(per valley)	0.35	0.33	0.22	0.38	0.067	0.014
DOS mass m_c/(total)	0.73	1.08	0.56	0.79	0.067	0.014
DOS mass m_v	0.60	0.81	0.34	0.83	0.53	0.43

expansion, and melting temperature. The thermal property is described in terms of the Debye temperature. The optical property is given by the maximum phonon energy at $k=0$ in the optical branch. The acoustic property simply lists the range of the longitudinal and transverse sound speeds in all possible crystal directions. The electrical properties of semiconductors are complex, and are related to the band structures.

The electrical properties of metals are described in terms of their conductivity σ or resistivity ρ. However, in semiconductors, the "mobility" of electrons μ_e and holes μ_h is the term used more often, because the electron density n and hole density p are variables under external conditions. In a semiconductor, following Eq. (6.6), current densities of electrons $\vec{j}_n$ and holes $\vec{j}_p$, total conductivity σ and mobilities are:

$$\vec{j}_n = -ne\vec{v}_n = ne\mu_e \vec{E}; \quad \vec{j}_p = pe\vec{v}_n = pe\mu_h \vec{E}; \tag{6.17}$$

$$\sigma = e(n\mu_e + p\mu_h); \quad \mu_e = e\tau_e/m_e^*; \mu_h = e\tau_h/m_h^*, \tag{6.18}$$

where m_e^* and m_h^* are the average effective masses m^* for electrons and holes as defined in Eq. (6.6) of conductivity. The relaxation times τ_e and τ_h in semiconductors are in the order of 10^{-13} s at room temperature, which is ten times that in metals, because the electron-phonon scattering is much weaker in semiconductors.

6.2.2 Carrier Density and Mobility

The conductivity in Eq. (6.18) is proportional to the carrier density n, p and the mobility μ_e, μ_h, therefore these are the most important quantities to control in semiconductors. The carrier density can vary vastly when affected by stimuli such as temperature, impurities, sound, and light; therefore semiconductors are the core functional materials in temperature-sensitive, acoustically-sensitive, and light-sensitive electronic devices.

Moreover, the active devices in integrated circuits can be prepared by doping different impurities in the adjacent areas on the silicon wafer. The semiconductor characteristics introduced in the last subsection are for pure semiconductors, and are called the intrinsic properties. The physical properties, especially the carrier densities in semiconductors doped with impurities, are usually called the extrinsic properties, and they are more important in solid electronic devices.

Law of Mass Action for Carrier Densities

The physical properties of solids have to be calculated based on the statistics of electrons. In conduction bands, the Fermi-Dirac statistics of electrons are $f_c(\varepsilon_e) = f_{F-D}(\varepsilon_e)$ and the energy density of states is labeled as $g_c(\varepsilon_e)$; in valence bands, the Fermi-Dirac statistics of holes are $f_v(\varepsilon_e) = 1 - f_{F-D}(\varepsilon_e)$ and the total energy density of states is labeled as $g_v(\varepsilon_e)$:

$$f_c(\varepsilon_e) = \frac{1}{e^{\beta(\varepsilon_e-\mu)} + 1} \simeq e^{-\beta(\varepsilon_e-\mu)};$$

$$g_c(\varepsilon_e) = \frac{1}{2\pi^2}\left(\frac{2m_c}{\hbar^2}\right)^{3/2}\sqrt{\varepsilon_e - \varepsilon_c};$$

$$f_v(\varepsilon_e) = \frac{1}{e^{\beta(\mu-\varepsilon_e)} + 1} \simeq e^{-\beta(\mu-\varepsilon_e)};$$

$$g_v(\varepsilon_e) = \frac{1}{2\pi^2}\left(\frac{2m_v}{\hbar^2}\right)^{3/2}\sqrt{\varepsilon_v - \varepsilon_e}, \qquad (6.19)$$

which are plotted versus the electron energy $E = \varepsilon_e$ in Figure 6.10. Most semiconductors satisfy the "non-degenerate conditions" $\varepsilon_c - \mu > 3k_B T$ and $\mu - \varepsilon_v > 3k_B T$, therefore $f_c(\varepsilon_e)$ and $f_v(\varepsilon_e)$ can be expressed in the former exponential form, which is similar to the classical Maxwell-Boltzmann statistics.

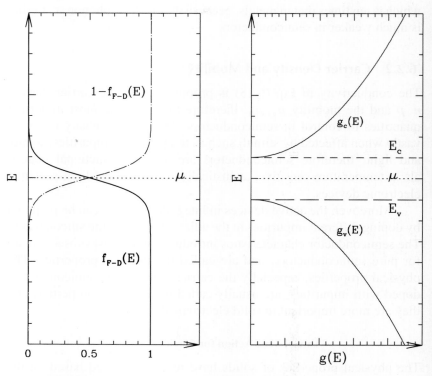

Figure 6.10 Fermi-Dirac statistics and energy density of states in semiconductors

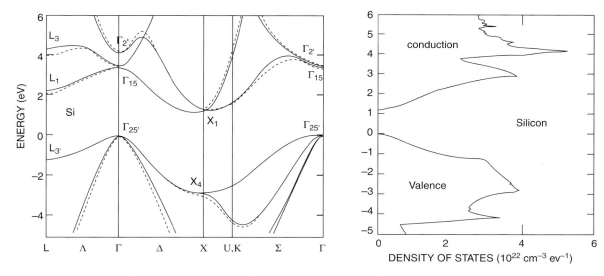

Figure 6.11 Energy density of states in valence and conduction bands of silicon

Near valance and conduction band edges, the EDOS $g_c(\varepsilon_e)$ and $g_v(\varepsilon_e)$ of a semiconductor have similar forms to that of free electron gas $C\sqrt{\varepsilon}$, as shown in Figure 6.11, because the band $\varepsilon_e(\vec{k})$ can be expanded to $\varepsilon_0 \pm C(\vec{k} - \vec{k}_0)^2$ near the extremum. The DOS effective masses m_c and m_v in Table 6.6 are related to the detailed effective masses describing the energy band structure in Table 6.5 as:

$$g_c(\varepsilon) = M_c g_c^0 = C' M_c ((m_L^* m_T^* m_T^*)^{1/3})^{3/2} \sqrt{\Delta\varepsilon};$$
$$m_c = (M_c^2 m_L^* m_T^* m_T^*)^{1/3} \quad (6.20)$$

$$g_v(\varepsilon) = g_v^{lh} + g_v^{hh} = C'(m_{lh}^{*\,3/2} + m_{hh}^{*\,3/2})\sqrt{-\Delta\varepsilon};$$
$$m_v' = (m_{lh}^{*\,3/2} + m_{hh}^{*\,3/2})^{2/3} \quad (6.21)$$

where g_c^0 is the EDOS with respect to one energy valley near the conduction band edge, M_c is the valley degeneracy, and g_v^{lh} and g_v^{hh} are the EDOS of the light-hole and heavy hole band respectively. It should be noted that, when the split-off band is included, the DOS mass of holes $m_v = (m_{lh}^{*\,3/2} + m_{hh}^{*\,3/2} + m_{so}^{*\,3/2})^{2/3}$ is larger.

The electron density n and hole density p can be found following Eq. (6.19):

$$n = \int_{\varepsilon_c}^{\infty} d\varepsilon_e g_c(\varepsilon_e) f_c(\varepsilon_e)$$
$$= \frac{1}{2\pi^2}\left(\frac{2m_c}{\hbar^2}\right)^{3/2} \int_{\varepsilon_c}^{\infty} d\varepsilon_e \sqrt{\varepsilon_e - \varepsilon_c}\, e^{-(\varepsilon_e - \varepsilon_F)/k_B T};$$

$$p = \int_{-\varepsilon_v}^{\infty} d\varepsilon_h g_c(\varepsilon_h) f_c(\varepsilon_h)$$

$$= \frac{1}{2\pi^2} \left(\frac{2m_v}{\hbar^2}\right)^{3/2} \int_{-\varepsilon_v}^{\infty} d\varepsilon_h \sqrt{\varepsilon_v + \varepsilon_h}\, e^{-(\varepsilon_F + \varepsilon_h)/k_B T}. \quad (6.22)$$

Then it is not difficult to find the general expression of carrier densities n and p:

$$n = N_c \exp(-(\varepsilon_c - \varepsilon_F)/k_B T); \quad N_c = 2\left(\frac{m_c k_B T}{2\pi \hbar^2}\right)^{3/2};$$

$$p = N_v \exp(-(\varepsilon_F - \varepsilon_v)/k_B T); \quad N_v = 2\left(\frac{m_v k_B T}{2\pi \hbar^2}\right)^{3/2}. \quad (6.23)$$

The Eq. (6.23) of carrier densities n and p are accurate for all semiconductors under the non-degenerate conditions. The constants N_c and N_v are the extreme electron and hole densities in the very heavily doped semiconductors, the metallic limit $\varepsilon_F \to \varepsilon_c$ or $\varepsilon_F \to \varepsilon_v$, i.e., the total DOS in the conduction and valence band respectively. The expressions of N_c and N_v can be written as:

$$N_c = 2.50 \left(\frac{m_c}{m_e}\right)^{3/2} \left(\frac{T}{300K}\right)^{3/2} \times 10^{19} \text{cm}^{-3};$$

$$N_v = 2.50 \left(\frac{m_v}{m_e}\right)^{3/2} \left(\frac{T}{300K}\right)^{3/2} \times 10^{19} \text{cm}^{-3}. \quad (6.24)$$

The law of mass action for carrier densities is an important relationship in non-degenerate semiconductors, and can be derived directly from Eq. (6.23):

$$np = N_c N_v\, e^{-E_g/k_B T}$$

$$= 6.25 \left(\frac{m_c m_v}{m_e^2}\right)^{3/2} \left(\frac{T}{300K}\right)^3 e^{-E_g/k_B T} \times 10^{38} \text{cm}^{-6} \quad (6.25)$$

where $E_g = \varepsilon_c - \varepsilon_v$ is the band gap. In an impurity-doped extrinsic semiconductor, the product of n and p only depends on the temperature and the band structure of the high purity intrinsic semiconductor (E_g, m_c, and m_v); i.e., when n is increased by a million times, p must be decreased by a million times as well. The law of mass action is true for both the intrinsic and extrinsic non-degenerate semiconductors.

In semiconductors, the comprehensive effective mass includes the DOS effective masses m_c and m_v used in Eq. (6.19) for the calculation of $g_c(\varepsilon)$ and $g_v(\varepsilon)$, and the conductivity effective mass m_e^* and m_h^*

Table 6.7 Semiconductor comprehensive effective masses (unit: m_e), room-T extreme carrier density N_c, N_v compared with atomic density N_0 (unit: 10^{19}cm^{-3}) and room-T relaxation time τ_e, τ_h for electrons and holes (unit: 10^{-13} s)

	m_c	m_v'	m_v	N_c	N_v	N_0	m_e^*	m_h^*	τ_e	τ_h
Si	1.080	0.52	0.81	2.810	1.82	5.00E3	0.260	0.37	2.1	0.95
Ge	0.560	0.29	0.34	1.050	0.39	4.44E3	0.120	0.21	2.7	2.30
GaAs	0.067	0.53	0.53	0.043	0.95	4.42E3	0.067	0.34	3.2	0.77

used in Eq. (6.18), which can be measured using the infrared reflection method (based on Lyden, 1964). In Table 6.7, the DOS effective masses m_c, m_v', m_v, conductivity effective masses m_e^*, m_h^*, room-temperature relaxation times τ_e, τ_h:

$$\tau_{e,h} = \frac{\mu_{e,h} m_{e,h}^*}{e} = \frac{\mu_{e,h}}{10^3 \text{cm}^2/\text{V/s}} \frac{m_{e,h}^*}{m_e} \times 5.69 \times 10^{-13}\text{s}, \quad (6.26)$$

N_c, N_v, and the atomic densities N_0 are listed in Table 6.7 for silicon, germanium, and gallium arsenide respectively.

Intrinsic and Extrinsic Carrier Density

In an intrinsic semiconductor, every electron in the conduction band must correspond to a hole in the valence band, i.e., $n = p$ must be true; then the intrinsic carrier density n_i can be easily found following the law of mass action in Eq. (6.25):

$$n_i = n = p = \sqrt{np} = \sqrt{N_c N_v} \exp(-E_g/2k_B T) \quad (6.27)$$
$$= 2.50 \left(\frac{\sqrt{m_v m_v}}{m_e}\right)^{3/2} \left(\frac{T}{300K}\right)^{3/2} e^{-E_g/2k_B T} \times 10^{19}\text{cm}^{-3}$$

The temperature dependence of the energy gap E_g can then be expressed as (based on Pearson and Bardon, and www.ioffe.rssi.ru):

$$\begin{aligned}
E_g(T) &= 2.396 - 6.0 \times 10^{-4} T^2/(T+1200)(\text{eV}) \quad &3C-\text{SiC} \\
E_g(T) &= 1.17 - 4.73 \times 10^{-4} T^2/(T+636)(\text{eV}) \quad &\text{Si} \\
E_g(T) &= 0.744 - 4.80 \times 10^{-4} T^2/(T+235)(\text{eV}) \quad &\text{Ge} \\
E_g(T) &= 1.519 - 5.405 \times 10^{-4} T^2/(T+204)(\text{eV}) \quad &\text{GaAs}
\end{aligned}$$

$$(6.28)$$

Table 6.8 Intrinsic carrier density n_i of main semiconductors (based on www.ioffe.rssi.ru)

Physical Properties	3C-SiC	Si	Ge	GaAs
DOS mass m_c/m_e	0.73	1.08	0.56	0.067
DOS mass m_v/m_e	0.55	0.81	0.34	0.53
Intrinsic $n_i/(\text{cm}^{-3})$, 1000 K	3.47E + 14	0.83E + 18	0.56E + 19	2.52E + 16
Intrinsic $n_i/(\text{cm}^{-3})$, 500 K	0.65E + 08	2.09E + 14	1.84E + 16	0.92E + 12
Intrinsic $n_i/(\text{cm}^{-3})$, 300 K	1.99E − 01	0.84E + 10	1.97E + 13	2.39E + 06
Intrinsic $n_i/(\text{cm}^{-3})$, 200 K	0.78E − 11	0.45E + 05	0.61E + 10	4.06E − 01
Slope $a/(\log_{10} \text{cm}^{-3})$	6.4	3.3	2.2	4.2
Intercept $b/(\log_{10} \text{cm}^{-3})$	20.64	20.93	20.65	20.35

The intrinsic carrier densities n_i at different temperatures are listed in Table 6.8; the corresponding $n_i(T)$ curves (solid lines) are plotted in Figure 6.12 and they approximately have the form $\log_{10} n_i = -a(1000K/T) + b$ (dashed lines). The slope a is proportional to the band gap E_g but the intercept b is similar for all materials.

In reality, intrinsic properties appear in a semiconductor when the impurity densities are much lower than the intrinsic carrier density n_i. Therefore, in intrinsic germanium, silicon, GaAs, and 3C-SiC, the atomic percentage of impurities must be purified to be below $n_i/N \sim 10^{-9}$, 10^{-13}, 10^{-16}, and 10^{-23} respectively. Among these semiconductors used in industry, germanium has the lowest requirement on

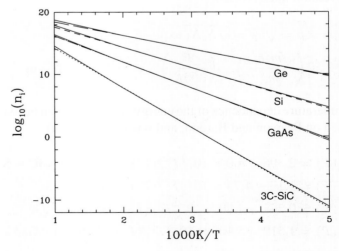

Figure 6.12 Intrinsic carrier density versus $1/T$ (dashed lines are linear fits)

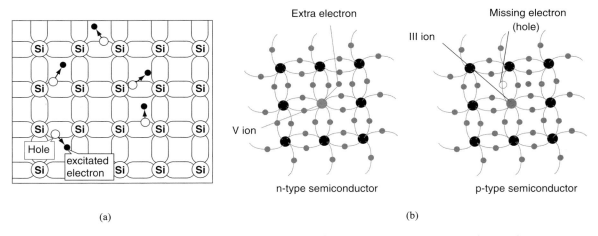

Figure 6.13 Schematics of chemical bonds in semiconductors: (a) intrinsic: pure silicon; (b) extrinsic: n-type and p-type semiconductors

atomic percentage of impurities, which is why germanium was used in first generation of semiconductor devices.

Extrinsic carrier densities are more complicated. In Figure 6.13(b), two types of extrinsic semiconductors are illustrated, which are implemented by doping impurities of family IV and III respectively. When the family V (P, As, Sb) atoms are doped, and the impurity atom (by a certain percentage) replaces a silicon atom with four covalent bonds around, an extra electron will appear in the n-type semiconductor crystal—that's why the impurities from family V are called donors. When the family III (B, Al, Ga, In) atoms are doped, and the impurity atom (by a certain percentage) replaces a silicon atom, an extra hole must appear in p-type semiconductor crystal—that's why the impurities from family III are called acceptors.

In n-type or p-type semiconductors, a donor level ε_D or an acceptor level ε_A can enter the intrinsic band structures, as illustrated in Figure 6.14:

$$\varepsilon_D = \varepsilon_c - E_d; \quad \varepsilon_A = \varepsilon_v + E_a, \qquad (6.29)$$

where E_d and E_a are the shallow ionized energies with respect to the donor level and acceptor level respectively. (Deep impurity levels are due to metal atoms). In intrinsic semiconductors, all electrons in the conduction band come from the valence band; in extrinsic semiconductors, the donors and the acceptors contribute extra electrons and holes to the conduction band and the valence band respectively.

Around donors and acceptors, the electron-V-ion or the hole-III-ion bound state is hydrogenic (an "exotic atom" state akin to that of a hydrogen atom); however, because of the screening effect, the

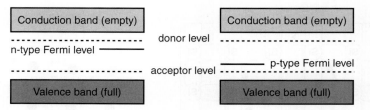

Figure 6.14 Donor level and acceptor level in extrinsic semiconductors

ionized energy E_d, $E_a = -E_1$ is much smaller and the orbit size r_1 is much larger than those in a hydrogen atom (the Rydeberg unit $\mathrm{Ry} = 2\pi^2 m e^4/h^2$, the Bohr radius $a_B = \hbar^2/me^2$, and the corresponding transformations are $e^2 \to e^2/\epsilon_r$ and $m \to m^*$). The ionized energies in silicon and germanium doped with different impurities are listed in Table 6.9 respectively.

At 0 K, the donor level in the n-type semiconductor or the acceptor level in the p-type semiconductor is half-filled, because there is only one carrier (spin up or down) in the hydrogenic atom centered at the impurity ion. At room temperature or high temperatures, most carriers contributed by impurities enter the conduction band or valence band, because the ionized energies E_d and E_a in the range 0.01 eV to 0.1 eV are quite small. The occupation rates by electrons/holes in the donor/acceptor level can be written as:

$$1 - \frac{N_D^+}{N_D} = \frac{1}{g_c^{-1}\exp[(\varepsilon_D - \varepsilon_F)/k_B T] + 1} = f_d$$
$$1 - \frac{N_A^-}{N_A} = \frac{1}{g_v^{-1}\exp[(\varepsilon_F - \varepsilon_A)/k_B T] + 1} = f_a \qquad (6.30)$$

where N_D^+/N_A^- are the densities of impurity ions whose electrons/holes are ionized and excited into the conduction/valence band respectively;

Table 6.9 Donor level and acceptor level analyzed by the hydrogenic states

Physical properties	Donor level	Acceptor level
Hydrogenic state	$e^- + (\text{V ion})^+ \sim$ atom	$h^+ + (\text{III ion})^- \sim$ atom
Energy levels	$E_n = -\frac{E_d}{n^2}$ $(n = 1, \ldots, \infty)$	$E_n = -\frac{E_a}{n^2}$ $(n = 1, \ldots, \infty)$
Ionized energy	$E_d \simeq 13.6\,eV\,(m^*/m_e)/\epsilon_r^2$	$E_a \simeq 13.6\,eV\,(m^*/m_e)/\epsilon_r^2$
Si E_d, E_a/(eV)	P:0.045, As:0.054, Sb:0.043	B:0.045, Al:0.072, Ga:0.074
Ge E_d, E_a/(eV)	P:0.013, As:0.014, Sb:0.010	B:0.011, Al:0.011, Ga:0.011
Orbital radius	$r_n = a_B n^2\, \epsilon_r/(m^*/m_e)$	$r_n = a_B n^2\, \epsilon_r/(m^*/m_e)$

the Fermi level ε_F locates in the upper/lower half of the band gap in n/p-type semiconductors; $g_c = 2$ and $g_v = 4$ (for both heavy holes and light holes) are the spin degeneracy and should be deducted from the original Fermi-Dirac statistics.

The carrier densities in extrinsic semiconductors can be found from the law of mass action together with the charge conservation law:

$$n + N_A^- = p + N_D^+ \qquad (6.31)$$

and the occupation rate at impurity levels in Eq. (6.30) should also be used to evaluate N_A^- and N_D^+:

$$n = \frac{1}{2}[\Delta N + \sqrt{\Delta N^2 + 4n_i^2}];$$
$$p = \frac{1}{2}[-\Delta N + \sqrt{\Delta N^2 + 4n_i^2}]; \qquad (6.32)$$

$$\rho = e^{-1}(n\mu_e + p\mu_h)^{-1};$$
$$\Delta N = N_D(1 - f_d) - N_A(1 - f_a), \qquad (6.33)$$

in which the Fermi energy ε_F has to be solved self-consistently with Eq. (6.23), and the resultant $\varepsilon_F(T)$ is plotted in Figure 6.15(b).

There are three distinct ranges in $n(T)$ or $p(T)$ curves of extensive semiconductors: the intrinsic range (>500 K), the saturation range or extrinsic range (150 K–450 K), and the freeze-out range (0–100 K), as shown in Figure 6.15(a). For example, in the n-type semiconductor ($N_D > N_A$), when the temperature is so high that $n_i \gg N_D$, the electron density n and the Fermi level ε_F in the intrinsic range are:

$$\begin{aligned} n = n_i &= \sqrt{N_c N_v} \exp(-E_g/2k_B T) \\ &= N_c \exp(-(\varepsilon_c - \varepsilon_F)/k_B T); \end{aligned} \qquad (6.34)$$

$$\varepsilon_F = \frac{1}{2}(\varepsilon_c + \varepsilon_v) + \frac{3}{4}k_B T \ln(m_v/m_c). \qquad (6.35)$$

When the temperature is around room temperature, almost all electrons in hydrogenic atoms around donors are excited to the conduction band ($N_D^+ \simeq N_D$); the donor density N_D is usually much larger than $n_i(T)$ (in Si, $n_i = 10^{10}$ cm^{-3} at 300 K and $n_i = 10^{14}$ cm^{-3} at 500 K). n and ε_F in the saturation range are

$$n \simeq \Delta N = N_c \exp(-(\varepsilon_c - \varepsilon_F)/k_B T); \qquad (6.36)$$

$$\varepsilon_F = \varepsilon_c - k_B T \ln(N_c/\Delta N) \quad \Delta N = N_D - N_A \simeq N_D. \qquad (6.37)$$

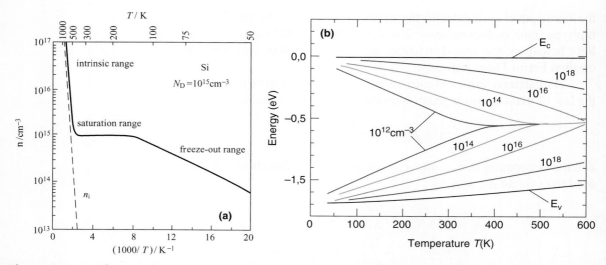

Figure 6.15 (a) Electron density n versus T in n-type silicon (based on Burns, 1985); (b) Fermi levels versus T at different doping levels (based on Grove, 1967)

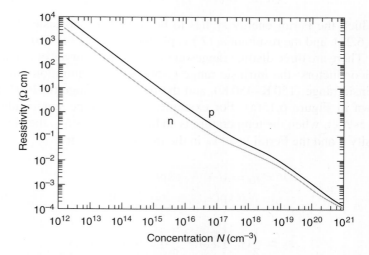

Figure 6.16 Resistivity of n/p-type silicon versus N_D/N_A (both labeled as N) (based on www.ioffe.rssi.ru)

Therefore, at room temperature, the resistivity $\rho = 1/(n\mu_e)$ should be inversely proportional to the impurity density N_D, as shown in Figure 6.16.

When the temperature is much lower than room temperature, only some of the electrons in the hydrogenic atoms around donors are excited to the conduction band ($N_D^+ < N_D$), thus the Fermi energy is trapped around the donor level; and n and ε_F in the freeze-out range have two slightly different forms with respect to different N_A:

$$n \simeq \sqrt{N_c N_D/2} \exp(-E_d/2k_B T); \quad N_D > n \gg N_A; \quad (6.38)$$

$$\varepsilon_F = \varepsilon_D + \frac{1}{2}E_d - \frac{1}{2}k_B T \ln(2N_c/N_D) \quad N_D > n \gg N_A; \quad (6.39)$$

$$n \simeq [(N_D - N_A)/2N_A] N_c \exp(-E_d/k_B T); \quad N_D > N_A \gg n; \quad (6.40)$$

$$\varepsilon_F = \varepsilon_D + k_B T \ln[(N_D - N_A)/2N_A]; \quad N_D > N_A \gg n. \quad (6.41)$$

To summarize the results in the three regions, in Figure 6.15(b), ε_F are plotted versus T in the n/p-type semiconductors with different doping (ε_F in upper/lower-half of E_g) respectively.

Hall Coefficient, Mobility, and Resistivity

The electrical properties of semiconductors have to be explained by both the carrier density property in Figure 6.15(a) and characteristics of the carrier mobility (i.e., the relaxation time). The conductivity in semiconductors can also be calculated using the language of energy bands, as illustrated in Eq. (6.6). For example, in the n-type semiconductors, the ratio of the conductivity σ (if the current density $\vec{j} = j_x \hat{e}_x$ is in the x-direction) versus the carrier density n is:

$$\frac{\sigma}{n} = \frac{\int_{\varepsilon_c}^{\infty} d\varepsilon_e g_c(\varepsilon_e) \left(-\frac{df_c}{d\varepsilon_e}\right) e^2 v_x v_x \tau}{\int_{\varepsilon_c}^{\infty} d\varepsilon_e g_c(\varepsilon_e) f_c(\varepsilon_e)}$$

$$= \frac{\int_0^{\infty} d\varepsilon \sqrt{\varepsilon} \left(-\frac{d}{d\varepsilon} e^{-\varepsilon/k_B T}\right) \frac{e^2}{3} \sqrt{\frac{2\varepsilon}{m_c}} l(\varepsilon)}{\int_0^{\infty} d\varepsilon \sqrt{\varepsilon} e^{-\varepsilon/k_B T}}$$

$$= \frac{\frac{e^2}{3k_B T} \sqrt{\frac{2}{m_c}} (k_B T)^2 \Gamma(2) \langle l(\varepsilon) \rangle}{(k_B T)^{3/2} \Gamma\left(\frac{3}{2}\right)} = \frac{4e^2 l_c}{3\sqrt{2\pi m_c k_B T}} = e\mu_e, \quad (6.42)$$

where $l_c = \langle l(\varepsilon) \rangle$ is the mean free path of electrons in the conduction band, which is treated as an independent value in semiconductors. This expression of conductivity in semiconductors was first obtained by Hendrik Lorentz.

The carrier densities n and p can be measured using the Hall coefficient, which was discussed in Section 5.2.4 regarding the Sommerfeld model of free electron Fermi gas in metals. The Hall coefficient for n-type (or p-type) semiconductors can also be calculated using the ratio of the longitudinal conductivity σ_{11} and transverse conductivity σ_{12} under the condition $\omega_c \tau = eH_z \tau / m^* c \ll 1$ (cgs unit, based on Seitz, 1940):

$$\begin{aligned}
R_H &= \frac{E_y}{j_x H_z} = \frac{1}{\sigma H_z} \frac{E_y}{E_x} = \frac{1}{\sigma H_z}\left(-\frac{\sigma_{12}}{\sigma_{11}}\right) \\
&= \frac{-1}{\sigma H_z} \frac{\int_{\varepsilon_c}^{\infty} d\varepsilon_e g_c(\varepsilon_e)\left(-\frac{df_c}{d\varepsilon_e}\right)(\omega_c \tau) e^2 v_x v_x \tau}{\int_{\varepsilon_c}^{\infty} d\varepsilon_e g_c(\varepsilon_e)\left(-\frac{df_c}{d\varepsilon_e}\right) e^2 v_x v_x \tau} \\
&= \frac{-\omega_c}{\sigma H_z} \frac{\int_0^{\infty} d\varepsilon \sqrt{\varepsilon}\, e^{-\varepsilon/k_B T} l^2(\varepsilon)}{\int_0^{\infty} d\varepsilon \sqrt{\varepsilon}\, e^{-\varepsilon/k_B T} \sqrt{\frac{2\varepsilon}{m_c}} l(\varepsilon)} \\
&= \frac{-\omega_c}{\sigma H_z} \frac{(k_B T)^{3/2}\Gamma(\frac{3}{2}) l_c^2}{\sqrt{\frac{2}{m_c}}(k_B T)^2 \Gamma(2) l_c} \\
&= -\frac{\omega_c l_c}{\sigma H_z}\frac{1}{4}\sqrt{\frac{2\pi m_c}{k_B T}} = -\frac{3\pi}{8}\frac{1}{nec}
\end{aligned} \quad (6.43)$$

$$\sigma = \frac{j_x}{E_x} = \frac{\sigma_{11}^2 + \sigma_{12}^2}{\sigma_{11}} = \frac{4}{3}\frac{ne^2 l_c}{\sqrt{2m_c k_B T}}, \quad (6.44)$$

which has an extra factor $-3\pi/8$ compared to the Hall coefficient in metals.

It was proved in Section 6.1 that, in metals, the relaxation time τ_F of electrons on the Fermi surface is proportional to the inverse temperature T^{-1}. In semiconductors, the carrier's group velocity $\vec{v}_{c,v}(\vec{k})$ is very small near band edges; and it is the mean free path l_c, l_v of carriers, not the relaxation time τ_e and τ_h, that is proportional to T^{-1}.

τ_e and τ_h are the relaxation times of electrons and holes at the conduction band edge $\vec{k}_c$ and the valence band edge $\vec{k}_v$ respectively, and they are much longer than τ_F in metals. In semiconductors, according to Matthiessen's rule, both the τ_d of carrier-defect scattering (dominant at low-T, related to the ultimate crystal quality) and the τ_{cr} of carrier-phonon scattering (dominant at high-T) contribute to the mean free path l_c for electrons or l_v for holes. The mean free path l_{cr} with respect to the carrier-phonon scattering in semiconductors can be estimated as:

$$\begin{aligned}
\frac{g - f_{F-D}}{l_{cr}} &= \iiint \frac{d^3\vec{k}'}{4\pi^3} P_{\vec{k},\vec{k}'} \frac{g(\vec{k}) - g(\vec{k}')}{\hbar|\vec{k} - \vec{k}_c|/m_c} \\
\Rightarrow \frac{1}{l_{cr}} &\propto \iiint \frac{d^3\vec{k}'}{4\pi^3} P_{\vec{k},\vec{k}'} \propto T
\end{aligned} \quad (6.45)$$

Therefore the electron and hole mobility in the saturation range and the intrinsic range (carrier-defect scattering negligible) can be found by Eq. (6.42):

$$\mu_{e,h} = \frac{e}{m^*_{e,h}} \tau_{e,h} = \frac{4el_{c,v}}{3\sqrt{2\pi m_{c,v} k_B T}} \propto \frac{1}{T^{3/2}} \quad (6.46)$$

This relation $\mu_{e,h} \propto T^{-3/2}$ has been proved by the measured mobility in the intrinsic range where the carrier-phonon scattering is dominant (based on Pearson and Bardeen, 1949). In extrinsic semiconductors with larger doping densities N_D and N_A, the mobility is lower due to stronger carrier-defect scattering, and the intrinsic range starts at a higher temperature. The universal slope $\log(\mu/\mu_0)/\log(T/T_0)$ in the intrinsic range is near the expected value $-3/2$, as shown in Figure 6.17.

The resistivity $\rho(T)$ curve can be found by combining the curves of density n, p and mobility μ_e, μ_h versus the temperature. For the n-type

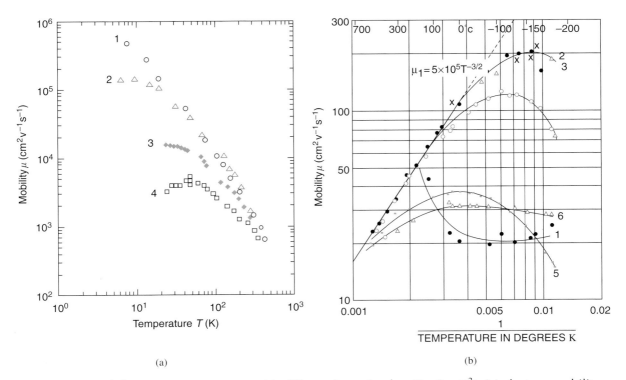

Figure 6.17 Mobility versus temperature with different impurity densities/(cm^{-3}): (a) electron mobility: N_D is 1) $< 1E12$, 2) $< 4E13$, 3) $1.75E16$ ($N_A = 1.48E15$), 4) $1.3E17$ ($N_A = 2.2E15$) (based on www.ioffe.rssi.ru); (b) hole mobility: N_A is 1) $1.5E16$ ($N_D = 0.9E16$), 2) $6.0E17$, 3) $1.3E18$, 4) $2.2E18$, 5) $5.3E18$, 6) $1.4E19$ (based on Pearson and Bardeen, 1949)

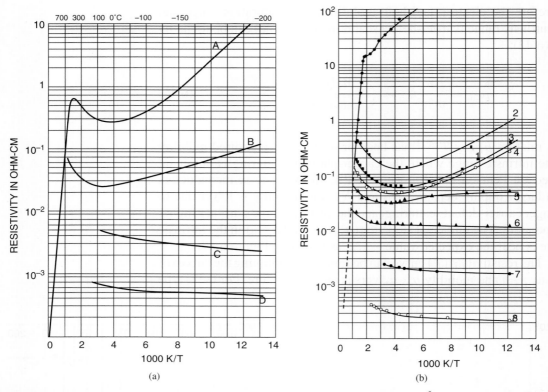

Figure 6.18 Resistivity versus temperature at different impurity densities/(cm^{-3}): (a) n-type silicon, the phosphorous density N_D is, from top to bottom, A:1.2E17(N_A = 1.5E16), B:1.25E18, C:1.7E19, D:2.6E20; (b) p-type silicon, the boron density N_A is, from top to bottom, 1) 1.5E16(N_D = 0.9E16), 2) 6.0E17, 3) 1.3E18, 4) 2.2E18, 5) 5.3E18, 6) 1.4E19, 7) 1.2E20, 8) 4.8E20 (based on Pearson and Bardeen, 1949)

semiconductor, n and ρ vary with the temperature in the three regions as:

$$\ln n = -\frac{E_g}{2k_B T} + b_i; \quad \ln \rho \simeq \frac{E_g}{2k_B T} + \ln A; \quad \text{intrinsic} \quad (6.47)$$

$$\ln n = \ln N_D; \quad \ln \rho = -\frac{3}{2} \ln \frac{T_0}{T} + \ln B; \quad \text{saturation} \quad (6.48)$$

$$\ln n = -\frac{E_d}{k_B T} + b_f; \quad \ln \rho \simeq \frac{E_d}{k_B T} + \ln C; \quad \text{freeze-out} \quad (6.49)$$

In the low-temperature freeze-out range, the resistivity (not very heavily doped) decreases as $\exp(E_d/k_B T)$; in the saturation range, the resistivity increases mildly with T, due to the mobility $\mu \propto T^{-\eta}$; in the very high-T intrinsic range, the resistivity decreases as $\exp(E_g/T)$, which is

Table 6.10 Electrical properties of extrinsic silicon in the saturation range

Physical Properties	n/p-type (n for n, p; m^* for m_c, m_v; l for l_c, l_v)
carrier density/(cm^{-3})	$n \simeq N_D$, $N_A \in (10^{14}, 10^{18})$
Hall coefficient/(cm^3/C)	$R_H = \mp 7.36 \times 10^{18}(n/\text{cm}^{-3})^{-1} \in (10^4, 10^0)$
Mobility/(cm^2/V/s)	$\mu = 1.39 \times 10^8 \left(\frac{l}{\text{cm}}\right)\left[\frac{m^*}{m_e}\frac{T}{300\,\text{K}}\right]^{-1/2}$; $l \sim 10^{-5}$ cm
Resistivity/($\Omega\cdot$cm)	$\rho = \frac{\lvert R_H\rvert}{\mu} = 5.29 \times 10^{10} \left(\frac{n}{\text{cm}^{-3}}\right)^{-1}\left(\frac{l}{\text{cm}}\right)^{-1}\left[\frac{m^*}{m_e}\frac{T}{300\,\text{K}}\right]^{1/2}$

universal for any doping condition and proved by the experiments using silicon by G. L. Pearson and J. Bardeen in Figure 6.18.

The extrinsic properties in the saturation range are the most important for applications, because the carrier density is relatively stable. The important electronic quantities in the saturation range are listed in Table 6.10.

As we have seen, John Bardeen made major contributions to the clarification of extrinsic semiconductor band structures. In 1928, he graduated from the University of Wisconsin with a B.S. in Electrical Engineering. For the next two years he undertook graduate studies, and it was during this period that he was first introduced to quantum theory by Professor John H. Van Vleck, the mentor of Walter Kohn. From 1928 to 1933, Bardeen worked on practical mathematical problems, such as geophysics, radiation in antennae, and oil prospecting using magnetic and gravitational surveys. From 1933 to 1936, Bardeen's interest in mathematical physics led him to become a graduate student of E. P. Wigner at Princeton University. From 1938 to 1945, Dr. Bardeen worked at the University of Minnesota and the Naval Ordnance Laboratory (NOL). From 1945 to 1951, he worked with the solid state research group at Bell Telephone Laboratories; then he went on to his last career position, as a professor at the University of Illinois. It was this combination of academic and industrial experience that led to Bardeen's achievements in semiconductor research.

6.2.3 Basic Concepts of Semiconductor Devices

The most important solid state electronic devices—diodes and triodes—are made by contacting the n-type and p-type semiconductors with the assistance of insulators (white area) and metals (black area) to form an IC active device, as illustrated in Figure 6.19.

In industry, the silicon wafer is already doped to be n-type or p-type, for the sake of easier processing of active devices. In the

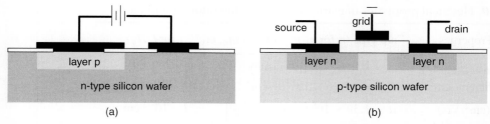

Figure 6.19 Diodes and triodes on silicon wafer: (a) p-n junction diode; (b) n-p-n transistor or n-channel MOSFET (metal-on-semiconductor field-effect-transistor)

Table 6.11 Wafer processing techniques in the fabrication of a modern electronic device

categories	processing technology
Removal	wet clean, wet etching, dry etching; and in the back end processing, chemical-mechanical planarization (CMP)
Modification	doping p-n junctions or transistors in the opening of masks by doping impurities in diffusion furnaces followed by furnace anneals, or ion implantation followed by rapid thermal anneal (RTA); also ultra-violet exposure to reduce ϵ in the low-k insulators (UV)
Deposition	physical vapor deposition (PVD), chemical vapor deposition (CVD), electrochemical deposition (ECD), molecular beam epitaxy (MBE), atomic layer deposition (ALD)
Patterning	photolithography, where photoresist are exposed to create masks, and remaining photoresist are removed by plasma ashing

n-type silicon wafer preparation process, an LnP_5O_{14} source wafer is pressed onto the pure single crystal silicon wafer. After 850° furnace treatment, the phosphorus is doped into the silicon wafer by the chemical reaction $2P_2O_5 + 5Si \rightarrow 4P + 5SiO_2$. In the p-type silicon wafer preparation process, a surface oxidized boron nitride (BN) source wafer is pressed onto the pure single crystal silicon wafer. After 935° furnace treatment, boron is doped into silicon by a chemical reaction $2B_2O_3 + 3Si \rightarrow 4B + 3SiO_2$.

The manufacture of modern IC chips involves several hundred sequenced steps. The first big step is the "front end processing," referring to the formation of the transistors directly on the silicon wafer. The next big step is the "back end processing," referring to the interconnecting of semiconductor devices by aluminum or copper wires to form the desired electrical circuits. The four categories of wafer processing techniques are listed in Table 6.11. After the IC preparation, a silicon wafer is cut into dies, and after testing, the dies are packaged into the final IC chips.

It should be emphasized that, in the modification process, the doped impurity density falls below the surface of the wafer, with a roughly Gaussian dependence on the depth $\exp(-y^2/\sigma^2)$, $[\sigma \in (10^0 - 10^2)\,\mu m]$. Therefore, in reality, the p-n junction is always a graded junction, i.e., the impurity density varies gradually with the depth.

P-N Junction

The p-n junction built on an n-type silicon wafer, as illustrated in Figure 6.19(a), is chosen as an example for analysis, and a step or abrupt junction is considered for the sake of simplicity. The band structures in a p-n junction can be analyzed by assuming the "before and after" contact strategy. Before the contact, the band edges ε_c, ε_v are the same in the n/p regions, because both regions are doped silicon. However the Fermi levels are different: ε_F in the n/p region is located in the upper/lower half of the band gap respectively, as shown in Figure 6.20(b). The chemical potential difference between n/p regions is

$$n_n = N_c\, e^{-(\varepsilon_c - \mu_n^0)/k_B T} \simeq N_D$$
$$p_p = N_v\, e^{-(\mu_p^0 - \varepsilon_v)/k_B T} \simeq N_A - N_D$$
$$\Rightarrow \mu_n^0 - \mu_p^0 = k_B T \ln\left(\frac{N_D(N_A - N_D)}{n_i^2}\right), \quad (6.50)$$

where the impurity densities are N_A (p-region) and N_D (p- and n-region) respectively; both regions are assumed to be in the saturation range at room temperature, therefore the majority carrier densities n_n and p_p are just constants.

After the contact, electrons in the n-region and holes in the p-region diffuse into the opposite region and recombine with each other; then a depletion layer or space charge region appears at the p-n junction, as illustrated in Figure 6.20(a). This diffusion-recombination process finally reaches equilibrium when a sufficiently large block-off field $E(x)$ is built up between the p and n regions, as seen in Figure 6.20(b). After equilibrium is reached, the Fermi energy (chemical potential) ε_F must be the same throughout the space; thus the band edges ε_c, ε_v must be higher in the p-region than in the n-region by an energy difference of

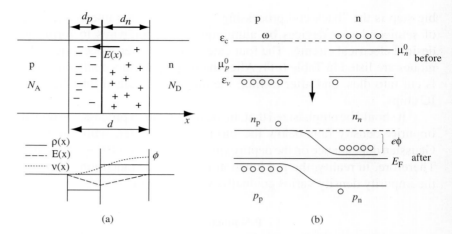

Figure 6.20 PN junction (a) in real space, the junction and the space charge density; (b) in x-dependent energy space, band structures before and after the contact

$e\phi = \mu_n^0 - \mu_p^0$. The build-in voltage ϕ is

$$
\begin{aligned}
0 &= j_{\text{diffusion}} + j_{\text{drift}} \\
0 &= -eD_e\left(-\frac{dn}{dx}\right) + ne\mu_e E \\
0 &= +eD_h\left(-\frac{dp}{dx}\right) + pe\mu_h E \\
E(x) &= -\frac{k_B T}{e}\frac{1}{n}\frac{dn}{dx} = +\frac{k_B T}{e}\frac{1}{p}\frac{dp}{dx}, \\
\Rightarrow \quad \phi &= -\int dx\, E = \frac{k_B T}{e}\ln\left(\frac{n_n}{n_p}\right) = \frac{k_B T}{e}\ln\left(\frac{p_p}{p_n}\right),
\end{aligned}
\qquad (6.51)
$$

where p_p/n_p are the majority/minority carrier densities in the p-region, n_n/p_n are the majority/minority carrier densities in the n-region; and the ratio of diffusion coefficients D_e, D_h over mobilities μ_e, μ_h satisfy the Einstein relation $D_e/\mu_e = D_h/\mu_h = k_B T/e$.

In the depletion layer, the space charge densities are $\rho(x < 0) = \rho_- = -e(N_A - N_D)$ and $\rho(x > 0) = \rho_+ = eN_D$, The electric field $E(x)$ and potential $V(x)$ are:

$$
\begin{aligned}
\frac{dE}{dx} &= 4\pi\rho(x)/\epsilon_r \\
-\frac{dV}{dx} &= E(x) \qquad V_- = 0, \quad V_+ = \phi, \\
E_\mp(x) &= \frac{4\pi\rho_\mp}{\epsilon_r}(x \pm d_\mp), \qquad (-: x < 0;\ +: x > 0) \\
\Rightarrow \quad V_\mp(x) &= V_\mp - \frac{2\pi\rho_\mp}{\epsilon_r}(x \pm d_\mp)^2, \qquad (d_- = d_p,\ d_+ = d_n)
\end{aligned}
$$

$$(6.52)$$

The depletion width $d = d_p + d_n$ can be found by the charge conservation, the continuous condition for the electric field $E(0^-) = E(0^+)$ and the potential $V(0^-) = V(0^+)$ (cgs unit):

$$+\rho_- d_p = -\rho_+ d_n$$

$$0 - \frac{2\pi\rho_-}{\epsilon_r} d_p^2 = \phi - \frac{2\pi\rho_+}{\epsilon_r} d_n^2$$

$$d = d_p + d_n = \left(\frac{\epsilon_r \phi}{2\pi} \frac{\rho_- + \rho_+}{\rho_- \rho_+}\right)^{1/2}$$

$$\Rightarrow \quad d_p = d\rho_+/(\rho_- + \rho_+) = dN_D/N_A$$
$$\quad d_n = d\rho_-/(\rho_- + \rho_+) = d(N_A - N_D)/N_A$$

(6.53)

The depletion capacitance per unit area (without external voltage) can be found by the depletion width: $C_0/A = \epsilon_0 \epsilon_r / d$. In Table 6.12, the numerical description of the built-in voltage ϕ, the depletion width d, and the depletion capacitance C_0/A at room temperature are listed for the p-n junctions built on silicon, germanium, GaAs, InP, and 3C-SiC wafers respectively. It should be emphasized that the intrinsic and extrinsic densities n_i, n, p vary rapidly with the temperature T, as shown in Figure 6.12 and Figure 6.15(a); therefore ϕ, d, and C_0 are only stable in the saturation region of extrinsic properties.

All the main applications of p-n junctions with respect to the diode or rectifier in electronics are related to their one-way conduction characteristics. When the conduction properties of a p-n junction are studied, the p-n junction must be "biased," i.e., an external potential difference V must be applied to the p-n junction. When the bias is applied, the total potential difference is the sum of the built-in voltage and the external

Table 6.12 Room temperature basic parameters of p-n junctions (assuming doping density $N_A \gg N_D$ in the n-type wafer or $N_D \gg N_A$ in the p-type wafer)

Wafer	E_g (eV)	Build-in voltage ϕ (V)	ϵ	Depletion layer d (μm)	C_0/A (nF/cm^2)
3C-SiC	2.30	$0.0258 \ln \frac{N_A N_D}{3.96 \times 10^{-2} \text{cm}^{-6}}$	9.7	$1.04 \sqrt{\frac{e\phi}{eV} \frac{(N_A+N_D)10^{15}\text{cm}^{-3}}{N_A N_D}}$	$8.6 \frac{\mu m}{d}$
Si	1.12	$0.0258 \ln \frac{N_A N_D}{0.70 \times 10^{20} \text{cm}^{-6}}$	11.7	$1.14 \sqrt{\frac{e\phi}{eV} \frac{(N_A+N_D)10^{15}\text{cm}^{-3}}{N_A N_D}}$	$10.3 \frac{\mu m}{d}$
Ge	0.66	$0.0258 \ln \frac{N_A N_D}{3.88 \times 10^{26} \text{cm}^{-6}}$	16.2	$1.34 \sqrt{\frac{e\phi}{eV} \frac{(N_A+N_D)10^{15}\text{cm}^{-3}}{N_A N_D}}$	$14.3 \frac{\mu m}{d}$
GaAs	1.43	$0.0258 \ln \frac{N_A N_D}{5.71 \times 10^{12} \text{cm}^{-6}}$	12.9	$1.19 \sqrt{\frac{e\phi}{eV} \frac{(N_A+N_D)10^{15}\text{cm}^{-3}}{N_A N_D}}$	$11.4 \frac{\mu m}{d}$
InP	1.35	$0.0258 \ln \frac{N_A N_D}{2.10 \times 10^{14} \text{cm}^{-6}}$	12.1	$1.16 \sqrt{\frac{e\phi}{eV} \frac{(N_A+N_D)10^{15}\text{cm}^{-3}}{N_A N_D}}$	$10.7 \frac{\mu m}{d}$

voltage ($V > 0$ means positive voltage is added in the p-region):

$$V_+(d_n) - V_-(-d_p) = \phi - V;$$

$$d = d_p + d_n = \left(\frac{\epsilon_r(\phi - V)}{2\pi} \frac{N_A + N_D}{N_A N_D}\right)^{1/2}$$

(6.54)

Therefore the depletion widths d_p and d_n vary with $\sqrt{\phi - V}$; i.e., with the positive/negative bias voltage V, the total potential difference $\phi - V$ must be lower/higher, then the depletion layer becomes thinner/thicker. The depletion capacitance of the biased p-n junction is related to the depletion width and the bias voltage:

$$\frac{C}{A} = \frac{\epsilon_0 \epsilon_r}{d} = \frac{C_0}{A}\sqrt{\frac{\phi}{\phi - V}}$$

(6.55)

Therefore the capacitance of the p-n junction increases with V.

The current passing through a p-n junction is realized by two processes: the diffusion of carriers over an energy barrier $e(\phi - V)$ and the recombination of electron-holes:

$$j = eD_e \frac{n_n e^{-e(\phi-V)/k_B T} - n_p}{d} - eD_h \frac{p_n - p_p e^{-e(\phi-V)/k_B T}}{d};$$

$$I = [(n_n e^{-e(\phi-V)/k_B T} - n_p)\bar{v}_e + (p_p e^{-e(\phi-V)/k_B T} - p_n)\bar{v}_h]eA$$

$$= 1.6 \times \frac{n_p \bar{v}_e + p_n \bar{v}_h}{10^6 \text{cm}^{-3} \cdot 10^7 \text{cm/s}} \frac{A}{\text{cm}^2}[\exp(eV/k_B T) - 1](\mu A),$$

(6.56)

where only the $e^{-e(\phi-V)/k_B T}$ percentage of majority carriers can overcome the electric field, diffuse into the opposite region, and be annihilated afterwards by electron-hole recombination. The drift velocities of electrons and holes are $\bar{v}_e \simeq D_e/d$ and $\bar{v}_h \simeq D_h/d$ respectively, which are both in the order of 10^7 cm/s.

The I-V curve in Figure 6.21(c) and Eq. (6.56) is the ideal rectifier property with carrier charge e, where the saturation current $I = -I_0 = -(n_p \bar{v}_e + p_n \bar{v}_h)eA$ at backward bias is the minority current. The I-V curve would break down at a very large backward bias; "-50V" labeled in Figure 6.21(d) approximately shows the breakdown voltage V_b. If the doping concentrations N_A and N_D are not very high, the mechanism of breakdown is due to the avalanche effect:

$$eE_b d \simeq 10^5 \text{eV/cm} \cdot 10^{-5} \text{cm} \simeq \text{eV}$$

(6.57)

When the backward voltage $|V|$ is so high that the minority carrier is accelerated over the depletion layer with the energy ~ 1 eV, one carrier

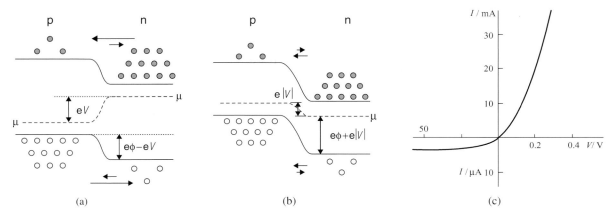

Figure 6.21 Biased p-n junction: (a) forward $V > 0$; (b) backward $V < 0$; (c) I-V curve

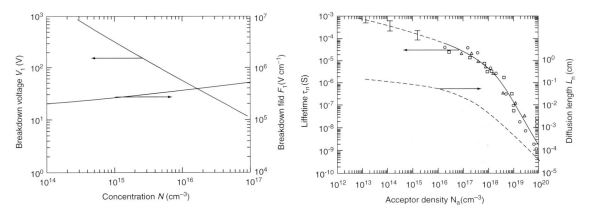

Figure 6.22 (a) Breakdown voltage V_b and breakdown field E_b versus doping concentration N (based on Sze, 1981); (b) lifetime and diffusion length of electrons versus N_A (based on del Alamo and Swanson, 1987)

will "hit and excite" another. Then the carrier density suddenly increases and the p-n junction breaks down. The breakdown voltage V_b decreases for higher doping concentrations N_A or N_D, because the depletion width $d \propto \sqrt{N_D^{-1}}$ is thinner. The breakdown field E_b slightly increases for higher doping, as seen in Figure 6.22(a).

The diffusion coefficient D of carriers can be found by the Einstein relation:

$$D = \frac{k_B T}{e}\mu = 26 \times \frac{T}{300\,K} \frac{\mu}{10^3\,\mathrm{cm}^2/\mathrm{V/s}} (\mathrm{cm}^2/\mathrm{s}), \qquad (6.58)$$

which proves that the drift velocity is $\bar{v} = D/d \sim 10^7$ cm/s in silicon. The relaxation time of carriers τ can also be found by the mobility $\mu = e\tau/m^*$. In most semiconductors such as silicon, germanium, GaAs, or 3C-SiC, τ is in the order of 10^{-13} s. There is another class of time scales for carriers: the life time τ_n and τ_p in the minority diffusion equation ($E = 0$) characterizing the electron-hole recombination:

$$\frac{\partial n}{\partial t} = D_e \frac{\partial^2 n}{\partial x^2} - \frac{n}{\tau_n} + G \quad \Rightarrow \quad n = G\tau_n e^{-x/L_n} \tag{6.59}$$

where G is the generation rate and $R = n/\tau_n$ is the recombination rate in the p-region. The life time τ_n and the diffusion length L_n of electrons in p-type silicon versus the doping concentration N_A is plotted in Figure 6.22(b). In silicon that is not very heavily doped, both τ_n of electrons and τ_p of holes are in the order of 10^{-3}–10^{-5} s, which is 10^{10}–10^8 times longer than the relaxation time τ_e or τ_h; the diffusion lengths L_n and L_p are between 0.1 cm and 0.01 cm, which is longer than the depletion width $d \sim 1\mu$m for a proper operation of the p-n junction.

Metal-Semiconductor Junction

Historically, the study of metal-semiconductor junctions preceded that of p-n junctions. In 1938, a German physicist, Walter Schottky, developed a theoretical explanation for the metal-semiconductor junction. Schottky was a Ph.D. student under Max Planck. After he obtained his Ph.D. at the University of Berlin in 1920, he initially worked at two universities; he then worked at Siemens AG from 1927 until he was in his late eighties. In 1919 he invented the tetrode, which was the first multi-grid vacuum tube. In his book, *Thermodynamik* (1929), he was the first person to point out the existence of electron holes in the valence-band structure of semiconductors. As we can see, Schottky played a very important role in semiconductor research in Germany.

The metal-semiconductor junction is of great practical importance in integrated circuit processing, because there are millions of connecting points between wires and active devices. There are three kinds of metal-semiconductor junctions: (1) The Schottky barrier, the I-V curve of which is analogous to a p-n junction; (2) The Ohmic contact, the I-V curve of which is similar to a resistor; (3) The MOS (Metal-oxide-semiconductor) contact, which will be discussed in the next subsection on MOSFET devices.

In the metal-n-semiconductor Schottky barrier in Figure 6.23(a), before the contact, the Fermi energy ε_F^m of the metal is between the Fermi energy ε_F of the n-type semiconductor and the valence band edge ε_v; after the contact, the Fermi energy ε_F is universal and the semiconductor surface is p-type because the local ε_F is at the lower half of the gap, while the in-depth semiconductor is still n-type. Compared to semiconductors,

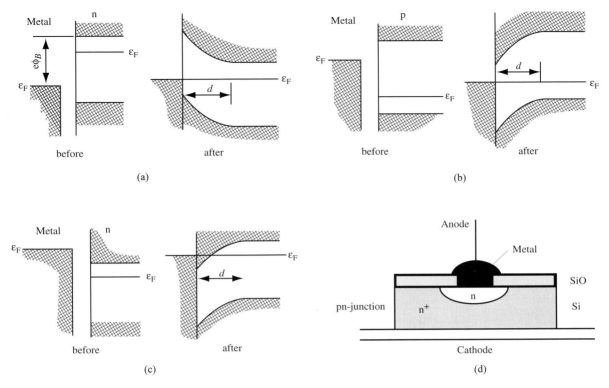

Figure 6.23 MS junction: (a) metal-(n-type semiconductor) Schottky barrier; (b) metal-(p-type semiconductor) Schottky barrier; (c) metal-(n-type semiconductor) Ohmic contact (based on Sze, 1981); (d) schematics of a Schottky diode (from www.at-mix.de)

in metals both the electron density and "hole density" are huge; therefore, due to the surface diffusion, the metal-n-semiconductor Schottky barrier becomes a metal-(p)-n junction.

In a metal-p-semiconductor Schottky barrier in Figure 6.23(b), before the contact, the Fermi energy ε_F^m of the metal is located between the conduction band edge ε_c and the Fermi energy ε_F of a p-type semiconductor; after the contact, the Fermi energy ε_F is universal and the semiconductor surface is n-type because the local ε_F is at the upper half of the gap, while the in-depth semiconductor is still p-type. The huge electron density in metals diffuses into semiconductors, therefore the metal-p-semiconductor Schottky barrier becomes a metal-(n)-p junction.

For a metal-n-semiconductor Ohmic contact in Figure 6.23(c), before the contact, the Fermi energy ε_F^m of the metal is higher than the conduction band edge ε_c; after the contact, the Fermi energy ε_F is universal. The semiconductor surface becomes metallic because the local ε_F is in the conduction band, while the in-depth semiconductor is still n-type. The high-density electrons in metals can diffuse into the surface of the semiconductor, therefore the metal-n-semiconductor Ohmic contact becomes a metal-(metal)-n junction and the electrons can flow through the junction quite freely, in a manner similar to a resistor.

Schottky diodes are commonly used in integrated circuits, and their I-V relationship is

$$I = I_s \left[\exp \frac{eV - IR_s}{k_B T} - 1 \right] \qquad (6.60)$$

where R_s is the series of resistance. The Schottky diode illustrated in Figure 6.23(d) is built with two different metals: metal I-GaAs forms the Schottky barrier, and metal II-GaAs forms the Ohmic contact ($\varepsilon_F^m > \varepsilon_c$ in metal II).

The metal wires in integrated circuits are aluminum or copper. Aluminum has a similar atomic number to silicon, therefore it is quite possible that the aluminum-silicon junction is a Schottky barrier. If an Ohmic contact is required, an interlayer of a heavier metal should be deposited between the wire and the semiconductor, such that the Fermi surface ε_F^m of the interlayer is higher than the ε_c of the semiconductor. Silicon alloys are used for Ohmic contacts to silicon, such as $Si_{1-x}M_x$ (the element M can be Co, Ti, W, Ta, Pt, Al); while gold alloys are used for Ohmic contacts to gallium arsenide, such as $Au_{1-x}M_x$ (the element M can be Zn, In, Ge, Si, Sn, Te).

MOS Transistor

The most basic element in integrated circuit design is the transistor. Since 1960, the type most commonly used transistor has been the MOSFET: the metal-oxide-semiconductor field effect transistor. A MOSFET contains two p-n junctions between the source and the drain: the p-n-p structure on the n-type wafer, and the n-p-n structure on the p-type wafer. More importantly, a gate for the control of signal amplification, equivalent to a grid in a vacuum tube triode, is directly deposited on an oxide layer of the wafer, as illustrated in Figure 6.24(a). Silicon dioxide is a very good insulator, so a very thin layer, typically only a few hundred molecules thick, is required for the oxide layer between the metal and semiconductor.

The NPN or PNP type of structure does not allow the flow of current between the source and the drain, because there is always a backward biased p-n junction in either direction of the current. When a gate voltage $V_G > V_t$ is applied to the NPN structure, holes in the underlying p-region are pushed away, $\varepsilon_c - \varepsilon_F$ is lower at the surface, and an n-channel appears for the electron current flow. When a gate voltage $V_G < -V_t$ is applied to the PNP structure, electrons in the underlying n-region are

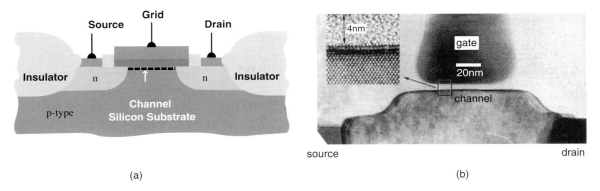

Figure 6.24 MOS transistor: (a) schematics of n-channel MOS; (b) TEM image of n-channel MOS (from http://www.chem.wisc.edu/)

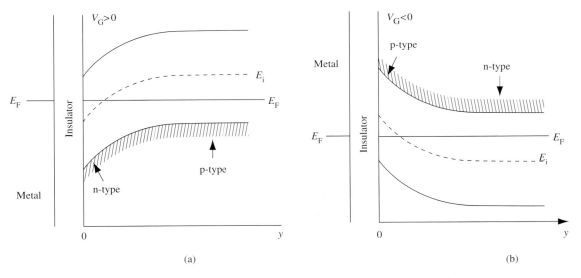

Figure 6.25 Band structures: (a) $V_G > 0$, n-channel MOS; (b) $V_G < 0$, p-channel MOS

pushed away, $\varepsilon_F - \varepsilon_v$ is lower at surface, and a p-channel appears for the hole current flow.

The typical amplification characteristics of MOSFETs are shown in Figure 6.26. The drain-source I_{DS}-V_{DS} curves have two regions for different gate-source voltages V_{GS}: (1) the linear region or triode region: the MOSFET behaves like a resistor, and the conductance $G = I/V$ increases linearly with V_{GS}; (2) the saturation region: the saturation current I increases for higher V_{GS}.

In a MOSFET, the source-drain current $I_{DS} = dQ/dt$ is related to the source-drain voltage V_{DS}, the threshold voltage V_t (when $V > V_t$ the

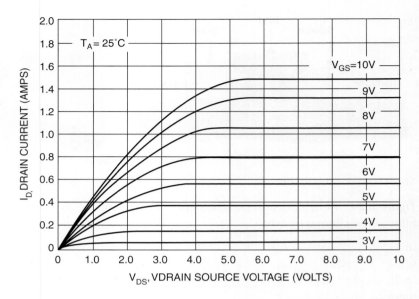

Figure 6.26 MOSFET amplifier: I-V curve under different gate voltage (from http://www.chem.wisc.edu/)

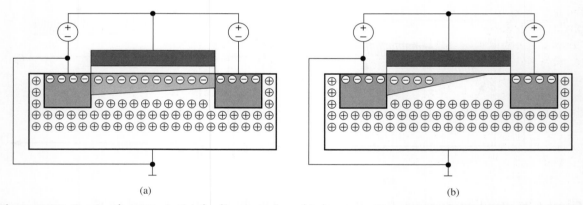

Figure 6.27 Carrier densities in (a) the linear region; (b) the saturation region of a MOSFET (from http://www.deas.harvard.edu/courses/es154/)

MOS conduct electricity), and the gate voltage V_{GS} as:

$$I_{DS} = \beta\left(V_{GS} - V_t - \frac{1}{2}V_{DS}\right)V_{DS}; \quad V_{DS} < V_0 \text{ linear} \quad (6.61)$$

$$I_{DS} = \frac{1}{2}\beta(V_{GS} - V_t)^2; \quad V_{DS} > V_0 \text{ saturation} \quad (6.62)$$

where $\beta = \mu C_{ox} W/L$ is determined by the mobility, oxide dielectrics, and MOS geometry. In the linear (triode) region of a nMOS, all regions underlying the gate become n-type, as illustrated in Figure 6.27(a); in the saturation region, part of underlying region is no longer n-type with too high V_{GS}, as seen in Figure 6.27(b), therefore, the current no longer increases. The pMOS behaves in a similar way to the nMOS.

6.3 Superconductors

Superconductors, the solids that have no resistance to the current flow, are one of the last great frontiers in science. Both the practical uses of superconductors and the theories that explain superconductivity are still imperfect. In the previous two sections of this chapter, it has been shown that the resistivity of matter is mainly due to carrier-phonon scattering; however phonons always exist in a crystal, therefore it is impossible to understand the appearance of superconductivity by the usual transport theories.

It had been known for many years that the resistance of metals fell linearly with the lowering of temperature, but the limiting value that the resistance would approach near 0 K was not known. In 1908, Dutch physicist Heike Kamerlingh-Onnes from Leiden University successfully liquefied helium gas, and with the use of liquid helium it was possible to study the extreme-low-temperature properties of matter. Onnes measured the resistivity of pure metals in this new temperature range below 4.2 K. In 1911, he observed superconductivity in mercury, and noticed that its resistance suddenly disappeared below a critical temperature $T_c \sim 4$ K, as shown in Figure 6.28(a). This was an exciting result that meant that, in superconductors, electricity could flow without any loss of energy. In 1913, Onnes won the Nobel Prize in Physics for these investigations.

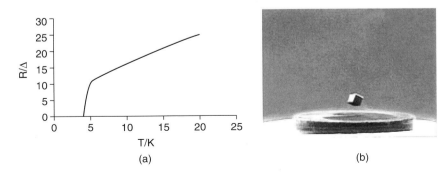

Figure 6.28 Superconductivity: (a) zero resistivity at low-T; (b) perfect diamagnetism

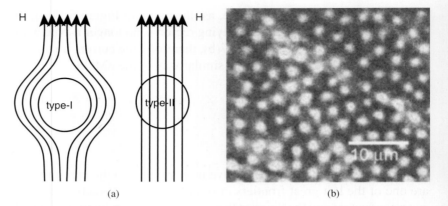

Figure 6.29 (a) Meissner effect in a Type I superconductor and flux penetration in a Type II superconductor; (b) Abrikosov lattice of vortices (from nobelprize.org)

In 1933, German physicists Walther Meissner and R. Ochsenfeld discovered that superconductors are more than perfect conductors of electricity. They found that the superconductor is a perfect diamagnet with a zero internal B-field; so when a magnet approaches a superconductor, it will experience a repulsive force from the superconductor, which, if strong enough, will cause the magnet to overcome the force of gravity and float above the superconductor, as illustrated in Figure 6.28(b). However, a large magnetic field above the critical field H_c destroys the superconductivity. This phenomenon is known as strong diamagnetism and is often referred to as the "Meissner effect" today.

The theory of superconductivity can be roughly classified into the phenomenological and microscopic theories. The phenomenological theory was first mooted in 1934 with the Gorter-Casimir model or the two-fluid model. In 1935, brothers Fritz and Heinz London proposed equations to explain the Meissner effect. In 1950, a new theoretical progression was made by Russian physicist Vitaly Ginzburg and the great theoretician, Lev Landau, called the Ginzburg-Landau theory. This explained superconductivity comprehensively and provided a derivation for the London equations. In 1957, Lev Landau's student, Alexei Abrikosov, simplified the Ginzburg-Laudau theory so that the hexagonal arrangement of magnetic flux tubes, known as the "Abrikosov lattice of vortex," could be explained. In 1962, Landau won the Nobel Prize in Physics, mainly for the general second-order phase transition theory; and in 2003, Ginzburg and Arikosov won the Nobel Prize in Physics for their contributions to superconductor theory.

In 1957, John Bardeen worked with two colleagues, Leon Cooper and Robert Schrieffer, to develop a microscopic theory for superconductivity, called the BCS theory. It states that a pair of electrons (Cooper pair) bound by electron-phonon interaction is the carrier of superconductor current. After years of experimentation, the BCS theory was proved to be true for all traditional superconductors (metals), and the three researchers were jointly awarded the 1972 Nobel Prize in Physics. This was the second Nobel Prize for Bardeen.

The next significant theoretical advancement came in 1962, when a young Englishman, Brian Josephson, predicted that current could flow through a S/M/S or S/I/S junction (S: superconductor; M: metal; I: insulator). This tunneling phenomenon, known as the "Josephson effect" today, has major applications in electronic devices such as the SQUID, and was also important for the verification of the BCS theory. Josephson won the 1973 Nobel Prize in Physics.

The most recent development in superconductivity came in 1986, when Georg Bednorz and Alex Muller, working at IBM in Zurich, Switzerland, created an oxide superconductor, LaBaCuO, that worked at the "high temperature" of 30 K. The superconductivity in high-Tc superconductors strongly depends on the detailed structures, therefore the BCS theory based on the picture of Fermi surfaces in metals no longer seems to be true. Bednorz and Muller won the 1987 Nobel Prize in Physics.

6.3.1 Characteristics of Superconductors

Superconductors can be classified into Type-I and Type-II by the order of their critical magnetic fields H_c, as listed in Table 6.13. Type-I superconductors are pure metals and their H_c are quite low. Type-II superconductors are alloys, where the Meissner effect only appears

Table 6.13 Superconductor basic physical properties: T_c/K is the critical temperature at zero magnetic field; H_c/Tesla is the critical magnetic field at 0K (based on Kittel, 1986, and www.superconductors.org)

Type	Name	T_c	H_c	Name	T_c	H_c	Name	T_c	H_c
I (2-4B)	Zn	0.85	–	Sc	0.05	–	Ti	0.40	–
I (2-4B)	Cd	0.56	–	Y	1.30	–	Zr	0.61	–
I (2-4B)	Hg	3.95	.04	La	6.00	.10	Hf	0.13	–
I (5-8B)	V	5.40	.14	Mo	0.92	–	Tc	7.80	.14
I (5-8B)	Nb	9.25	.20	W	0.015	–	Re	1.70	.02
I (5-8B)	Ta	4.47	.08	Ru	0.49	–	Os	0.66	–
I (3-4A)	Al	1.18	.01	In	3.41	.03	Sn	3.72	.03
I (3-4A)	Ga	1.08	–	Tl	2.38	.02	Pb	7.20	.08
II	Nb$_3$Ge	23.2	38	Nb$_3$Al	18.7	32	V$_3$Si	16.7	2.4
II	Nb$_3$AlGe	20.7	44	NbN	15.7	1.5	V$_3$Ga	14.8	2.1
II	Nb$_3$Sn	18.0	24	NbTi	10.0	15	PbMoS	14.4	6.0
High-Tc	LaBaCuO	30	–	YBaCuO	92	50	TlBaCuO	125	–

below a lower critical field H_{c1}. In the range $H_{c1} < H < H_{c2}$, the magnetic field penetrates through flux tubes in the form of an Abrikosov lattice, below the critical field $H_c \sim H_{c2}$. All high-Tc superconductors are Type-II superconductors.

All practical superconductor materials were Type-II, in which a large current could be carried without breaking superconductivity. In 1941, niobium nitride was found to superconduct. In 1953, vanadium silicon was also shown to display superconductive properties. In 1962, Westinghouse developed the first commercial NbTi superconducting wire. Since the 1960s, powerful superconductor electromagnets have been utilized in accelerators, such as those in the Rutherford-Appleton Laboratory in the UK and the Fermi Laboratory in the US. Recently, high-Tc superconducting wires, especially YBaCuO wires, have been developed.

The discovery of high-Tc superconductors is one of the most exciting events in physics in the late 20th century. In 1972, under the guidance of H. J. Scheel, a young German physicist, Johannes Georg Bednorz, began the experimental part of his diploma work on crystal growth and the characterization of $SrTiO_3$. In 1982, Bednorz formally joined IBM in Zurich, after he had completed his work on the crystal growth of perovskite-type solid solutions and the related structural, dielectric, and ferroelectric properties. From 1983 to 1986, Bednorz worked with the Swiss physicist, Alex Müller, to search for a high-Tc superconducting oxide. Finally, they realized the importance of the discovery of LaBaCuO in 1986, and were awarded the Nobel Prize in Physics in 1987.

The phase diagram of a LaSrCuO high-Tc superconductor is shown in Figure 6.30(a). Lanthanum strontium cuprate $(La_{1-x}Sr_x)CuO_4$ is a perovskite solid solution made by doping lanthanum copper oxide (La_2CuO_4) with strontium. La_2CuO_4 itself is an anti-ferromagnetic insulator; when 3–21% La is replaced by Sr, $(La_{1-x}Sr_x)_2CuO_4$ becomes a superconductor below T_c. The phase diagram of lanthanum barium cuprate $(La_{1-x}Ba_x)_2CuO_4$ is similar to Figure 6.30(a), except that the superconducting phase appears in two nearby ranges, 5–13% and 13–24%, of barium-doping.

The mechanism of high-Tc superconductivity is not yet fully understood, but most scientists agree that the carriers in high-Tc superconductors are the holes, not the electrons. In $(La_{1-x}Ba_x)_2CuO_4$, lanthanum is a IIIB element and barium is a IIA element, therefore there must be a hole created after the replacement of a lanthanum atom by a barium atom. In yttrium barium copper oxide $YBa_2Cu_3O_{7-\delta}$, the CUC of which is shown in Figure 6.30(b), the total positive charge is $3 + 2 \times 2 + 3 \times 2 = 13$, and the total negative charge is $-2 \times (7 - \delta) = -14 + 2\delta$; therefore there must be positive-charge holes conducting in Cu(2)-O(3) surfaces in Figure 6.30(b) to balance the total charge in the solid.

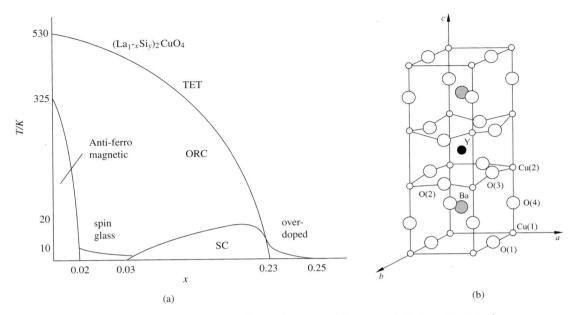

Figure 6.30 High-Tc superconductors: (a) phase diagram of $(La_{1-x}Sr_x)_2CuO_4$, "SC" is the superconducting phase; (b) atomic structure of $YBa_2Cu_3O_{7-\delta}$ (based on Han, 1998)

The first step in achieving an understanding of superconductivity came with the development of the Gorter-Casimir model in 1934. After Meissner and Ochsenfeld's discovery of the expulsion of the magnetic field from a superconductor, Dutch physicists Cornelis Gorter and Hendrik Casimir formulated the thermodynamic model of superconductivity. In this two-fluid model, it is assumed that an x fraction of electrons are in the normal fluid and the rest of the electrons are in the super fluid. The critical magnetic field H_c and the specific heat C_s can be found by the total free energy at equilibrium:

$$\mathcal{F}(x, T) = x^{1/2} f_n(T) + (1-x) f_s(T)$$

$$= x^{1/2} \left[-\frac{1}{2}\gamma T^2 \right] + (1-x)[-\beta]$$

$$0 = \left. \frac{\partial \mathcal{F}(x, T)}{\partial x} \right|_{x=x_0} ; \Rightarrow x_0(T) = \left(\frac{\gamma T^2}{4\beta} \right)^2 = \left(\frac{T}{T_c} \right)^4 ;$$

(6.63)

$$\frac{H_c^2(T)}{8\pi} = \bar{\mathcal{F}}_n - \bar{\mathcal{F}}_s = -\frac{1}{2}\gamma T^2 - \mathcal{F}(x_0, T)$$

$$= \beta \left[1 - \left(\frac{T}{T_c} \right)^2 \right]^2$$

(6.64)

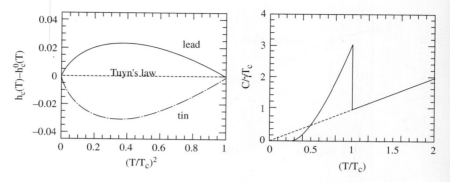

Figure 6.31 (a) $H_c - T$ curve's deviation from the two-fluid model (dotted) for tin (dot-dashed) and lead (solid); (b) specific heat versus temperature (based on Schrieffer, 1983)

$$C_s = -T\frac{d^2 \mathcal{F}(x_0, T)}{dT^2} = \frac{3\gamma^2 T^3}{4\beta} = 3\gamma T_c \left(\frac{T}{T_c}\right)^3 \quad (6.65)$$

where $C_n = \gamma T$ is the specific heat of "normal" metals, as discussed in Chapter Five; and $-\gamma T^2/2$ and $-\beta$ are the characteristic energies of normal and super fluids.

The real $H_c(T)$ and $C(T)$ of superconductors, plotted in Figure 6.31, slightly deviate from Eqs. (6.64) and (6.65) of the two-fluid model:

$$h_c(T) = \frac{H_c(T)}{H_c(0)} \simeq 1 - \left(\frac{T}{T_c}\right)^{2+\eta} \quad \eta = 0 \text{ for two-fluid model}$$

(6.66)

$$C_s(T) \simeq a \exp(-\Delta(0)/k_B T) \quad \text{for } T \ll T_c \quad (\Delta(0) \simeq 1.76 k_B T_c)$$

(6.67)

where η is about $+0.13$ and -0.16 for lead (Pb) and tin (Sn) respectively. $h_c^0(T)$ from the Gorter-Casimir model is also called Tuyn's law. The measured $C_s(T)$ at near 0 K shows the existence of an energy gap $\Delta(0)$, which is analogous to the Einstein model of phonons.

The existence of the energy gap Δ in the superconducting phase can be directly observed in several experimental methods. One method is the interaction of superconductors with the AC electromagnetic field. At near 0 K, it is found that the supercurrent only exists below a critical frequency ω_g:

$$\hbar \omega_g = 3.5 k_B T_c = 2\Delta(0), \quad (6.68)$$

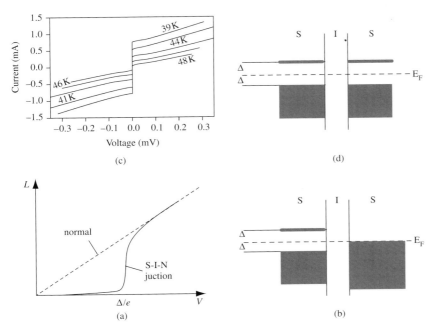

Figure 6.32 (a) I-V curve of S-I-N junction (based on de Gennes, 1989); (b) band structure of S-I-N junction; (c) I-V curve of Josephson junction (from http://www.ifp.fzk.de/); (d) band structure of Josephson junction

i.e., in an electromagnetic field with a frequency above ω_g, the resistivity is no longer zero in a superconductor. This phenomenon is related to carrier-photon scattering. Ultrasonic waves can also be used to measure $\Delta(0)$.

Another method for superconductor energy gap measurement is the S-I-N junction (superconductor-insulator-normal metal junction), as illustrated in Figure 6.32(a) and (b). When a positive voltage V is applied to the S-region in a SIN junction, the electron energy in the N-region is lifted by eV. If $eV > \Delta(0)$ is satisfied, the electrons at the Fermi level in the N-region can tunnel into the S-region and cause a sharp increase in the current; therefore the gap $\Delta(0) = eV_t$ can be found. The structure of the S-I-N junction is analogous to the S-I-S Josephson junction shown in Figure 6.32(c) and (d). The I-V curve of the Josephson junction has a large jump at $V = 0$, which is the main characteristic leading to its application as a switch.

6.3.2 Phenomenological Theories

The Gorter-Casimir or two-fluid model was actually the first phenomenological theory of superconductors. In 1935, Fritz and Heinz

London formulated equations to explain the Meissner effect, and predicted the depth to which a magnetic field could penetrate into a superconductor.

The picture of the London theory is still based on the two fluid model. The total free energy of a superconductor includes two terms, the kinetic energy of the superconducting electrons and the energy associated with the magnetic field:

$$\mathcal{F} = F_0 + K + E_{\text{em}} = F_0 + \iiint d^3\vec{r} \left[\frac{1}{2} m\vec{v}^2 n_s + \frac{\vec{H}^2}{8\pi} \right] \quad (6.69)$$

where n_s is the superconducting electron density. Based on the Maxwell equation (Ampere's law), the two energy terms are related (cgs unit):

$$\vec{\nabla} \times \vec{H} = \frac{4\pi}{c} \vec{j}_s = \frac{4\pi}{c}(-n_s e \vec{v}) \Rightarrow n_s \vec{v} = \frac{c}{4\pi e} \vec{\nabla} \times \vec{H} \quad (6.70)$$

$$\mathcal{F} = F_0 + \frac{1}{8\pi} \iiint d^3\vec{r}[\lambda_L^2 (\vec{\nabla} \times \vec{H})^2 + \vec{H}^2]; \quad \lambda_L^2 = \frac{mc^2}{4\pi n_s e^2} \quad (6.71)$$

λ_L is called the London length. The differential equation of the magnetic field $\vec{H}(\vec{r})$ in superconductors can be found by the variation of $\mathcal{F}$ versus $\vec{H}$:

$$\delta \mathcal{F} = \iiint \frac{d^3\vec{r}}{4\pi}[\lambda_L^2 (\vec{\nabla} \times \vec{H}) \cdot (\vec{\nabla} \times \delta\vec{H}) + \vec{H} \cdot \delta\vec{H}]$$
$$\Rightarrow \lambda_L^2 \vec{\nabla} \times \vec{\nabla} \times \vec{H} + \vec{H} = 0 \quad \text{or}$$
$$\vec{\nabla} \times \vec{j}_s = -\frac{n_s e^2}{mc} \vec{H} \quad \text{or} \quad \vec{j}_s = -\frac{n_s e^2}{mc} \vec{A}. \quad (6.72)$$

This is the famous London equation. The most outstanding achievement of the London theory lies in its explanation of the Meissner effect. If the condition $\vec{\nabla} \cdot \vec{H} = 0$ is used, the London equation can be written as:

$$-\lambda_L^2 \vec{\nabla}^2 \vec{H} + \vec{H} = 0 \quad \rightarrow \quad d^2\vec{H}/dz^2 = \vec{H}/\lambda_L^2$$
$$\rightarrow \vec{H}(z) = \vec{H}(0) \exp(-z/\lambda_L) \quad (6.73)$$

Therefore, near a surface $z = 0$, the magnetic field must decrease exponentially inside the superconductor, which is the Meissner effect. The London length λ_L is the penetration depth δ, and is proportional to $n_s^{-1/2}$ in the London theory.

If the ratio $(1 - x)$ of the superconducting phase in the two-fluid model is related to n_s, the temperature dependence of the London length can be found:

$$\frac{n_s}{n} = 1 - x = 1 - \left(\frac{T}{T_c}\right)^4$$

$$\rightarrow \lambda_L(T) = \frac{\sqrt{mc^2/4\pi n e^2}}{\sqrt{1-(T/T_c)^4}} = \frac{\lambda_L(0)}{\sqrt{1-(T/T_c)^4}}. \quad (6.74)$$

Therefore the London length $\lambda_L \to \infty$ and sc-density $n_s \to 0$ when $T \to T_c^-$.

Fritz London offered a quantum mechanical interpretation of the London Eq. (6.72), which led to an interesting conclusion about flux quantization in superconductors. In quantum mechanics, the current $\vec{j} = -en\vec{v} = -en\vec{p}/m$ can be derived from the many-body electron wavefunction $\Psi(\vec{r}_1, \vec{r}_2, \ldots, \vec{r}_N)$:

$$\vec{j} = -\frac{e}{2m}\sum_{j=1}^{N}\int d\tau[\Psi^*(-i\hbar\vec{\nabla}_j)\Psi + \Psi(i\hbar\vec{\nabla}_j)\Psi^*]\delta(\vec{r}_j - \vec{r}) \quad (6.75)$$

The London Eq. (6.72) describes the relationship between the super current $\vec{j}_s$ and the vector potential $\vec{A}$ ($\vec{\nabla} \times \vec{A} = \vec{B}$), so the many-body wavefunction Ψ_s in the superconducting phase must be related to the vector potential in the following manner:

$$\vec{j}_s = -\frac{e}{2m}\sum_{j=1}^{N}\int d\tau[\Psi_s^*(-i\hbar\vec{\nabla}_j)\Psi_s + \Psi_s(i\hbar\vec{\nabla}_j)\Psi_s^*]\delta(\vec{r}_j - \vec{r})$$

$$= -\frac{n_s e^2}{mc}\vec{A}$$

$$(6.76)$$

$$\Psi_s = e^{i\sum_j \phi_j}\Psi_0; \quad \vec{\nabla}_j\phi_j(\vec{r}_j) = \frac{e}{\hbar c}\vec{A}(\vec{r}_j). \quad (6.77)$$

Therefore there is a gauge transformation between the wavefunction Ψ_s and Ψ_0 with and without a magnetic field respectively. If a constant magnetic field B through a hole (radius r) in a cylindrical superconductor is considered, the vector potential must be directly related to the flux Φ as:

$$\Phi = \iint d^2s\vec{B} = \oint d\vec{l}\cdot\vec{A} \to A_\theta = \frac{\Phi}{2\pi r}, A_r = A_z = 0 \quad (6.78)$$

$$\phi_j = \frac{e\Phi}{2\pi\hbar c}\theta_j = \frac{\Phi}{\Phi_0}\theta_j, \to \Phi = n\Phi_0'; \quad \Phi_0' = \frac{hc}{e} = 2\Phi_0 \quad (6.79)$$

The flux Φ through a superconductor must be an integer n times the flux quantum Φ_0', because the wavefunction Ψ_s must be identical when $\theta_j \to \theta_j + 2\pi$.

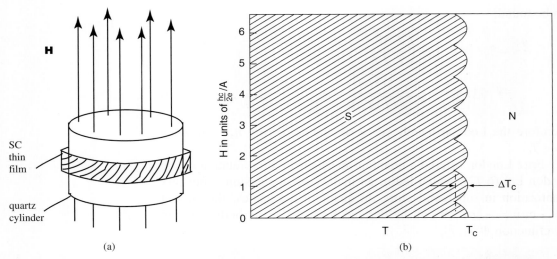

Figure 6.33 (a) Schematics of an experimental settlement to measure the flux quantum; (b) proof of $\Phi_0 = hc/2e$ by the quantization of the phase diagram (based on Little and Parks, 1962)

In 1951, Lars Onsager pointed out that the true flux quantum should be half of Φ'_0; and in 1962, the flux quanta $\Phi_0 = hc/2e \simeq 2 \times 10^{-7} \text{G} \cdot \text{cm}^2$ was proved by measuring T_c versus the scaled external magnetic field $H/(\Phi_0/A)$, as shown in Figure 6.33(b).

In fact the London equation is not true for all superconductors, but only for Type-II superconductors with few defects. In Type-I superconductors, the penetration depth δ is related to the correlation length ξ_0. The explanation of ξ_0 was given by Alfred Brian Pippard in 1953, which was a nonlocal generalization for the London theory. The Pippard equation and the Pippard length are given as:

$$\vec{j}_s(\vec{r}) = -\frac{3n_s e^2}{4\pi m c \xi_0} \iiint d^3\vec{r}' \frac{[\vec{A}(\vec{r}') \cdot (\vec{r} - \vec{r}')] \cdot (\vec{r} - \vec{r}')}{|\vec{r} - \vec{r}'|^4} e^{-|\vec{r} - \vec{r}'|/\xi} \tag{6.80}$$

$$\xi^{-1} = \xi_0^{-1} + \eta l^{-1}; \quad \xi_0 = \frac{\hbar v_F}{\pi \Delta} \quad l = \infty \text{ for pure materials} \tag{6.81}$$

where the Pippard correlation length $\xi_0 \simeq \hbar \delta p = \hbar v_F / \delta E$ can be viewed as the size of the free electron wave packet in the energy range ($\varepsilon_F - \Delta$, $\varepsilon_F + \Delta$) near the Fermi surface of a metal, and l is the mean free path with respect to impurities.

For Type-I element superconductors, the London length $\lambda_L \sim 100\text{Å}$ is relatively short, while the Pippard length $\xi_0 \sim 1\ \mu\text{m}$ (the Fermi velocity $v_F \sim 10^8$ cm/s) is quite long; the Pippard penetration depth

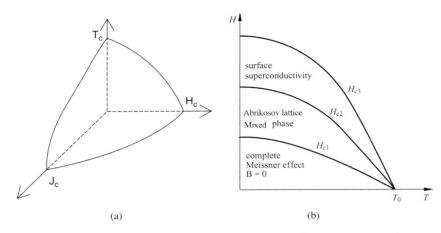

Figure 6.34 Phase diagram of (a) a Type-I superconductor or Pippard superconductor and (b) a Type-II superconductor or London superconductor

is $\delta = (\lambda_L^2 \xi_0)^{1/3}$, much longer than λ_L. For Type-II alloy superconductors, the Pippard length $\xi_0 \sim 100$Å is short (the Fermi velocity $v_F \sim 10^6$ cm/s due to complicated band structures), while the London length $\lambda_L \sim 1000$Å is quite long (the effective mass $m^*/m_e \gg 1$); the penetration depth is λ_L for pure samples and $\delta = \lambda_L (\xi_0/l)^{1/2}$ in impure materials (based on Landau, Liftshitz, and Pitaevskii, 1980).

The last breakthrough in the phenomenological theory was the Ginzburg-Landau theory, established in 1950, together with A. A. Abrikosov's work on the flux vortex in 1957. The phase diagram of Type-II superconductors is shown in Figure 6.34(b): between H_{c1} and H_{c2}, the magnetic field gradually penetrates the superconductor to form thin threads of magnetic flux surrounded by vortex currents, and the superconductor is at the s/n mixed state; between H_{c2} and H_{c3}, only a thin surface layer is superconducting, and the ratio $H_{c3} : H_{c2} = 1.7 : 1$.

The Ginzburg-Landau theory was based on Landau's theory for second-order phase transition, in which the order parameter of superconductivity is chosen as:

$$|\Psi(\vec{r})|^2 = n_s/n; \quad \Psi(\vec{r}) = \sqrt{n_s/n}\, e^{i\Phi} \qquad (6.82)$$

Ψ can be viewed as the many-body wavefunction of superconducting electrons. The total free energy is given as the function of Ψ:

$$\mathcal{F} = F_{n0} + \int dv \left\{ \frac{1}{2m^*} \left|\left(-i\hbar \vec{\nabla} - \frac{e^*}{c}\vec{A}\right)\Psi\right|^2 \right.$$
$$\left. + a|\Psi|^2 + \frac{1}{2}b|\Psi|^4 + \frac{\vec{B}^2}{8\pi} \right\} \qquad (6.83)$$

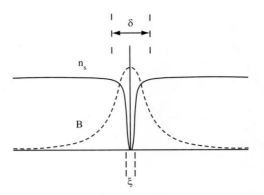

Figure 6.35 Flux tube characteristics of a Type-II superconductor (based on Landau and Liftshitz, 1987)

where the first term is the quantum kinetic energy of electrons in a magnetic field (the form of which will be discussed further in Chapter Seven). The effective mass $m^* = 2m$ and the effective charge $e^* = 2e$ are influenced by the microscopic BCS theory.

The Ginzburg-Landau equation can be found by the standard functional method:

$$\frac{1}{2m^*}\left(-i\hbar\vec{\nabla} - \frac{e^*}{c}\vec{A}\right)^2 \Psi + a\Psi + b|\Psi|^2\Psi = 0;$$
$$\vec{j}_s = \frac{e^*\hbar}{m^*}|\Psi|^2\vec{\nabla}\Phi - \frac{(e^*)^2}{m^*c}|\Psi|^2\vec{A} \qquad (6.84)$$

which is a complex non-linear differential equation. When the vector potential $\vec{A}$ is zero and the superconducting carrier density n_s is uniform, Eq. (6.84) is simplified and a and b can be related to the two-fluid model and the London theory:

$$a + b|\Psi_e|^2 = 0 \Rightarrow |\Psi_e|^2 = -\frac{a}{b} = \frac{\lambda_L^2(0)}{\lambda_L^2(T)};$$
$$\bar{\mathcal{F}}_s - F_{n0} = -\frac{1}{2}\frac{a^2}{b} = -\frac{H_c^2}{8\pi}$$
$$\Rightarrow a(T) = -\frac{H_c^2}{4\pi}\frac{\lambda_L^2(T)}{\lambda_L^2(0)}, \quad b(T) = -\frac{H_c^2}{4\pi}\frac{\lambda_L^4(T)}{\lambda_L^4(0)};$$
$$\delta = \left(\frac{m^*c^2 b}{8\pi e^*|a|}\right)^{1/2} \qquad (6.85)$$

where $a \sim \alpha(T - T_c)$ is true near T_c. Abrikosov linearized the Ginzburg-Landau equation by assuming a small fluctuation $\psi = \Psi - \Psi_e$ from the equilibrium near $\vec{r} = 0$:

$$-\frac{\hbar^2}{2m^*}\vec{\nabla}^2\psi + 2b|\Psi_e|^2\psi = 0 \Rightarrow \frac{\hbar^2}{2m^*}\vec{\nabla}^2\psi + 2a\psi = 0 \quad (6.86)$$

If ψ only depends on the radius ρ of a cylinder, the solution to the former equation is simply a Bessel function $K_0(\rho/\xi)$, which decays exponentially for large ρ:

$$\psi \sim e^{-\rho/\xi};$$
$$\xi = \left(-\frac{\hbar^2}{4m^*a}\right)^{1/2} = \left(\frac{\pi\hbar^2}{m^*H_c^2}\right)^{1/2}\frac{\lambda_L(0)}{\lambda_L(T)} \sim \frac{\xi_0}{(1-T/T_c)^{1/2}}$$

(6.87)

Therefore Abrikosov found that the size of the flux tube is in the order of the correlation length ξ. The flux quantum hc/e^* can be found by $\oint d\vec{l} \cdot \vec{j}_s = 0$.

6.3.3 Microscopic BCS Theory

The BCS theory was developed in 1957, six years after John Bardeen left Bell Laboratories and one year after Shockley, Bardeen, and Brattain won the Nobel Prize for their work on transistors. In 1957, Leon Cooper was a research associate at the University of Illinois, having obtained his Ph.D. at Columbia University in 1954. In the same year, John Robert Schrieffer began his graduate studies at Illinois under Professor John Bardeen, having just graduated from MIT with a B.S. obtained while working in Professor John C. Slater's research group.

The BCS theory originated from a discovery by Cooper in 1956. He proposed the idea of pairs of "bound" electrons to explain the ground state of materials at 0 K. These pairs are known as Cooper pairs. A Cooper pair contains two Bloch electrons $|\vec{k}\rangle$ and $|-\vec{k}\rangle$ near the Fermi surface of metal superconductors. In a superconducting current, the two electrons are at $|\vec{k}+\vec{q}/2\rangle$ and $|-\vec{k}+\vec{q}/2\rangle$ states respectively, and have a net drift velocity $\vec{v}_d = \hbar\vec{q}/m^*$ ($m^* = 2m$). The electrons will certainly interact with the phonons. However the drift velocity $\vec{v}_d$ is conserved for the Cooper pair at the superconducting phase: one electron will distort the lattice around it, creating an area of positive charge density around itself; another electron at some distance away in the lattice is then attracted to this positive charge, and is thus indirectly attracted to the first electron via the phonon, as illustrated in Figure 6.36.

The condensation of Cooper pairs is the foundation of the BCS theory. First, let us consider the one-pair problem, for which the

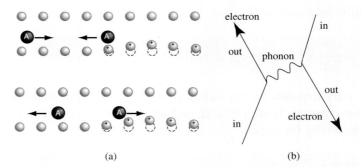

Figure 6.36 Cooper pair of two electrons with opposite velocities: (a) conservation of drift velocity with electron-phonon interactions; (b) abstract Feynman diagram (from www.hyperphysics.phy-astr.gsu.edu)

Schrodinger equation is:

$$\left(-\frac{\hbar^2}{2m}\vec{\nabla}_1^2 - \frac{\hbar^2}{2m}\vec{\nabla}_2^2 + V(\vec{r}_1, \vec{r}_2)\right)\Psi(\vec{r}_1, \vec{r}_2) = (2\varepsilon_F + E)\Psi(\vec{r}_1, \vec{r}_2)$$

(6.88)

where $\varepsilon_F = \hbar^2 k_F^2 / 2m$ is the Fermi energy, and E is the binding energy of the Cooper pair. In a pair with zero drift velocity, the wave function is

$$\Psi(\vec{r}_1, \vec{r}_2) = \sum_{\vec{k}} a_{\vec{k}} \exp[i\vec{k}\cdot(\vec{r}_1 - \vec{r}_2)] = \sum_{\vec{k}} a_{\vec{k}} \exp[i\vec{k}\cdot\vec{R}] \quad (6.89)$$

Then the Schrodinger equation for the Cooper pair can be rewritten as:

$$(E + 2\varepsilon_F - 2\varepsilon_{\vec{k}})a_{\vec{k}} = \frac{1}{\Omega}\iiint d^3\vec{R}\, V(\vec{r}_1, \vec{r}_2)\Psi(\vec{r}_1, \vec{r}_2)e^{-i\vec{k}\cdot\vec{R}}$$

$$= \sum_{\vec{k}'} V_{\vec{k}\vec{k}'} a_{\vec{k}'} \quad (6.90)$$

$$V_{\vec{k}\vec{k}'} = \frac{1}{\Omega}\iiint d^3\vec{R}\, e^{i\vec{k}'\cdot\vec{R}} V(\vec{r}_1, \vec{r}_2) e^{-i\vec{k}\cdot\vec{R}}$$

$$= \begin{cases} -\frac{V}{\Omega} & \varepsilon_{\vec{k}}, \varepsilon_{\vec{k}'} - \varepsilon_F \in \hbar\omega_D(-1, 1) \\ 0 & \text{otherwise} \end{cases} \quad (6.91)$$

where Ω is the total volume. A self-consistent equation can then be found:

$$C_{\vec{k}} = \sum_{\vec{k}'} V_{\vec{k}\vec{k}'} a_{\vec{k}'} = \sum_{\vec{k}'} V_{\vec{k}\vec{k}'} \frac{C_{\vec{k}}}{E + 2\varepsilon_F - 2\varepsilon_{\vec{k}}};$$

$$\Rightarrow 1 \simeq g(\varepsilon_F) \int_{-\hbar\omega_D}^{\hbar\omega_D} d\zeta \, \frac{V}{2\zeta + |E|} \quad (6.92)$$

where $\zeta = \varepsilon_{\vec{k}} - \varepsilon_F$ is the energy deviated from the Fermi surface, and $g(\varepsilon_F)$ is the EDOS of free electrons at the Fermi surface. The interaction potential $V_{\vec{k}\vec{k}'}$ and the binding energy E of the Cooper pair are less than zero; therefore $|E|$ is used in the equation above. The energy gap $\Delta(0)$ can then be obtained:

$$1 \simeq 2g(\varepsilon_F)V \int_0^{\hbar\omega_D} \frac{d\zeta}{2\zeta + |E|} = g(\varepsilon_F)V \ln\left(\frac{2\hbar\omega_D + |E|}{+|E|}\right)$$

$$\Delta(0) = |E| = \frac{2\hbar\omega_D}{\exp[1/g(\varepsilon_F)V] - 1} \simeq 2\hbar\omega_D \exp\left(-\frac{1}{g(\varepsilon_F)V}\right)$$

(6.93)

The critical magnetic field can also be found:

$$F_n - F_s = \frac{1}{2} g(\varepsilon_F) \Delta^2(0) = \frac{H_c^2(0)}{8\pi}; \Rightarrow H_c(0) = 2[\pi g(\varepsilon_F)]^{1/2} \Delta(0)$$

(6.94)

It can be seen that the isotope effect is naturally included in the BCS theory, because the Debye frequency $\omega_D \propto \sqrt{1/M}$ depends on the atomic number.

The theory for calculating the critical temperature has to deal with the creation and annihilation of Cooper pairs, therefore the second quantization method has to be utilized. This is too complex for readers of this book, so only the final equation for the critical temperature T_c is given here:

$$\frac{1}{g(\varepsilon_F)V} = \int_0^{\hbar\omega_D} \frac{d\zeta}{\zeta} \tanh\left(\frac{\zeta}{2k_B T_c}\right)$$

$$\Rightarrow k_B T_c = 1.14\hbar\omega_D \exp\left(-\frac{1}{g(\varepsilon_F)V}\right); \quad (6.95)$$

The derivation of the temperature dependence of the band gap $\Delta(T)$ is also complex. The band gap $\Delta(T)$, shown in Figure 6.37, is measured by SIN junction. The measured energy gap $2\Delta(0)$, critical temperature T_c,

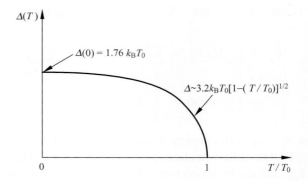

Figure 6.37 Energy gap versus temperature in the BCS theory

Table 6.14 Important parameters of the BCS theory of Type-I superconductors

	$2\Delta(0)/k_B$/K	T_c/K	$2\Delta(0)/k_B T_c$	Θ_D/K	$[g(\varepsilon_F)V]_{BCS}$	$[g(\varepsilon_F)V]_{exp}$
Zn	3.17	0.85	3.72	235	0.174	0.18
Cd	1.8	0.56	3.21	164	0.172	0.18
Hg	18.2	3.95	4.61	70	0.333	0.35
V	18	5.40	3.33	390	0.227	–
Nb	32	9.25	3.47	275	0.284	–
Al	4.3	1.18	3.64	375	0.170	0.18
In	12	3.41	3.52	109	0.284	0.29
Tl	8.45	2.38	3.55	100	0.258	0.27
Sn	13	3.72	3.49	195	0.244	0.25
Pb	29	7.20	4.02	96	0.367	0.39

and scaled interaction parameter $[g(\varepsilon_F)V]_{exp}$ for the Type-I superconductors are listed in Table 6.14. The BCS predictions $2\Delta(0) = 3.52 k_B T_c$ and $[g(\varepsilon_F)V]_{BCS}$ calculated by Eq. (6.95) agree quite well with the measured values in Table 6.14.

$2\Delta(0)/k_B T_c$ of mercury and lead deviate from the expected value 3.52 in the BCS theory, because they are strong-correlated superconductors. However, the isotope effect still holds for mercury, as seen in Figure 6.38. In fact, the strong-correlated condensed matter theory, based on the quantum field theory, can explain the superconducting behavior of mercury and lead very accurately.

This brings us to the end of this chapter. However, there are other interesting transport phenomena that have not been discussed in detail in this chapter. For example, the Mott insulators, such as La_2CuO_4 used in the preparation of high-Tc superconductors, would be expected to be

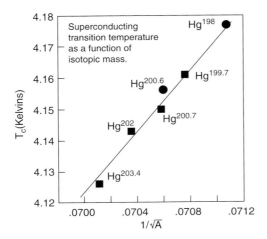

Figure 6.38 Isotope effect: the critical temperature is proportional to the $1/\sqrt{A}$, where A is the atomic mass (Maxwell, 1950, Reynolds et al., 1950)

half-filled conductors under conventional band theories, but in reality they are insulators. This phenomenon was partially explained by Sir Nevill Francis Mott: the band gap in a Mott insulator is due to electron-electron interactions, which are not considered properly in band theories or DFT. This is another example of strong-correlation condensed matter physics.

Summary

This chapter has analyzed the electrical properties of conductors, semiconductors, and superconductors, based on the band theories introduced in Chapter Five. The structure, processing, and electronic properties of semiconductor devices, including diodes and triodes, have also been introduced.

1. Conductivity: electric current is caused by the electrons in the half-filled energy bands. Therefore the metals, semiconductors, and insulators can be classified with zero, small, and large energy gaps at the Fermi surface. The conductivity of metals is mainly determined by the electron-phonon scattering, except that at very low temperatures, impurities also play a role.
2. Semiconductors (basics): semiconductors usually have diamond or zinc blende structures, with 0 to 3 eV energy gaps. The carriers in semiconductors are electrons and holes, and their effective masses are 0.01 to 1.1 times the bare electron mass. The resistivity of intrinsic or doped semiconductors depends on both the majority carrier

density and the mobility. Near room temperature, the majority carrier density in the doped semiconductor is almost a constant, but the mobility is proportional to $T^{-3/2}$.

3. Semiconductors (devices): in a p-n junction, there is a depletion layer with zero carrier density, and the ionized electric charges form a capacitor-like structure. The I-V curve of the p-n junction is influenced by both the majority carriers (V > 0) and the minority carriers (V ≪ 0). Depending on the band structure, a metal-semiconductor junction can be an Ohm-resistor or a p-n junction. The metal-oxide-semiconductor transistor is an amplifier, and the gate voltage controls the carrier densities beneath.

4. Superconductors: the basic phenomena of the superconducting phase are zero-resistance, zero magnetic field, and the existence of flux tubes. The traditional superconductors are metals, and among them, the niobium alloys are outstanding practical superconductor materials; the high-Tc superconductors are the doped perovskite-type ceramics, but the theory is not yet fully developed. The most famous superconductor theories are the two-fluid model, the London theory, the Ginzberg-Landau theory, and the Bardeen-Cooper-Schrieffer theory. The traditional superconductors are well-explained by the G-L theory and the BCS theory, which state that the paired electrons are the carriers moving without resistance even with the electron-phonon scattering.

References

Ashcroft, N. W., and Mermin, N. D. (1976). *Solid State Physics*, Holt, Rinehart and Winston, New York.
Burns, G. (1985). *Solid State Physics*, Academic Press, Orlando.
Chelikowsky, R., and Cohen, M. L. (1976). "Nonlocal Pseudopotential Calculations for the Electronic Structure of Eleven Diamond and Zinc-blende Semiconductors," *Phys. Rev. B*, Vol. 14, 556–582.
Chu, J. H. (2005). *Narrow Band Semiconductor Physics* (in Chinese), Science Press, Beijing.
de Gennes, P. G. (1989). *Superconductivity of Metals and Alloys*, Addison-Wesley, New York.
del Alamo, J. A., and Swanson, R. M. (1987). "Modeling of Minority Carrier Transport in Heavily Doped Silicon Emitters," *Solid State Electron*, Vol. 30, 1127–1136.
Dexter, R. N., Zeiger, H. J., and Lax, B. (1956). "Cyclotron Resonance Experiments in Silicon and Germanium," *Phys. Rev.*, Vol. 104, 637–644.
Feng, D., and Jin, G. (2003). *Condensed Matter Physics* (in Chinese), Higher Education Press, Beijing.
Grove, A. S. (1967). *Physics and Technology of Semiconductor Devices*, Wiley, New York.
Han, R. S. (1998). *High-Tc Superconductor Physics* (in Chinese), Beijing University Press, Beijing.

Huang, K., and Han, R. (1988). *Solid State Physics* (in Chinese), Higher Education Press, Beijing.
Khoshenevisan M., Pratt Jr. W. P., Schroeder P. A., and Steenwyk S. D. (1979). "Low-Temperature Resistivity and Thermoelectric Ratio of Copper and Gold ," *Phys. Rev. B*, Vol. 19, 3873–3878.
C. Kittel, C. (1986). *Introduction to Solid State Physics*, Wiley, New York.
Landau, L. D., Liftshitz, E. M., and Pitaevskii, L. P. (1980), translated by Sykes, J. B., and Kearsley, M. J. *Statistical Physics, Part 2*, Pergamon Press, Oxford.
Little, W. A., and Parks, R. D. (1962). "Observation of Quantum Periodicity in the Transition Temperature of a Superconducting Cylinder," *Phys. Rev. Lett.*, Vol. 9, 9–12.
Lyden, H. A. (1964). "Measurement of the Conductivity Effective Mass in Semiconductors using Infrared Reflection," *Phys. Rev.*, Vol. 134, A1106–A1112.
Maxwell, E. (1950). "Isotope Effect in the Superconductivity of Mercury". *Phys. Rev.*, Vol. 78, 477.
Pearson, G. L., and Bardeen, J. (1949). "Electrical Properties of Pure Silicon and Silicon Alloys containing Boron and Phosphorus," *Phys. Rev.*, Vol. 75, 865–883.
Persson, C., and Lindefelt, U. (1996). "Detailed Band Structure for 3C-, 2H-, 4H-, 6H-SiC, and Si around the Fundamental Band Gap," *Phys. Rev. B*, Vol. 54, 10257–10260.
Reynolds C. A., Serin, B., Wright, W. H., and Nesbitt, L. B.(1950). "Superconductivity of Isotopes of Mercury". *Phys. Rev.*, Vol. 78, 487.
Seitz, F. (1940). *The Modern Theory of Solids*, McGraw-Hill, New York.
Shockley, W. (1953). "Cyclotron Resonances, Magnetoresistance, and Brillouin Zones in Semiconductors," *Phys. Rev.*, Vol. 90, 491.
Schrieffer, J. R. (1983). *Theory of Superconductivity*, Addison-Wesley, New York.
Sze, S. M. (1981). *Physics of Semiconductor Devices*, John Wiley and Sons, New York.
"Cooper Pairs." Nave, C. R. Department of Physics and Astronomy, Georgia State University, City of Atlanta, USA. 8 April 2006. <http://hyperphysics.phy-astr.gsu.edu/hbase/solids/coop.html>
"The Nobel Prize in Physics 2003." Nobel Prize Organization, City of Stockholm, Sweden. 31 March 2006. <http://nobelprize.org/nobel_prizes/physics/laureates/2003/public.html>
"Nanostructures in our Daily Life." Breitzer, J. Department of Chemistry, University of Wisconsin, City of Madison, USA. 25 March 2006. <http://www.chem.wisc.edu/courses/801/Spring00/Ch1_3.html>
"Electronic Devices and Circuits." Yang, W. Department of EECS, Harvard University, City of Boston, USA. 25 March 2006. <http://www.deas.harvard.edu/courses/es154/>
"Exploring Materials Engineering: Semiconductor Materials". Pizzo, P. P. Department of Chemical and Materials Engineering, San Jose State University, City of San Jose, USA. 20 March 2006. <http://www.engr.sjsu.edu/WofMatE/Semiconductors.htm>
"Semiconductors on NSM." Physico-Technical Institute of the Russian Academy of Sciences, City of St Petersburg, Russia. 3 February 2006. <http://www.ioffe.rssi.ru/SVA/NSM/Semicond/>
"Semiconductor Materials." Ruzyllo, J. Department of Electrical Engineering and Materials Research Institute, Pennsylvania State University, City of University Park, USA. 15 January 2006. <http://www.semi1source.com/materials/>
"Superconductor." Eck, J. electronics engineer, USA. 15 April 2006. <http://www.superconductors.org/>
"Hotline Applied Superconductivity." Institut für Festkörperphysik, Technische Universitat München, City of München, Germany. 28 March 2006. <http://wwwifp.fzk.de/ISAS/Hottline/oct98/jj_hts_other.htm>

Exercises

6.1 According to Matthiessen's rule, the resistivity of metals is mainly caused by electron-phonon scattering and defects. Find (from other data sources) the resistivity of Cu at 0K and at room temperature, and use this data to estimate the defect-related mean free path l_d and the phonon-related mean free path l_F at 0K and room temperature respectively.

6.2 Using the tight-binding energy band expression in Chapter Five and Eq. (6.6) in this chapter,

  a. calculate the average mean effective mass m^* in a 1D tight-binding band.
  b. analyze the relationship between m^* and the band width.

6.3 Using the thermal conductivity data in Table 6.1, the Debye temperature data in Table 6.6, and Eq. (6.13), calculate the phonon mean free path l of Si, Ge, and GaAs.

6.4 The valence band of a semiconductor can be described by the form $E = -10^{-37}(k/m^{-1})^2 \text{J}$. If an electron at state $\vec{k} = 10^9 \text{m}^{-1} \hat{e}_x$ is taken away from the fully-filled valence band,

  a. effective mass of the hole
  b. effective wavevector of the hole
  c. hole velocity
  d. hole energy.

6.5 The band gap of Cu_2O is similar to GaP or 3C-SiC. Why is it that GaP and 3C-SiC are regarded as semiconductor materials, but Cu_2O is not?

6.6 Comparing the conductivity effective mass in Eq. (6.18) and the DOS effective mass listed in Table 6.6, which one is larger? Explain your conclusion using the energy bands of semiconductors shown in Figure (6.8).

6.7 In Table 6.8, the coefficients describing intrinsic semiconductor density are listed. What are the main physical quantities affecting the coefficients a and b?

6.8 In Table 6.9, the ionized energy of doped silicon is listed for different impurities. Use the formulae for the ionized energy given in Table 6.9 to calculate the effective mass m^*. What are the main physical quantities that affect m^*? Compare m^* with other effective masses of silicon.

6.9 Using the basic physical properties of InSb listed in Table 6.1 and Table 6.6, Estimate the:

  a. ionized energy E_d
  b. ground state orbital radius

c. donor density N_D when the ground state orbitals start to overlap with each other. What are the ratios of N_D/n_i and N_D/N_0?

6.10 If there is only one kind of impurity atom with $N_D = 10^{15}$ cm^{-3} from family V in an n-type silicon, and the measured carrier density at 40 K is $n = 10^{12}$ cm^{-3}, estimate the ionized energy E_d.

6.11 a. If phosphorous-doped silicon is purified to $N_D = 10^{12}$ cm^{-3}, estimate the temperatures at the border of intrinsic/saturation region and saturation/freeze-out region.
   b. If phosphorous-doped silicon is highly doped to $N_D = 10^{19}$ cm^{-3}, estimate the temperatures at the border of intrinsic/saturation region and saturation/freeze-out region.

6.12 There are three As-doped germanium ($E_d = 0.0127$ eV) and the impurity densities N_D/cm^{-3} are 5.5E16, 1.7E15, 1.4E14 respectively. Draw the $\log n - 1/T$ curve for the three extrinsic semiconductors by computer.

6.13 In Figure 6.18, the four impurity densities N_D/cm^{-3} are 1E17, 1.25E18, 1.7E19, and 2.6E20 respectively. Calculate the carrier density n, the Hall coefficient R_H, the electron mean free path l_c, and the resistivity ρ at room temperature for the first two impurity densities.

6.14 At room-T, if the n-type silicon wafer has an impurity density of $N_D = 10^{16}$ cm^{-3}, and the doping density in the p-layer is $N_A = 10^{18}$ cm^{-3}, find the intrinsic potential ϕ, the depletion layer width d, and the depletion capacitance C_0/A.

6.15 In a p-n junction, if the negative current is 3 μA and the negative bias field is 0.15 V, calculate the positive current with the positive bias field 0.15 V.

6.16 Use the data for carrier lifetimes in Figure 6.22 to estimate the diffusion lengths L_n, L_p in Si, Ge, and As. What is the importance of the diffusion length in the design of p-n junction devices?

6.17 Use Eq. (6.57) to draw the I-V curve of the Schottky diode, and label the characteristic values in the plot.

6.18 Why is there a threshold voltage V_b in the I-V amplification curve of transistor, i.e., what is the physical meaning of this voltage?

6.19 In the two-fluid model of superconductors, the balance between the free-electron specific heat and the superconducting specific heat is considered. However, the situation for lead is somewhat different:
   a. Estimate its free electron specific heat near T_c using the data in Table (5.7).
   b. Estimate the atomic vibration specific heat near T_c using the Debye model.
   c. Which term, a. or b., is larger? Could Eq. (6.62) be directly used for lead?

6.20 Using the metal electron density n listed in Table (5.1) of Chapter Five:

    a. Give the relationship between the London length λ_L and the normalized superconducting density $|\Psi|^2 = n_s/n$. Note: real data and units should be given for all parameters.

    b. When the London length is in the range of $100 - 1000\text{Å}$, what is the range of n_s/n?

6.21 Using the data for Al given in Table (6.14), and the results of the Sommerfeld model in Chapter Five, calculate the attraction potential constant V for Al.

Magnetic Properties of Solids

7

> **ABSTRACT**
> - Spin: quantum mechanical origin of magnetism
> - Landau levels: electron gas in a magnetic field
> - Magnetism: four classes based on spin orientations
> - Interactions: fundamental particles interacting with spins

The magnetic properties of solids have been recognized since ancient times. One of the earliest uses of magnetic materials was in compasses, which first appeared in China around 100 B.C. during the Han Dynasty, and which have been used in maritime navigation since the 12th century.

In the late 16th century, Sir William Gilbert wrote a book, *De Magnete*, in which he described the earth itself as a huge magnet. In 1832, the great mathematician Carl Friedrich Gauss proposed the concept of a "system of units," following his analysis of geomagnetic fields by examining the mechanical forces acting on a compass. The Gauss units, or cgs units, subsequently came to form an important part of the basis for future scientific research.

Electric and magnetic phenomena were thought to be totally independent until the early 19th century. In 1820, Hans Christian Oersted, a professor of science at Copenhagen University, arranged a science demonstration for his friends and students at home. Unexpectedly, he found that a compass placed near an electric wire would turn at a right angle of the current. Thus the correlation between electricity and magnetism was discovered, and it was subsequently expressed mathematically in Ampere's law in 1822. In 1831, Michael Faraday found that an electric current could be induced by the variation of a magnetic field. Coulomb's law, Ampere's law, and Faraday's induction law were summarized in the Maxwell equations in 1864 and Hertz's differential form in 1890.

Magnetism in matter is a very complex phenomenon. In 1778, Anton Brugmans found that bismuth was repelled by a magnet, which

258 Chapter 7 Magnetic Properties of Solids

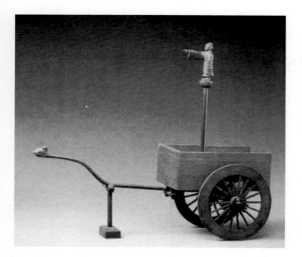

Figure 7.1 The original Chinese compass and a ship's compass

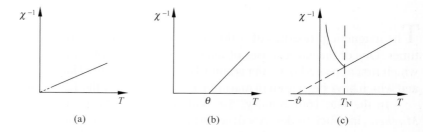

Figure 7.2 Inverse susceptibility versus temperature: (a) paramagnetism; (b) ferromagnetism; (c) antiferromagnetism

was effectively the discovery of diamagnetism (although this term was not used until later). In 1845, Faraday investigated the magnetization of non-metallic materials, and coined the terms "paramagnetic" and "diamagnetic" to differentiate the forces of attraction and repulsion to and from a magnet. He also found that, among all the known elements at that time, only iron, cobalt, and nickel were ferromagnetic. This was the first attempt to classify magnetism in matter. About 100 years later, a French physicist, Louis Néel, clearly distinguished ferromagnetism, antiferromagnetism, and ferrimagnetism (ferrites), thus completing the classification of magnetism in matter, and was awarded a Nobel Prize in 1970.

In 1895, Pierre Curie found that diamagnetism was independent of temperature; however, in paramagnetic materials or ferromagnetic materials in the paramagnetic phase, the inverse susceptibility χ^{-1} was proportional to temperature T. In 1905, sixty years after Faraday's classification, Paul Langevin used these statistics to explain paramagnetism and diamagnetism. He thought that diamagnetism was related to the current loop, and that paramagnetism was due to a randomly oriented permanent magnet of molecules, which explained Curie's law. In 1907, Pierre-Ernest Weiss, (Louis Néel's mentor), suggested the concept of an "intrinsic magnetic field" to explain ferromagnetism; this assumption seemed to be arbitrary at the time, but was later proved by the spin of electrons in solids.

The quantum mechanical explanation of magnetism is based on the intrinsic spin of fundamental particles. In 1896, Pieter Zeeman, an assistant of Kamerlingh Onnes and Hendrik Lorentz at Leyden University, discovered the magnetic splitting of spectral lines. In 1916, Arnold Sommerfeld modified Bohr's orbits to elliptical orbits, explained the Zeeman splitting, and paved the way for the discovery of spin quantum numbers l and m. In 1921, Otto Stern and Walther Gerlach proved that the basic quantum of atomic spin was simply the Planck constant $\hbar$. In 1925, Samuel Goudsmit and George E. Uhlenbeck found that the spin of an electron was $\hbar/2$, and Wolfgang Pauli formulated his Exclusion Principle; after which the number of elements in the periodic table could naturally be explained. In 1928, Werner Heisenberg explained the Weiss field by the exchange interaction, a concept borrowed from the Heitler-London theory, and Paul Dirac proposed the relativistic quantum theory of electrons. Zeeman, Heisenberg, Dirac, Stern, and Pauli won Nobel Prizes in Physics in 1902, 1932, 1933, 1943, and 1945 respectively.

Another quantum base of magnetism is the Landau level of free electrons in a magnetic field, proposed by Lev Landau in 1930, which explained the Shubnikov-de Haas oscillation, the de Haas-van Alphen effect, and the quantum Hall effect.

A new branch of magnetism concerns the interactions of fundamental particles with spins. In 1944, electron spin resonance (ESR) was discovered by E. K. Zavoisky at Kazan State University in the former Soviet Union. In 1945, Felix Bloch from Stanford University and E. M. Purcell's group from Harvard University developed nuclear spin resonance (NMR), a pure electromagnetic method for the study of nuclear moments in solids, liquids, or gases. NMR is one of the most accurate experiments in physics, and Bloch and Purcell won the Nobel Prize in Physics in 1952 for their contributions. Neutron elastic and inelastic scattering, developed by C. G. Shull and B. N. Brockhouse respectively, are also basic methods for analyzing magnetic structure and low temperature spin waves or magnons.

In industrial applications, (ferro)magnetic materials are classified as permanent magnets, soft magnetic materials, and magnetic recording materials, which provide static magnetic fields, magnetic field paths (circuits), and the correspondence between the average magnetization and 0–1 signals respectively. Micromagnetic theory explains the magnetic domain, hysteresis, and characteristics of magnetic devices. The micromagnetic theory was first proposed by Lev Landau in 1935, but it is not discussed in this book.

7.1 Quantum Mechanical Origin of Magnetism

The quantum mechanical theory of magnetism originated from the "basic" magnetic moment in the Bohr model. In 1821–1822, Andre-Marie Ampere proposed the idea of "molecular current" to explain the ferromagnetic moment. In 1905, Paul Langevin pointed out that a circular current would perform diamagnetically or non-magnetically in an external magnetic field; therefore the ferromagnetic "permanent magnet" could not be the circular current. In 1916, Arnold Sommerfeld modified the circular Bohr orbits into the elliptical form by introducing quantum number l, where $l\hbar = mvr$ is the angular momentum; as a result, the magnet moment of a quantum orbit could be found from Ampere's idea:

$$\mu = \frac{IA}{c} = \frac{\pi r^2}{c} \frac{-e}{2\pi r/v} = -\frac{e(mvr)}{2mc} = -\frac{e\hbar}{2mc} l = -\mu_B l \quad (7.1)$$

where the Bohr magneton $\mu_B = 9.27 \times 10^{-21}$ erg/G is the quantum of moment. In an external magnetic field, the energy level would split due to the Zeeman energy:

$$E = -\vec{\mu} \cdot \vec{H} = \mu_B H l_z; \quad (l_z = l, l-1, \ldots, -l) \quad (7.2)$$

Therefore the Zeeman splitting of spectra lines was simply proportional to $\mu_B H$. It is interesting to note that both the normal Zeeman effect with respect to odd number splitting levels (integer l) and the anomalous Zeeman effect with respect to even number splitting levels (half-integer l) are found, as shown in Figure 7.3.

The quantum numbers in the Zeeman effect have to be explained by Schrodinger's equation in the presence of a magnetic field. The Hamiltonian of a single charged-particle in Schrodinger's equation can be derived from the covariant energy of a relativistic particle in the electromagnetic potential $(\phi, \vec{A})$:

$$(E - q\phi)^2 = m^2 c^4 + c^2 \left(\vec{p} - \frac{q}{c}\vec{A}\right)^2 \quad (7.3)$$

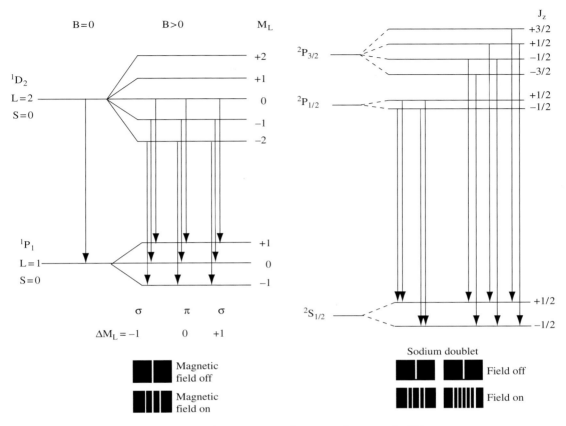

Figure 7.3 Zeeman splitting: (a) normal l = integer; (b) anomalous l = half-integer

In the non-relativistic limit, the energy of an electron can be expressed as:

$$E \simeq mc^2\left(1 + \frac{\left(\vec{p} + \frac{e}{c}\vec{A}\right)^2}{2m^2c^2}\right) - e\phi = mc^2 + \frac{\left(\vec{p} + \frac{e}{c}\vec{A}\right)^2}{2m} - e\phi$$

(7.4)

The Zeeman energy related to electron spin is naturally given by Paul Dirac's relativistic quantum mechanics, developed in 1928:

$$\Delta E = -\vec{\mu}_s \cdot \vec{H} = g_0 \mu_B \vec{s} \cdot \vec{H} = g_0 \mu_B H s_z \quad (g_0 = 2.002319304386)$$

(7.5)

where $s_z = \pm\frac{1}{2}$ is the electron spin component along the external magnetic field.

The total multi-electron Hamiltonian can be found following Eqs. (7.4) and (7.5):

$$\mathcal{H} = \sum_i \left[\frac{(\vec{p}_i + \frac{e}{c}\vec{A}(\vec{r}_i))^2}{2m} + v(\vec{r}_i) + \frac{1}{2}\sum_{j\neq i} \frac{e^2}{r_{ij}} + g_0 \mu_B \vec{s}_i \cdot \vec{H} \right]$$

(7.6)

This Eq. (7.6) is called the Pauli Hamiltonian or Schrodinger-Pauli Hamiltonian, as described by Wolfgang Pauli in 1927. The eigenstates of the Pauli Hamiltonian are complex. Its applications at different limits are the key to explaining the quantum mechanical origin of magnetism.

7.1.1 Single Atom Approximation: Atomic Moments

The magnetic moment of an atom or an ion in solids can be found using the single atom approximation of the Pauli Hamiltonian in Eq. (7.6). In an atom, the total Hamiltonian can be expanded into the zeroth, first, and second order of the uniform magnetic field $\vec{H} = H\hat{e}_z$ using the Landau vector potential $\vec{A}$:

$$\begin{aligned}
\mathcal{H} &= \sum_i \left[\frac{(\vec{p}_i - \frac{e}{2c}\vec{r}_i \times \vec{H})^2}{2m} + v(\vec{r}_i) + \frac{1}{2}\sum_{j\neq i}\frac{e^2}{r_{ij}} + g_0\mu_B \vec{s}_i \cdot \vec{H} \right] \\
&= \sum_i \left[\frac{\vec{p}_i^2}{2m} + v(\vec{r}_i) + \frac{1}{2}\sum_{j\neq i}\frac{e^2}{r_{ij}} \right] + \sum_i \mu_B(\vec{l}_i + g_0\vec{s}_i)\cdot \vec{H} \\
&\quad + \sum_i \frac{e^2}{8mc^2}(\vec{r}_i \times \vec{H})^2 \\
&= \mathcal{H}^{(0)} + \Delta \mathcal{H}^{(1)} + \Delta \mathcal{H}^{(2)}
\end{aligned}$$

(7.7)

where the Landau vector potential $\vec{A}$ in the uniform magnetic field is utilized:

$$\begin{aligned}
\vec{A} &= -\frac{1}{2}\vec{r} \times \vec{H}; \quad (\vec{\nabla} \times \vec{A})_\alpha = \epsilon_{\alpha\beta\gamma}\partial_\beta A_\gamma \\
&= -\frac{1}{2}\epsilon_{\alpha\beta\gamma}\epsilon_{\gamma\kappa\eta}\partial_\beta(r_\kappa H_\eta) = H_\alpha.
\end{aligned}$$

(7.8)

In Eq. (7.7), $v(\vec{r}_i) = -e\phi(\vec{r}_i)$ is the single-electron potential; $\vec{l}_i = m\vec{r}_i \times \vec{p}_i$ is the orbital angular momentum operator of the i'th electron; $\mathcal{H}^{(0)}$ is the total Hamiltonian of the atom in zero $\vec{H}$ field; $\Delta\mathcal{H}^{(1)}$ is the first-order perturbation induced by a magnetic field with the magnitude of

$\mu_B H \sim 10^{-4}$ eV for $H = 1$ T; and $\Delta\mathcal{H}^{(2)}$ is the second-order perturbation, whose order of magnitude is $(\mu_B H)^2/(e^2/a_B) \sim 10^{-9}$ eV for $H = 1$ T.

The first order perturbation Hamiltonian $\Delta\mathcal{H}^{(1)}$ is directly related to the atomic magnetic moment operator $\vec{\mu}_J = -\mu_B \sum_i (\vec{l}_i + g_0 \vec{s}_i)$. In the ground state, if the atomic magnetic moment is not zero, the second order perturbation $\Delta\mathcal{H}^{(2)}$ can obviously be neglected; however, in some atoms or ions, $\langle \vec{\mu}_J \rangle$ does vanish, in which case the second order perturbation $\Delta\mathcal{H}^{(2)}$ is determinant for the magnetism. The eigen-values of $\Delta\mathcal{H}^{(1)}$ should just be the Zeeman energy in Eq. (7.2):

$$\Delta E^{(1)} = \langle 0|\Delta\mathcal{H}^{(1)}|0\rangle = -\langle 0|\vec{\mu}_J|0\rangle \cdot \vec{H} = g_J \mu_B H J_z \quad (7.9)$$

where $|0\rangle$ stands for the multi-electron ground state of an atom with $H = 0$, and the corresponding wavefunction of $|0\rangle$ is in the form of the Slater determinant given by Eq. (2.3) in Chapter Two. $\langle 0|\vec{\mu}_J|0\rangle$ is just the permanent atomic magnet first mentioned by Paul Langevin in 1905.

To determine the ratio between the atomic magnetic moment and the Bohr magneton—the gyromagnetic factor g_J in Eq. (7.9)—the following equivalence should be used:

$$\vec{J} = \vec{L} + \vec{S}; \quad \vec{L} = \sum_i \vec{l}_i; \quad \vec{S} = \sum_i \vec{s}_i \quad (7.10)$$

$$\langle 0|(\vec{L} + 2\vec{S}) \cdot \vec{J}|0\rangle = \langle 0|g_J \vec{J} \cdot \vec{J}|0\rangle = g_J J(J+1)$$

$$\langle 0|(\vec{J} + \vec{S}) \cdot \vec{J}|0\rangle = \langle 0|\vec{J}^2 + \frac{1}{2}[\vec{S}^2 + \vec{J}^2 - (\vec{J} - \vec{S})^2]|0\rangle$$

$$= J(J+1) + \frac{1}{2}[S(S+1) + J(J+1) - L(L+1)]$$

$$\Rightarrow g_J = 1 + \frac{S(S+1) + J(J+1) - L(L+1)}{2J(J+1)}.$$

(7.11)

g_J is also called the Landé g-factor, as it was first described by Alfred Landé in 1921. In an atom or ion, when the total orbital quantum number $L = 0$, the Landé g-factor reaches the limit $g_J = g_0 = 2$; when the total spin $S = 0$, the Landé g-factor $g_J = 1$; and when $L > S = J$, the Landé g-factor is minimized.

The quantum numbers L, S, and J for the ground state of an atom or ion can be determined using Hund's rule. In 1926, the German scientist Friedrich Hund applied Heisenberg's new quantum mechanics to molecules. Hund's rule of maximum multiplicity, (simplified as Hund's rule), states that a larger total angular momentum quantum number L usually makes the atom or molecule more stable (i.e., in a lower energy state). The reason for Hund's rule is that the various occupied spatial

$\{l_i\}$-orbits create a larger average distance between electrons to reduce electron-electron repulsion energy. Therefore, following Hund's rule, the sequence of electrons filling in an atomic suborbital shell $|n, l\rangle$ is:

$$s_z = -\frac{1}{2}; \quad l_z = -l, -l+1, \ldots 0, \ldots, l-1, l \tag{7.12}$$

$$s_z = +\frac{1}{2}; \quad l_z = -l, -l+1, \ldots 0, \ldots, l-1, l \tag{7.13}$$

At the ground state, the orbital and spin quantum numbers are defined as $L = |\sum_i l_z^i|$ and $S = |\sum_i s_z^i|$ respectively, and the atomic spin quantum number J is determined as:

$$J = |L - S|; \ (m \le 2l+1); \quad J = |L + S|; \ (m > 2l+1); \tag{7.14}$$

where m is the total number of electrons in the l'th sub-shell. In a fully-filled shell, $L = S = J = 0$ is certainly true; in a half-filled shell, the symbol term is $^{2S+1}X_J$, where the label X is written as S, P, D, F, G, H, I respectively for $L = 0, 1, 2, 3, 4, 5, 6$. The magnetic moment of an atom is determined by the half-filled shells.

In an insulator, all the anions and cations in families 1A, 2A, 3A, 1B, and 2B have full-shells, thus they have zero magnetic moments (of electrons); the cations of transition metals and rare-earth elements have a half-filled d-shell and f-shell respectively, and are the main contributors of atomic moments, as listed in Table 7.1.

In an atom or ion, the Zeeman splitting is influenced by both the ground state $|0\rangle$ and the first excited state $|1\rangle$. The angular momentum of a photon is $\pm\hbar$, and there is a selection rule between $|0\rangle$ and $|1\rangle$ to satisfy the conservation of angular momentum. The Zeeman splitting in the spectrum and the selection rules are

$$\begin{aligned}\Delta E &= g_J^1 \mu_B H J_z^1 - g_J^0 \mu_B H J_z^0 \\ &\simeq \left(\frac{g_J^1 J_z^1 - g_J^0 J_z^0}{2}\frac{H}{1\text{T}}\right) \times 0.11\,\text{meV}\end{aligned} \tag{7.15}$$

$$\Delta J = 0, \pm 1; \quad \Delta J_z = 0, \pm 1; \quad \Delta S = 0; \quad \Delta L = \pm 1. \tag{7.16}$$

where $\Delta J_z = 0$ is not allowed when the incident light is parallel to $\vec{H}$.

In the normal Zeeman splitting in Figure 7.3(a), both the excited state and the ground state have $S = 0$ and $J = L$, so the selection rules are $\Delta J_z = \Delta L_z = 0, \pm 1$, and the normal Zeeman splitting only has three values:

$$\Delta E_n = 0, \pm 0.06(H/1\text{T})\,\text{meV} \tag{7.17}$$

In the anomalous Zeeman splitting of sodium in Figure 7.3(b), both the excited state and the ground state have $S = 1/2$ (one electron), but the excited state $S = 1/2$, $L = 1$ splits into $J = L + S$ and $J =$

7.1 Quantum Mechanical Origin of Magnetism

Table 7.1 Symbol terms of transition metal and rare-earth ions

d-shell ($l = 2$)

m	$m_l = 2$,	1,	0	−1,	−2,	S	L	J	Symbol	Ion
1	↓					1/2	2	3/2	$^2D_{3/2}$	Ti^{3+}
2	↓	↓				1	3	2	$^3F_2$	V^{3+}
3	↓	↓	↓			3/2	3	3/2	$^4F_{3/2}$	Cr^{3+}
4	↓	↓	↓	↓		2	2	0	$^5D_0$	Cr^{2+}
5	↓	↓	↓	↓	↓	5/2	0	5/2	$^6S_{5/2}$	Fe^{3+}, Mn^{2+}
6	↓↑	↑	↑	↑	↑	2	2	4	$^5D_4$	Fe^{2+}
7	↓↑	↓↑	↑	↑	↑	3/2	3	9/2	$^4F_{9/2}$	Co^{2+}
8	↓↑	↓↑	↓↑	↑	↑	1	3	4	$^3F_4$	Ni^{2+}
9	↓↑	↓↑	↓↑	↓↑	↑	1/2	2	5/2	$^2D_{5/2}$	Cu^{2+}
10	↓↑	↓↑	↓↑	↓↑	↓↑	0	0	0	$^1S_0$	

f-shell ($l = 3$)

m	$m_l = 3$,	2,	1,	0,	−1,	−2,	−3	S	L	J	Symbol	g_j	Ion
0								0	0	0	$^1S_0$	0	La^{3+}
1	↓							1/2	3	5/2	$^2F_{5/2}$	6/7	Ce^{3+}
2	↓	↓						1	5	4	$^3H_4$	4/5	Pr^{3+}
3	↓	↓	↓					3/2	6	9/2	$^4I_{9/2}$	8/11	Nd^{3+}
4	↓	↓	↓	↓				2	6	4	$^5I_4$	3/5	pM^{3+}
5	↓	↓	↓	↓	↓			5/2	5	5/2	$^6H_{5/2}$	2/7	Sm^{3+}
6	↓	↓	↓	↓	↓	↓		3	3	0	$^7F_0$	−	Eu^{3+}
7	↓	↓	↓	↓	↓	↓	↓	7/2	0	7/2	$^8S_{7/2}$	2	Gd^{3+}
8	↓↑	↑	↑	↑	↑	↑	↑	3	3	6	$^7F_6$	3/2	Tb^{3+}
9	↓↑	↓↑	↑	↑	↑	↑	↑	5/2	5	15/2	$^6H_{15/2}$	4/3	Dy^{3+}
10	↓↑	↓↑	↓↑	↑	↑	↑	↑	2	6	8	$^5I_8$	5/4	Ho^{3+}
11	↓↑	↓↑	↓↑	↓↑	↑	↑	↑	3/2	6	15/2	$^4I_{15/2}$	6/5	Er^{3+}
12	↓↑	↓↑	↓↑	↓↑	↓↑	↑	↑	1	5	6	$^3H_6$	7/6	Tm^{3+}
13	↓↑	↓↑	↓↑	↓↑	↓↑	↓↑	↑	1/2	3	7/2	$^2F_{7/2}$	8/7	Yb^{3+}
14	↓↑	↓↑	↓↑	↓↑	↓↑	↓↑	↓↑	0	0	0	$^1S_0$	0	Lu^{3+}

$|L − S|$ states due to the spin-orbital interaction. In this case, the selection rules are $\Delta L = \pm 1$, $\Delta J_z = 0, \pm 1$. $g_J^1 \neq g_J^0$, thus the anomalous Zeeman splitting has $(2J^1 + 1)(2J^0 + 1) = 4$ lines when $J^1 = J^0 = 1/2$, and has $(2J^1 + 1)(2J^0 + 1) − 2 = 6$ lines when $J^1 = 3/2$, $J^0 = 1/2$.

It should be mentioned that the nucleus also has a magnetic moment:

$$\mu_z^n = g(e\hbar/2m_p c)J_z = g\mu_n J_z \tag{7.18}$$

Table 7.2 Quantum numbers and gyromagnetic factors in isolated Na atoms

States	S	L	J	g_J	E in magnetic field $\vec{H}$	$2J+1$
$^2S_{1/2}$	1/2	0	1/2	2	J_z^0 $(H/1\text{T}) \times 0.11$ meV	2
$^2P_{1/2}$	1/2	1	1/2	2/3	J_z^1 $(H/1\text{T}) \times 0.04$ meV	2
$^2P_{3/2}$	1/2	1	3/2	4/3	J_z^1 $(H/1\text{T}) \times 0.08$ meV	4

where a nuclear magneton μ_n is 1/1863 of the Bohr magneton μ_B. For protons and neutrons, the spin $J = 1/2$, and the g factor equals 5.5856912 and -3.8260837 respectively, which were only partially understood by QCD (quantum chromo-dynamics).

7.1.2 Free Electron Approximation: Landau Levels

In insulators, the former single atom approximation of the Pauli Hamiltonian gives a reasonably good description of the atomic moments. In metals, the behavior of the valence electron gas in a magnetic field obviously cannot be explained by the single atom approximation, so a new approach to the Pauli Hamiltonian—the free electron approximation—is used. In a magnetic field, the energy of the free electron Fermi gas is different from the cases in the Sommerfeld model; the relevant model here is the Landau levels, which were first formulated by Lev Landau in 1930.

Following Eq. (7.7), Schrodinger's equation for a free electron in the magnetic field is

$$\left[\frac{\vec{p}^2}{2m} + \frac{\mu_B H}{\hbar}(xp_y - yp_x) + \frac{e^2 H^2}{8mc^2}(x^2+y^2) + g_0 \mu_B H s_z\right]\psi = E\psi \quad (7.19)$$

If the wavefunction ψ is rewritten in the form

$$\psi(x, y, z) = \phi(x, y, z) \exp\left(i\frac{eH}{2c\hbar}xy\right) \quad (7.20)$$

Eq. (7.19) can be simplified as:

$$\left[\frac{\vec{p}^2}{2m} + 2\frac{\mu_B H}{\hbar}xp_y + \frac{e^2 H^2}{2mc^2}x^2 + g_0 \mu_B H s_z\right]\phi = E\phi \quad (7.21)$$

The modified wavefunction ϕ only feels the potential in the x-direction, therefore there should only be plane waves in the y-direction and

z-direction:
$$\phi(x, y, z) = \lambda(x) \exp(i(k_y y + k_z z)) \tag{7.22}$$

Eq. (7.19) is finally simplified into a one-dimensional form:

$$\left[\frac{p_x^2}{2m} + \frac{1}{2m}\left(\hbar k_y + \frac{eH}{c}x\right)^2 + \frac{\hbar^2 k_z^2}{2m} + g_0 \mu_B H s_z\right] \lambda(x) = E \lambda(x), \tag{7.23}$$

which is analogous to Schrodinger's equation for a harmonic oscillator centered at $x_0 = -(\hbar c/eH)k_y$. The eigen-energies and the eigen-wavefunction of a free electron in a magnetic field must have the forms:

$$E_n(k_z) = (n + \frac{1}{2} + s_z)\hbar \omega_c + \frac{\hbar^2 k_z^2}{2m}; \quad \omega_c = \frac{eH}{mc} \tag{7.24}$$

$$\lambda_n(x) = \frac{1}{\pi^{1/4}(2^n n! a_H)^{1/2}} \exp\left(-\frac{(x-x_0)^2}{2a_H^2}\right) H_n\left(\frac{x-x_0}{a_H}\right), \tag{7.25}$$

where ω_c is the classical cyclotron frequency of a charged particle in the uniform magnetic field, $\hbar \omega_c = g_0 \mu_B H$, and $a_H = \sqrt{\hbar/m\omega_c}$. It should be noted that the form of the wavefunction depends on the choice of the gauge for $\vec{A}$.

In a two-dimensional free electron gas, the perpendicular momentum $\hbar k_z$ is zero, so the eigen-energies really become the Landau levels:

$$E_n = (n + \frac{1}{2} + s_z)\hbar \omega_c \quad s_z = \pm\frac{1}{2} \tag{7.26}$$

The relationship between Landau levels and the magnetic field H can be plotted visually as the "Landau fan," as shown in Figure 7.4(a). The number of quantum states N_L and the filling factor v of a Landau level are:

$$N_L = \frac{L_x}{\Delta x_0} = \frac{L_x}{(\hbar c/eH)\Delta k_y} = L_x L_y \frac{eH}{hc};$$
$$v = \frac{N}{N_L} = \frac{Nhc}{eHL_x L_y}; \quad \left(\Delta k_y = \frac{2\pi}{L_y}\right). \tag{7.27}$$

where L_x and L_y are sizes of the thin film. In an increasing magnetic field, valence electrons in the free electron gas will fully fill ..., 4, 3, 2, 1 Landau levels, as seen in Figure 7.4(a). Thus the resistivity will have periodicity: for example, in Figure 7.4(c), the resistivity ρ_{xx} has periodicity with respect to the Landau-level filling factor v.

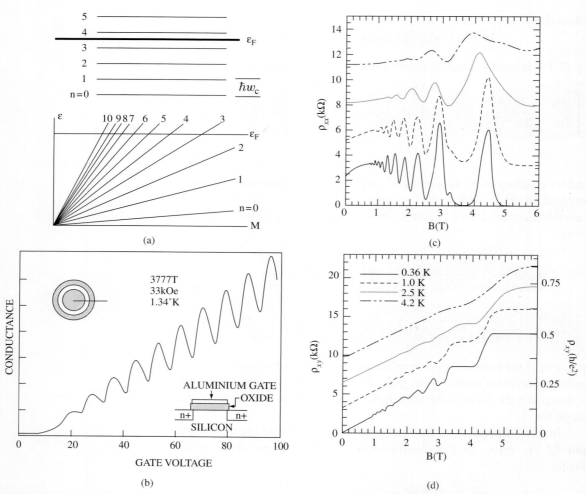

Figure 7.4 Shubnikov-de Haas Oscillation: (a) "Landau fan"; (b) conductivity versus gate voltage or ε_F in silicon (based on Fowler, Fang, Howard, and Stiles, 1966); (c)(d) ρ_{xx} and ρ_{xy} versus H-field in metals (based on Knobel, Samarth, Harris, and Awschalom, 2002)

This periodicity of ρ_{xx} is also called the Shubnikov-de Haas oscillation, as mentioned in Chapter Six for the measurement of effective mass in semiconductors:

$$\varepsilon_F = \left(n + \frac{1}{2} + s_z\right)\hbar\omega_c;$$

$$\rightarrow \quad m^* = \frac{e\hbar H}{c(\varepsilon_F^n - \varepsilon_F^{n-1})} \quad \text{or} \quad \frac{e\hbar}{\varepsilon_F c(H_n^{-1} - H_{n-1}^{-1})} \quad (7.28)$$

The measured effective mass m^* depends on the crystal orientation with respect to $\vec{H}$. In semiconductors, the Fermi energy ε_F is controlled by the carrier density or the gate voltage in MOS, as seen in Figure 7.4(b), such that $m^* \perp \vec{H}$ can be measured. In metals, ε_F is fixed, therefore the magnetic field H must be varied.

The transverse resistivity ρ_{xy} in Figure 7.4(d) shows plateaus, which are a result of the quantum Hall effect, and are also related to the filling factor of Landau levels. In 1978, when German physicists Klaus von Klitzing and T. Englert measured the Hall voltage rather than the Hall current, they found the flat Hall plateaus. The Hall conductance quantum e^2/h was verified in 1980; five years later, Klaus von Klitzing was awarded the Nobel Prize in Physics for the discovery of the quantum Hall effect. The conductance contributed by electrons in a fully-filled Landau level is

$$G_{xy} = \frac{\partial}{\partial H}\frac{j_x H}{E_y} = \frac{\partial}{\partial H}\frac{-1}{R_H} = ec\frac{\partial}{\partial H}\frac{N_L}{L_x L_y} = ec\frac{e}{ch} = \frac{e^2}{h} \quad (7.29)$$

This conductance quantum is universal in quantum systems, and first appeared in the Landauer formula for mesoscopic conductance, formulated by Rolf Landauer in 1957.

In 1982, Daniel C. Tsui, Horst L. Stoermer, and A. C. Gossard discovered the existence of Hall steps with respect to the rational fractional filling factor $\nu = p/q$, which is now known as the fractional quantum Hall effect:

$$G_{xy} = \frac{p}{q}\frac{e^2}{h}; \quad \nu = \frac{p}{q}, \quad (7.30)$$

Tsui, Stoermer, and Robert B. Laughlin shared the Nobel Prize in 1997. The theory of Laughlin is beyond the scope of this book, but it has a similar form to Eq. (7.25).

7.2 Categories of Magnetism

The categories of magnetism, namely diamagnetism, paramagnetism, ferromagnetism, antiferromagnetism, and ferrimagnetism, can be traced back to Anton Brugmans's exploration of diamagnets in 1778 and Michael Faraday's classification in 1845.

The mechanisms of magnetism can be universally understood by the multi-electron Hamiltonian in Eq. (7.6); however it is almost impossible to solve the Hamiltonian accurately as a whole, so a series of approximations for ions and electrons in solids has to be used to understand magnetism. Diamagnetism appears in zero-spin atoms or ions; diamagnetism and ferromagnetism correspond to different space orientations of non-zero atomic magnetic moments; and valence electron gas also contributes to magnetism.

7.2.1 Diamagnetism

A diamagnet is matter with negative susceptibility, which is repulsed by a magnet. In a "normal" diamagnet, all atoms or ions have zero magnetic moment; however in a "perfect diamagnet," i.e., a superconductor, the mechanism of diamagnetism is totally different, as discussed in Chapter Six. Ionic crystals, covalent crystals, and atomic and molecular solids are diamagnetic.

In a magnetic field $\vec{H} = H\hat{e}_z$, the second-order perturbation energy of an ion at the ground state can be found by Eq. (7.7):

$$\Delta E_0^{(2)} = \frac{e^2}{8mc^2}\left\langle 0\left|\sum_i (x_i^2 + y_i^2)\right|0\right\rangle H^2 + \sum_{n\neq 0}\frac{|\langle n|L_z + g_0 S_z|0\rangle|^2}{E_0 - E_n}H^2 \tag{7.31}$$

The second term in Eq. (7.31) is zero in most cases, so it will not be considered for the moment. In a weak magnetic field, the ground state of a diamagnet is invariant with $\vec{H}$, therefore the Larmor theorem of diamagnetic moments can be found by the first term in Eq. (7.31):

$$\mu_d = -\frac{\partial \Delta E_0^{(2)}}{\partial H} = -\eta H;$$

$$\eta = \frac{e}{2mcH}m\omega_L\left\langle 0\left|\sum_i(x_i^2 + y_i^2)\right|0\right\rangle;$$

$$\omega_L = \frac{eH}{2mc} \tag{7.32}$$

where ω_L is called the Larmor frequency. The diamagnetic susceptibility of a material with the atomic density $n = N/V$ can be found by Eq. (7.31):

$$\chi_d = \frac{n\mu_d}{H} = -\frac{ne^2}{4mc^2}\left\langle 0\left|\sum_i(x_i^2 + y_i^2)\right|0\right\rangle \simeq -\frac{ne^2}{6mc^2}\left\langle 0\left|\sum_i r_i^2\right|0\right\rangle \tag{7.33}$$

At the ground state, the fully-filled shells are rotationally symmetric in total; thus the former replacement $\sum_i x_i^2 + y_i^2 \to \frac{2}{3}\sum_i r_i^2$ is reasonable.

In compounds, the Larmor diamagnetic susceptibility per mole of ions is:

$$\chi_d^{\text{mole}} = -\frac{N_A a_B^3}{6}\left(\frac{e^2}{\hbar c}\right)^2 Z\langle (r/a_B)^2\rangle$$

$$= -0.79 Z\langle (r/a_B)^2\rangle \times 10^{-6}\,\text{cm}^3/\text{mol} \tag{7.34}$$

The diamagnetic susceptibility χ_d^{mole} is roughly proportional to the atomic number $A \simeq Z$, which is in the order of $10^{-6} - 10^{-5}\,\text{cm}^3/\text{mol}$,

Table 7.3 Estimated molar diamagnetic susceptibility of ions (Based on Mayers, 1952)

Alkali/Nobel ions	Li$^+$	Na$^+$	K$^+$	Rb$^+$	Cs$^+$	Cu$^+$	Au$^+$
$\chi_d^{mole}/(10^{-6} cm^3/mol)$	−0.7	−8.2	−15.9	−23.6	−34.8	−14	−48
Halogen/Oxygen ions	F$^-$	Cl$^-$	Br$^-$	I$^-$		O^{2-}	
$\chi_d^{mole}/(10^{-6} cm^3/mol)$	−8.6	−22.6	−32.9	−47.7		−12	
Alkali-earth ions		Mg^{2+}	Ca^{2+}	Sr^{2+}	Ba^{2+}		
$\chi_d^{mole}/(10^{-6} cm^3/mol)$		−7.4	−10.5	−18.8	−29.9		

as listed in Table 7.3. Among the neighboring elements in the same period, the heavier cations/anions have less diamagnetic susceptibility due to the smaller cation/anion radii.

It should be emphasized that, for different ionic crystals, the local electron densities are different for the same kind of ions, therefore the values listed in Table 7.3 may vary from one solid to another. For example, $-\chi_d^{mole}/(10^{-6} cm^3/mol)$ of F$^-$ could range from 7 to 17; $-\chi_d^{mole}/(10^{-6} cm^3/mol)$ of Na$^+$ could range from 3.7 to 12.5; and $-\chi_d^{mole}/(10^{-6} cm^3/mol)$ of Mg^{2+} could range from 3 to 14.

Diamagnetism of Free Electron Gas

The previous discussions on diamagnetism are suitable for discretely-distributed ions. In metals, the valence electrons have totally different behaviors from ions. Interestingly, the free electron gas has both diamagnetism and paramagnetism. The paramagnetism of free electron gas will be discussed later.

In Eq. (7.24), the Landau levels of 3D free electron gas $E_n(k_z)$ are a function of the wavevector k_z along the magnetic field. The partition function of 3D free electron Fermi gas (with N electrons) in the landau levels $\varepsilon = \left(n + \frac{1}{2}\right)\hbar\omega_c + \frac{\hbar^2 k_z^2}{2m^*}$ is (based on Seitz, 1940):

$$\ln Z = 2 \sum_{n=0}^{n_L} N_d \int_{-\infty}^{\infty} dk_z \ln[1 + e^{-(\varepsilon - \mu)/k_B T}];$$

$$\simeq 2 \sum_{n=0}^{n_L} N_d \int_{-\infty}^{\infty} dk_z \exp\left\{\frac{\mu}{k_B T} - \frac{1}{k_B T}\left(\left(n + \frac{1}{2}\right)\hbar\omega_c + \frac{\hbar^2 k_z^2}{2m^*}\right)\right\}$$

$$= 2 \sum_{n=0}^{n_L} N_d \frac{4\sqrt{2m^* k_B T}}{3h} \left(\frac{\mu}{k_B T} - \frac{\hbar\omega_c}{k_B T}\left(n + \frac{1}{2}\right)\right)^{3/2}$$

$$\simeq \frac{n_L N_d}{k_B T}\left[\frac{2\pi \hbar^2 k_F^2}{5m^*} - \frac{\hbar^2 \omega_c^2}{8\hbar^2 k_F^2/m^*}\right] \qquad (7.35)$$

where μ is the chemical potential or Fermi energy, and the relation $(\mu - \varepsilon)/k_B T \gg 1$ has been used in the former derivation; $N_d = N_L = L_x L_y(eH/hc)$ is the degeneracy of a Landau level; and $n_L = \nu = \text{int}(\mu/\hbar\omega_c)$ is the total number of filled Landau levels. The diamagnetic susceptibility of free electron gas can thus be found:

$$\chi_d^e = \frac{\partial^2(k_B T \ln Z)}{\partial H^2} = -n_L N_d \frac{(\hbar\omega_c/2H)^2}{\hbar^2 k_F^2/m^*} = -n_L N_d \left(\frac{m_e}{m^*}\right)^2 \frac{\mu_B^2}{2\varepsilon_F} \tag{7.36}$$

The order of $-\chi_d^e$ of electrons is also around $N_A \mu_B^2/\varepsilon_F \sim 10^{23-40+11} \sim 10^{-6}$ cm^3/mol. In the next section, we will see that the magnitude of the paramagnetic susceptibility in metals is greater than the magnitude of the diamagnetic susceptibility in Eq. (7.36).

7.2.2 Paramagnetism

In 1905, Paul Langevin used the Maxwell-Boltzman statistics to calculate the paramagnetic susceptibility of a randomly orientated permanent magnet $\vec{\mu}_a$:

$$\begin{aligned}\bar{\mu} &= \frac{\int d\phi \int d\theta \sin\theta \mu_a \cos\theta \exp(\mu_a H \cos\theta/k_B T)}{\int d\phi \int d\theta \sin\theta \exp(\mu_a H \cos\theta/k_B T)} \\ &= \mu_a \frac{\int_{-1}^{1} du\, u \exp[(\mu_a H/k_B T)u]}{\int_{-1}^{1} du \exp[(\mu_a H/k_B T)u]} = \mu_a \mathcal{L}\left(\frac{\mu_a H}{k_B T}\right),\end{aligned} \tag{7.37}$$

where $\mathcal{L}(x) = \coth(x) - \frac{1}{x}$ is the Langevin function. At low-field or high-temperature limits, one can expand the Langevin function as $\mathcal{L}(x) \simeq x/3$ ($x \ll 1$), and obtain

$$M = n\bar{\mu} \simeq \frac{n\mu_a^2}{3k_B T} H; \quad \chi = \frac{n\mu_a^2}{3k_B T} \tag{7.38}$$

which explains Curie's law for paramagnetic susceptibility $\chi^{-1} \propto T$.

In 1927, Léon Brillouin modified the Langevin theory based on the quantized Zeeman energy $\Delta E^{(1)} = g_J \mu_B H J_z$ and quantum statistics (Brillouin, 1927), where the partition function per atom/ion in a paramagnet is:

$$Z = \sum_{J_z=-J}^{J} \exp(-g_J \mu_B H J_z/k_B T) = \frac{\sinh[g_J \mu_B H(2J+1)/(2k_B T)]}{\sinh[g_J \mu_B H/(2k_B T)]}$$

$$\tag{7.39}$$

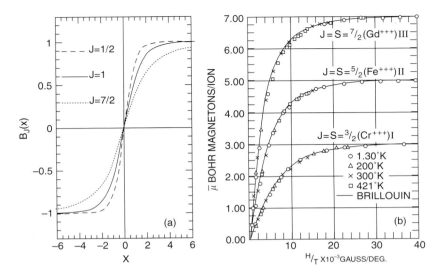

Figure 7.5 (a) Brillouin function $B_J(x)$ with different quantum number J; (b) measured magnetic moment M/n of ions versus H/T compared with $B_J(x)$ (based on Henry, 1953)

If there are N atoms or ions in a paramagnet, the average magnetization is:

$$M = \frac{N}{V} \frac{\partial(k_B T \ln Z)}{\partial H} = n g_J \mu_B J\, B_J(x); \quad x = \frac{g_J \mu_B J H}{k_B T} \quad (7.40)$$

$$B_J(x) = \frac{2J+1}{2J} \coth\left(\frac{2J+1}{2J}x\right) - \frac{1}{2J}\coth\left(\frac{1}{2J}x\right) \quad (7.41)$$

where $M_s = n g_J \mu_B J$ is the saturation magnetization; and $B_J(x)$ is the Brillouin function. It is a quantum mechanically modified version of the Langevion function $\mathcal{L}(x)$.

The Brillouin function can be expanded in the region $|x| \ll 1$ as:

$$\begin{aligned} B_J(x) &\simeq \frac{2J+1}{2J}\left(\frac{2J}{(2J+1)x} + \frac{1}{3}\frac{2J+1}{2J}x\right) \\ &\quad - \frac{1}{2J}\left(\frac{2J}{x} + \frac{1}{3}\frac{1}{2J}x\right) = \frac{J+1}{3J}x \end{aligned} \quad (7.42)$$

Then the quantum mechanical version of the paramagnetic susceptibility is

$$\begin{aligned} \chi = \frac{M}{H} &= n g_J \mu_B J \frac{B_J(x)}{H} \simeq n g_J \mu_B J \frac{J+1}{3J} \frac{g_J \mu_B J}{k_B T} \\ &= \frac{n(g_J \mu_B)^2 J(J+1)}{3 k_B T} \end{aligned} \quad (7.43)$$

If we compare Eq. (7.38) with Eq. (7.43), the quantum atomic paramagnetic moment is $\mu_a = g_J \mu_B \sqrt{J(J+1)}$, which is different from the atomic magnetic moment derived from the saturation magnetization $M_s/n = g_J \mu_B J$.

The atomic paramagnetic moment $\mu_a = g_J \mu_B \sqrt{J(J+1)}$ is a measurable quantity via the susceptibility; therefore it is possible to check the correctness of the Brillouin theorem of paramagnetism and Hund's rule. The magnetic paramagnetic behavior of a substance is mainly due to the ions with half-filled orbits.

In Table 7.4, the theoretical and experimental atomic paramagnetic moments are compared. For the rare-earth ions, the measured μ_a^{exp}/μ_B agrees very well with $g_J \sqrt{J(J+1)}$, except for Pm^{3+}, Sm^{3+}, and Eu^{3+}. For the transition metal ions, the measured μ_a^{exp}/μ_B basically agrees with $2\sqrt{S(S+1)}$, and the contribution of the orbital momentum L seems to

Table 7.4 Calculated and measured paramagnetic moments of ions (based on Kittel, 1986, and Stapleton and Jeffries, 1961)

Ion	$|0\rangle$	g_J	$g_J\sqrt{J(J+1)}$	$2\sqrt{S(S+1)}$	μ_a^{exp}/μ_B
Ti^{3+}, V^{4+}	$^2D_{3/2}$	0.80	1.55	1.73	1.7
V^{3+}	$^3F_2$	0.67	1.63	2.83	2.8
V^{2+}, Cr^{3+}, Mn^{4+}	$^4F_{3/2}$	0.40	0.77	3.87	3.8
Cr^{2+}, Mn^{3+}	$^5D_0$	–	0	4.90	4.9
Mn^{2+}, Fe^{3+}	$^6S_{5/2}$	2.00	5.92	5.92	5.9
Fe^{2+}	$^5D_4$	1.50	6.71	4.90	5.4
Co^{2+}	$^4F_{9/2}$	1.33	6.63	3.87	4.8
Ni^{2+}	$^3F_4$	1.25	5.59	2.83	3.2
Cu^{2+}	$^2D_{5/2}$	1.20	3.55	1.73	1.9
La^{3+}	$^1S_0$	–	0	0	0
Pr^{3+}	$^2F_{5/2}$	0.86	2.54	1.73	2.4
Pr^{3+}	$^3H_4$	0.80	3.58	2.83	3.6
Nd^{3+}	$^4I_{9/2}$	0.73	3.63	3.87	3.6
Pm^{3+}	$^5I_4$	0.60	2.68	4.90	1.7
Sm^{3+}	$^6H_{5/2}$	0.28	0.83	5.92	1.5
Eu^{3+}	$^7F_0$	–	0	6.93	3.4
Gd^{3+}	$^8S_{7/2}$	2.00	7.94	7.94	8.0
Tb^{3+}	$^7F_6$	1.50	9.72	6.93	9.5
Dy^{3+}	$^6H_{15/2}$	1.33	10.64	5.92	10.6
Ho^{3+}	$^5I_8$	1.25	10.61	4.90	10.4
Er^{3+}	$^4I_{15/2}$	1.20	9.58	3.87	9.5
Tm^{3+}	$^3H_6$	1.17	7.56	2.83	7.3
Yb^{3+}	$^2F_{7/2}$	1.14	4.53	1.73	4.5

disappear somehow, except that μ_a^{exp}/μ_B of Fe^{2+}, Co^{2+}, and Ni^{2+} stays between the two theoretical limits $g_J\sqrt{J(J+1)}$ and $2\sqrt{S(S+1)}$.

The quenching of the orbital angular momentum in a transition metal ion is caused by spin-orbit coupling, such that the total orbital quantum number L for an ion is no longer a "good" one. When L is a good quantum number, the orbits at ground state $|0\rangle$ must be invariant under a simple rotation of the multi-electron wavefunction. In a free ion, the electrons can be transferred between the orbits in a sub-shell, which are all degenerate, but there will still be partial orbital quenching as the probability distributions in different orbits are not identical. In transition metals, the electrons in the d_{xy}, d_{yz}, d_{zx}, $d_{x^2-y^2}$, and d_{z^2} orbits cannot be transferred from one orbit to another by a simple rotation, as the electron-electron repulsive interaction is not identical for different orbits under certain filling situations of d-shells. This process is also called the quenching of L.

Pauli Paramagnetism of Free Electron Gas

In the section on diamagnetism, it was stated that free electron gas has both diamagnetism and paramagnetism. The diamagnetism is partly due to the fully-filled Landau levels. The paramagnetism is caused by the Zeeman splitting of the spin up/down electrons, which was first explored by W. Pauli.

The electron moment is in the opposite direction to the electron spin. The total energy of a free electron in a magnetic field $\vec{H} = H\hat{e}_z$ is,

$$E = K - \vec{\mu} \cdot \vec{H} = \frac{\hbar^2 \vec{k}^2}{2m} + 2\mu_B H s_z; \quad s_z = \pm\frac{1}{2}, \quad (7.44)$$

The energy of spin-up electrons increases by $\mu_B H$, therefore the total number of moment-down electrons must be decreased; the energy of spin-up electrons is reduced by $\mu_B H$, therefore the total number of

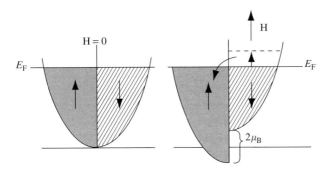

Figure 7.6 Zeeman splitting of the free electron energy (the arrows in bands refer to the magnetic moments of electrons)

moment-up electrons must be increased, as seen in Figure 7.6. The Pauli paramagnetism is caused by more spin-down electrons below the Fermi surface ε_F:

$$\chi_p^e = V\frac{M}{H} = V\frac{\mu_B \times (2\mu_B H)g(\varepsilon_F)/2}{H} = N\frac{3\mu_B^2}{2\varepsilon_F}; \quad (7.45)$$

The total specific susceptibility of the valence electron gas can be found by summarizing the Pauli paramagnetic susceptibility in Eq. (7.45) and the diamagnetic susceptibility of electrons in the fully-filled Landau levels in Eq. (7.36) (it should be noted that if the magnetic field is not very large, $n_L N_d \simeq N$ is usually satisfied):

$$\chi_e = \frac{\chi_p^e + \chi_d^e}{V} = \left[1 - \frac{n_L N_d}{3N}\left(\frac{m_e}{m^*}\right)^2\right]\frac{n/N_A}{\varepsilon_F/10\,\text{eV}} \times 4.85 \times 10^{-6}$$

(7.46)

If this formula is used to explain the susceptibility of solids (which is not accurate due to the free electron assumption), in normal metals, the valence electron gas must be paramagnetic; while in semiconductors, the carrier "gas" might be diamagnetic, because m^*/m_e is quite small in compound semiconductors.

To explain the susceptibility of simple metals, both the χ_e of valence electrons and the diamagnetic susceptibility of inner core electrons in Eq. (7.34) should be considered:

$$\chi_{tot} = \chi_e + \chi_i = \chi_e + \frac{n_a}{N_A}\chi_d^{mole} \quad (7.47)$$

If we utilize the Fermi energy of free electrons in Table 5.1, and let $m^* = m_e$, the specific susceptibility can be found for simple metals, as listed in Table 7.5.

It is obvious that, if the theory separately considers the valence and inner-core susceptibilities, it will not totally agree with the experiment,

Table 7.5 Calculated and measured specific susceptibility/(10^{-6}) of metals (based on Seitz, 1940, Stapleton and Jeffies, 1961, and Bitter, 1930)

Metal	χ_e	χ_i	χ^{exp}	Metal	χ_e	χ_i	χ^{exp}
Li	0.53	−0.05	0.27–2.04	Cu	0.64	−1.9	−0.76
Na	0.44	−0.36	0.49–0.63	Ag	0.57	−3.0	−2.1
K	0.35	−0.37	0.35–0.55	Au	0.57	−4.7	−2.9
Rb	0.33	−0.45	0.14–0.34	Mg	0.65	−0.22	0.95
Cs	0.31	−0.53	−0.19–0.41	Ca	0.53	−0.43	1.7

because the electron-electron repulsion and exchange are not considered. For the alkali and the alkali-earth metals, it seems that the susceptibility of the inner ion is "quenched," and only the valence electron is important. For the noble metals, the susceptibility of inner core electrons is important, in that the total susceptibility is diamagnetic.

de Haas-van Alphen Effect

Within the low-temperature/high-field limit $k_B T \leq \omega_c$, the oscillation of susceptibility in metals is called the de Haas-van Alphen oscillation, named after the two scientists who first observed the phenomenon in bismuth in 1931. This is a very important theory for determining the Fermi surface characteristics of metals. The oscillation period of susceptibility is proportional to the inverse magnetic field H^{-1}, which is related to Landau levels:

$$\varepsilon_F = i_{dHvA}\hbar\omega_c; \rightarrow i_{dHvA} = \frac{\varepsilon_F}{\hbar\omega_c} = \frac{\varepsilon_F c}{\hbar e}m_\perp^* H^{-1} \tag{7.48}$$

When i_{dHvA} is an integer, the susceptibility is minimized. The de Haas-van Alphen effect is anisotropic in crystals, because the perpendicular effective mass $m_\perp^* = \sqrt{m_1^* m_2^*} \perp \vec{H}$ varies with the orientation of $\vec{H}$.

The de Haas-van Alphen oscillation usually has one period, but sometimes two periods are found, as shown in Figure 7.7. These two periods can be used to determine the detailed structures of Fermi surfaces, such as the "neck" of the Fermi surface of the noble metals copper, silver, and gold, or the Fermi surfaces of the transition metals.

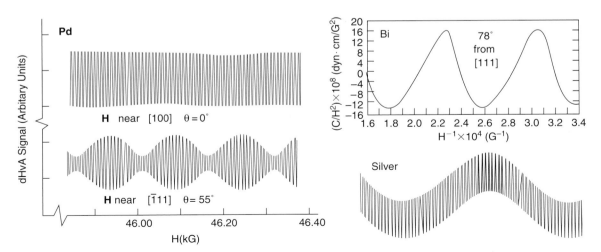

Figure 7.7 de Haas-van Alphen effect: normal oscillation (Bi,Pd) and oscillation with two periods (Pd,Ag) (based on Weiner, 1962, and Vuillemin, 1966)

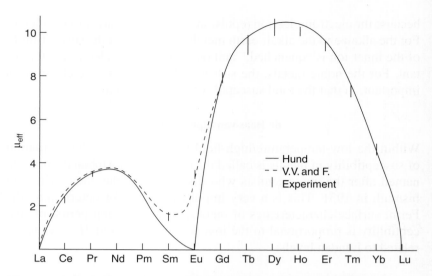

Figure 7.8 Paramagnetic moment of rare-earth ions with the Van Vleck paramagnetic correction (from Van Vleck's Nobel lecture in 1977)

Within the low-temperature or high-magnetic field limits, besides the oscillations of the resistivity (Shubnikov-de Haas oscillations) and the susceptibility (de Haas-van Alphen oscillation), other physical properties, like specific heat, also oscillate. All these effects can be explained by the periodic filling of Landau levels.

Van Vleck Paramagnetism

The measured paramagnetic susceptibility of rare-earth ions Pm^{3+}, Sm^{3+}, and Eu^{3+} deviates from the results of the Brillouin theory and Curie's law. John H. Van Vleck used the second order perturbation of $\Delta H^{(1)}$ to explain this phenomenon:

$$M_z = -\frac{\partial \Delta E}{\partial H} = -N\frac{\partial}{\partial H}\frac{|\langle 0|\mu_z H|1\rangle|^2}{E_0 - E_1} = 2N\frac{|\langle 0|\mu_z|1\rangle|^2}{E_1 - E_0}H = \chi_v H$$

(7.49)

The paramagnetic susceptibility χ_v is independent of temperature, and is called the Van Vleck paramagnetism. The deviations in the paramagnetic susceptibility of Sm^{3+} and Eu^{3+} ions are well explained, as shown in Figure 7.8.

7.2.3 Ferromagnetism

Ferromagnetism, which was first observed a long time ago, has only been explained quite recently. In fact it is one problem in physics which is still not fully understood.

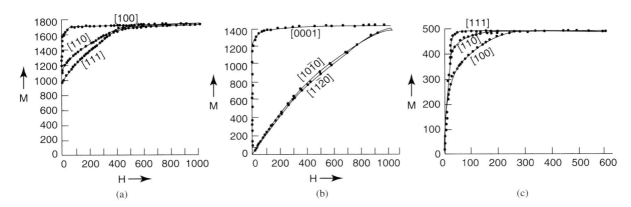

Figure 7.9 Magnetization process in a single crystal ferromagnet (based on Kaya, 1928)

In a ferromagnetic crystal, the spins of electrons spontaneously stay in a certain direction, which is called the easy axis $\hat{k}$. In iron, cobalt, and nickel single crystals, the magnetization processes along different crystal directions are shown in Figure 7.9, where the easy axes $\hat{k}$ are along [100], [0001], [111] directions respectively. Iron and nickel have cubic anisotropy (i.e., three easy axes) with positive/negative anisotropic energy, while the HCP cobalt has uniaxial anisotropy along the c-axis.

In polycrystal magnetic materials, the easy axis varies from one region to another, or from one crystal grain to another, therefore the magnetic properties can be very complex. Micromagnetics is the only theory available for analysis of domain structure and hysteresis of ferromagnetic materials, and the characteristics and the size-correlation of magnetic devices. However, as it is an "extended" rather than a "classical" solid state theory, it is not introduced here.

The classical theory of ferromagnetism is the Weiss theory, devised by Pierre-Ernest Weiss in 1907. Weiss introduced a large (unknown-source) intrinsic field to explain the spontaneous order of spins, which is known as a mean field theory.

Following the establishment of quantum mechanics, in 1928, Werner Heisenberg explained the Weiss field by exchange interaction; the Heisenberg model is now widely used in condensed matter physics. However, the exchange coupling constant in the Heisenberg model has different values in magnetic neutron scattering and low-temperature magnetization curves (to be discussed below), so it is not flawless.

Weiss Theory

For ferromagnetic materials, Weiss introduced an intrinsic molecular magnetic field H_E to align the spins below the Curie temperature. The Wiess field H_E is a mean field, where the Zeeman energy of $H_E(0)$ at

0 K should be equivalent to the thermal energy at T_c:

$$H_E = \lambda_E M; \quad gS\mu_B H_E(0) = k_B T_c \quad (7.50)$$

where S is used instead of J due to the quenching of the orbital momentum in transition metals. For example, the Curie temperature of iron is $T_c = 1043\,K$ and the average magnetic moment μ_a of an iron atom is $2.2\,\mu_B$. The Weiss field at 0 K can be estimated as:

$$H_E(0) = \frac{k_B T_c}{gS\mu_B}$$
$$= \frac{8.62 \times 10^{-5}\,(eV/K) \times 1043\,K}{2.2 \times 5.79 \times 10^{-9}\,(eV/G)} = 0.70 \times 10^7\,\text{Oe} \quad (7.51)$$

This Weiss field is huge compared to the dipolar magnetic field of the atomic moment $\mu_B/a^3 \simeq 10^{-20+24-1} = 10^3$ Oe, therefore the physical picture of the spontaneous ordering of ferromagnetism by interacting atom moments cannot be true.

Weiss modified the Langevin theory of paramagnetism in Eq. (7.38) by adding the Weiss field, thus obtaining the Curie-Weiss law for ferromagnetism:

$$\frac{M}{H_{tot}} = \frac{M}{H_{ext} + H_E} = \frac{M}{H_{ext} + \lambda_E M} = \frac{n\mu_a^2}{3k_B T}; \quad T > T_c$$
$$\chi = \frac{M}{H_{ext}} = \frac{n\mu_a^2}{3k_B(T-\theta)}; \quad \theta = \lambda_E \frac{n\mu_a^2}{3k_B} \quad (7.52)$$

where θ is the Weiss temperature or paramagnetic Curie temperature. The Weiss and Curie temperatures of ferromagnets are listed in Table 7.6.

The Weiss theory is quite successful at explaining the susceptibility of ferromagnetic materials in the paramagnetic phase, as shown in

Table 7.6 Characteristics of important ferromagnets (based on Seitz, 1940, and Kittel, 1986)

	Fe	Co	Ni	Gd	Dy	Ni$_3$Fe	CoFe
M_s/(emu/cc) at 0 K	1752	1446	512	2060	2920	–	–
M_s/(emu/cc) at 300 K	1707	1400	485	–	–	1007	1950
$\mu_a/(\mu_B)$ at 0 K	2.2	1.71	0.606	7.63	10.2	0.68/3	2.5
Easy axis	$\langle 100 \rangle$	[001]	$\langle 111 \rangle$	–	–	$\langle 111 \rangle$	$\langle 100 \rangle$
T_c/K	1043	1388	627	293	85	890	1256
θ/K	1037	1504	634	–	–	–	–

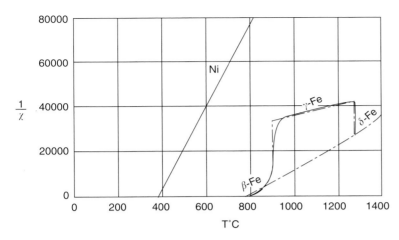

Figure 7.10 Inverse susceptibility of Ni and Fe (based on Kaya, 1928)

Figure 7.10. In nickel, when the temperature is above T_c, the inverse susceptibility $\chi^{-1} \propto (T - \theta)$ simply follows the Curie-Weiss law; in iron, there are two $\beta - \gamma$ and $\gamma - \delta$ phase transitions when the temperature is above T_c, but χ^{-1} of the β phase and δ phase are on the same $\xi(T - \theta)$, which shows the correctness of the Weiss theory. It should be emphasized that below the Curie temperature, the ferromagnetic susceptibility is multi-valued and cannot be well-defined; usually the initial permeability, the ratio B/H near the demagnetized state $M \simeq 0$, is used instead. The initial susceptibility is related to the chemical composition, microstructure, and geometry of the sample, and is totally different from the susceptibility at the paramagnetic phase described by the Curie-Weiss law.

Table 7.6 shows that the saturation magnetization M_s decreases at higher temperatures. This phenomenon can also be explained by the Weiss theory. In Eq. (7.40) of the Brillouin theory of paramagnetism, if the magnetic field H is the Weiss field H_E, the spontaneous magnetization M at ferromagnetic phase can be solved self-consistently:

$$M = (ng\mu_B J) B_J \left(\frac{g\mu_B J (\lambda_E M)}{k_B T} \right) \Rightarrow B_J(y) = \frac{k_B T}{n(g\mu_B J)^2 \lambda_E} y$$

(7.53)

When $T < T_c$, Eq. (7.53) has a non-zero solution $B_J(y_0) = (\eta T) y_0$. When $T \geq T_c$, Eq. (7.53) only has a zero trivial solution $y = 0$, as shown in Figure 7.11(b).

It was found by F. Tyler in 1930 that in the temperature range $T/T_c \in (0.1, 1)$, all the measured $M(T)$ curves of iron, cobalt, and nickel agreed well with the $M(T)$ solved by the Brillouin function with

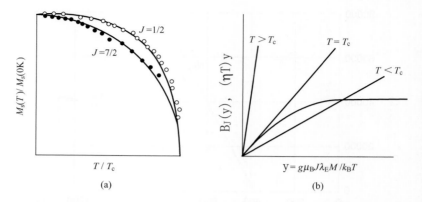

Figure 7.11 (a) $M(T)$ function for Fe,Co,Ni ($J = 1/2$) and for Gd,Dy ($J = 7/2$); (b) find $M(T)$ by the Brillouin function (based on Burns, 1985)

Table 7.7 Atomic spin S in ferromagnetic elements (based on Seitz, 1940, and Kittel, 1986)

	Fe	Co	Ni	Gd	Dy
μ_a^{exp}/μ_B at 0 K	2.2	1.71	0.606	7.63	10.2
S	1	1/2	1/2	7/2	5
$2S$	2	1	1	7	10
$2\sqrt{S(S+1)}$	2.83	1.73	1.73	7.93	10.95

$J = 1/2$; and the measured $M(T)$ of gadolinium and dysprosium agreed with the $M(T)$ calculated by $B_J(y)$ with $J = 7/2$. The atomic spin quantum number S could also be found from the measured average atomic magnetic moment μ_a^{exp}/μ_B, as listed in Table 7.7, where the spin quantum numbers S of iron, cobalt, and nickel did not totally agree with $J = 1/2$, but the spin quantum number S of gadolinium and dysprosium were not far from $J = 7/2$ in the Weiss theory.

In the ferromagnetic elements iron, gadolinium, and dysprosium, the experimental atomic moment is between the two values $2\mu_B S$ and $2\mu_B\sqrt{S(S+1)}$; in cobalt, μ_a^{exp} equals $2\mu_B\sqrt{S(S+1)}$; and in nickel, none of the integer or half-integer quantum numbers S fit μ_a^{exp} very well and the closest is $2\mu_B S$ with $S = 1/2$.

The disagreement between the theoretical and experimental μ_a shows the imperfectness of the Weiss theory. If the DFT-LDA band computation method is used, the average atomic moment of nickel is found to be $\mu_a = 0.54 \, \mu_B$, which is due to holes in the spin-down bands of nickel plotted in Figure 7.12.

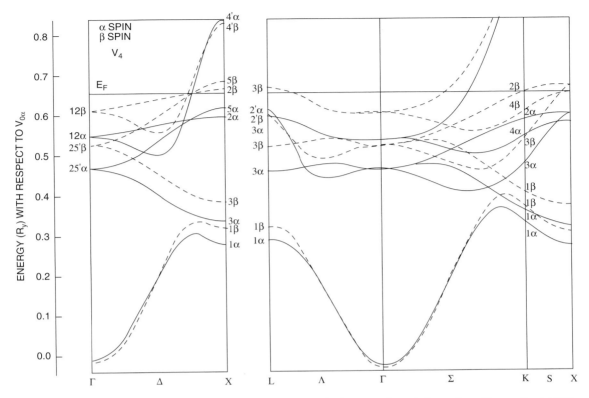

Figure 7.12 Calculated bands of nickel (solid: spin-up; dash: spin-down) (based on Connolly, 1967)

Heisenberg Model

In 1928, Werner Heisenberg used the concept of the exchange interaction in the Heitler-London theory of bi-electron states in the covalent bond to explain the quantum mechanical origin of the Weiss field.

The anti-symmetric eigenstates of a pair of electrons can be written as:

$$|\Psi_\eta\rangle = \frac{1}{2}(|\alpha_1\beta_2\rangle + \eta|\beta_1\alpha_2\rangle)\sigma_{12} \qquad (7.54)$$

where $\eta = 1$ stands for the spatially symmetric state and $\eta = -1$ the spatially anti-symmetric state. The perturbation energy with respect to the zeroth-order single-electron hamiltonian $h_l = \vec{p}_l^2/2m - Ze^2/|\vec{r}_l - \vec{R}|$ ($l=1,2$; $\vec{R}$ the position of nucleus) is

$$V = \frac{Z^2 e^2}{r_{\alpha\beta}} + \frac{e^2}{r_{12}} - \frac{Ze^2}{r_{1\beta}} - \frac{Ze^2}{r_{2\alpha}} \qquad (7.55)$$

Table 7.8 Exchange coupling constant versus $k_B T_c$ by the Heisenberg model

	Fe	Co	Ni	Gd	Dy
structure	BCC	HCP	FCC	HCP	HCP
z	8	12	12	12	12
S	1	1/2	1/2	7/2	5
$J_e/k_B T_c$	0.093	0.167	0.167	0.0079	0.0042

The spatially symmetric and anti-symmetric state would have the eigen-energy:

$$\varepsilon_+ = \langle \Psi_+ | \mathcal{H}_0 + V | \Psi_+ \rangle$$
$$= \frac{1}{2}(\varepsilon_\alpha + \varepsilon_\beta) + \bar{V} + J_e; \quad S = 0, S(S+1) = 0$$
$$\varepsilon_- = \langle \Psi_- | \mathcal{H}_0 + V | \Psi_- \rangle$$
$$= \frac{1}{2}(\varepsilon_\alpha + \varepsilon_\beta) + \bar{V} - J_e; \quad S = 1, S(S+1) = 2 \quad (7.56)$$

where the Heitler-London approximation $\langle \alpha_1 \beta_2 | \beta_1 \alpha_2 \rangle \simeq 0$ has been used; the exchange term is $J_e = \langle \alpha_1 \beta_2 | V | \beta_1 \alpha_2 \rangle = \langle \beta_1 \alpha_2 | V | \alpha_1 \beta_2 \rangle$.

Heisenberg used the eigen-values of the total spin operator $\vec{S} = \vec{s}_1 + \vec{s}_2$ to rewrite Eq. (7.56) as a single equation:

$$\varepsilon = \frac{1}{2}(\varepsilon_\alpha + \varepsilon_\beta) + \bar{V} - J_e(\vec{S}^2 - 1) = \varepsilon_0 - 2J_e \vec{s}_1 \cdot \vec{s}_2 \quad (7.57)$$

where $\vec{s}_1$ and $\vec{s}_2$ are the normalized spin operators of the first and second electrons. In covalent bonds, the exchange coupling J_e is negative, therefore the ground states of the two spins are antiparallel. In ferromagnetic materials, the exchange coupling J_e is positive, so the ground states of the two spins are parallel.

In a metal, if each atom has z neighbors and spin S, the total exchange energy by Eq. (7.57) must be equivalent to the Zeeman energy of the Weiss field at 0 K:

$$E_{ex} = -2z J_e S^2 \simeq -g\mu_B H_e(0) S \simeq -g\mu_B S(\lambda_E g \mu_B S) \quad (7.58)$$

Then the Heisenberg exchange constant J_e can be directly related to the Weiss temperature or the Curie temperature:

$$J_e = \lambda_E \frac{n(g\mu_B)^2}{2z} = \frac{3k_B T_c}{n(g\mu_B)^2 S(S+1)} \frac{n(g\mu_B)^2}{2z} = \frac{3k_B T_c}{2z S(S+1)}$$
$$(7.59)$$

This relationship is not always quantitatively correct if the J_e measured by neutron diffraction or spin-wave is used. For example, $J_e/k_B T_c$ of iron is 0.22 from the spin wave experiment, but the $J_e/k_B T_c$ in Eq. (7.59) is 0.093 ($S = 1, z = 8$) or 0.25 ($S = 1/2, z = 8$). This reflects the fact that current ferromagnetic theories cannot explain all the properties of ferromagnets consistently.

7.2.4 Antiferromagnetism and Ferrimagnetism

Louis Néel predicted the existence of antiferromagnetism as early as 1932, when he was working at Professor Weiss's laboratory in Strasbourg. This prediction predated the experimental discovery of antiferromagnetic manganese oxide (MnO) in 1938. Néel formulated the theory of ferrimagnetism in 1954, and investigated the magnetic properties of materials such as magnetite and garnet, which had traditionally been treated as ferromagnetic materials. These contributions won him the Nobel Prize in Physics in 1970.

Among the element crystals, only chromium is antiferromagnetic. The susceptibility in the single crystal antiferromagnetic phase is anisotropic: its parallel-to-spin component $\chi_\parallel$ is zero at 0 K; but its perpendicular-to-spin component $\chi_\perp$ is a constant at low temperatures. In polycrystalline chromium, the susceptibility is almost a constant versus temperature: $\chi = 3.4 \times 10^{-6}$ at 80 K and $\chi = 3.7 \times 10^{-6}$ at $T_N \simeq 300$ K.

Above the Néel temperature T_N, the magnetic susceptibility of an antiferromagnetic material has an explanation similar to the Weiss theory. A negative molecular field is added to the Langevin theory of paramagnetism in Eq. (7.38):

$$\frac{M}{H_{tot}} = \frac{M}{H_{ext} - H_A} = \frac{M}{H_{ext} - \lambda_A M} = \frac{n\mu_a^2}{3k_B T}$$

$$\chi = \frac{M}{H_{ext}} = \frac{n\mu_a^2}{3k_B(T + \theta)}; \quad \theta = \lambda_A \frac{n\mu_a^2}{3k_B} \quad (7.60)$$

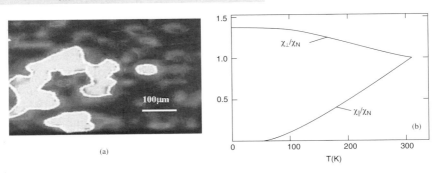

Figure 7.13 (a) antiferromagnetic domain of chromium measured by XRD; (b) calculated anisotropic susceptibility of single crystal chromium (based on Moyer, Arajs, and Hedman, 1976)

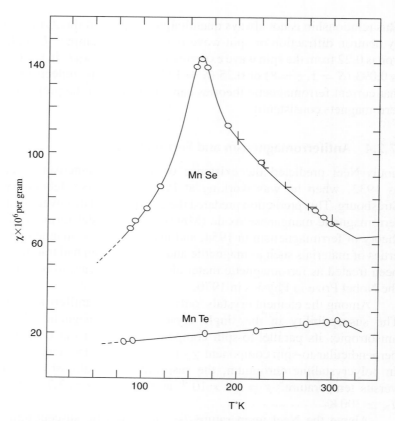

Figure 7.14 Susceptibility versus temperature for MnSe and MnTe (based on Squire, 1939)

where $-\theta$ is the negative Curie-Weiss temperature of the linear inverse susceptibility $\chi^{-1} \propto (T + \theta)$ at the paramagnetic phase. $-\theta$ can be explained by the negative exchange constant J_e in the Heisenberg model.

Antiferromagnetic materials are relatively uncommon. Some, including metals such as chromium, alloys such as iron manganese (FeMn), and oxides such as nickel oxide (NiO), are listed in Table 7.9. In the giant magnetoresistive (GMR) or tunnelling magnetoresistive (TMR) read head in hard disk drives, the antiferromagnetic alloy iron-manganese (FeMn) or iridium-manganese (IrMn) is used to "pin" the ferromagnetic-thin film permalloy (NiFe) or cobalt-iron (CoFe) for a linear read out. This practical FeMn/NiFe/Cu/NiFe spin-valve structure was devised by B. Dieny and his colleagues at the IBM Almaden center. (Dieny, Soeruisy, Parkin, Gurney, Wilhoit, and Mauri, 1991.)

The magnetite Fe_3O_4 is the earliest known magnetic material; however it was only after Néel's work in the 1950s that it became apparent that magnetite was not at a "pure" ferromagnetic phase, but was actually ferrimagnetic. In Figure 7.15(a), the structure of magnetite is shown:

Table 7.9 Characteristics of antiferromagnetic materials: T_N the Néel temperature; $-\theta$ the Curie-Wiess temperature (based on Kittel, 1986, and Squire, 1939)

Matter	T_N/K	$-\theta$/K	Structure	Matter	T_N/K	$-\theta$/K	Structure
NiO	525	−2000	FCC	MnTe	307	−690	HEX
Cr	311	–	BCC	MnSe	160	–	FCC
CoO	293	−330	FCC	MnS	140	−528	FCC
FeO	198	−570	FCC	MnO	116	−610	FCC
NiMn	1070	–	FCC	IrMn	520	–	FCC
FeMn	500	–	FCC	CrMn	393	–	FCC

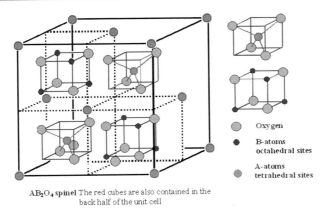

AB_2O_4 spinel The red cubes are also contained in the back half of the unit cell

Figure 7.15 Magnetite AB_2O_4 structure (www.tf.uni-kiel.de)

Fe^{3+} at the tetrahedral A-sites has a magnetic moment $\mu_{3+} = gS\mu_B = 5\mu_B$; Fe^{2+} at the octahedral B-sites has a magnetic moment $\mu_{2+} = gS\mu_B = 4\mu_B$ (orbital quench is considered). The measured magnetic moment per Fe_3O_4 molecule is 4.1 to $4.2\mu_B$, therefore the only possible explanation is that the Fe^{3+} ions at the A-sites are antiparallel-aligned, but the Fe^{2+} ions at the B-sites are parallel-ordered, as shown in Figure 7.15(b). If the Fe^{2+} ions at the B-sites in magnetite are replaced by metallic M^{2+} ions, the resultant ferrite is still ferrimagnetic, as listed in Table 7.10. The different ferromagnetic or antiferromagnetic couplings in separate sublattices are characteristic of ferrimagnetic materials.

Ferrimagnetism is exhibited by ferrites and magnetic garnets. On the one hand, below the Néel temperature T_N, ferrimagnets behave like anti-ferromagnets; on the other hand, below the Curie temperature T_c, ferrimagnets behave like ferromagnets in that there is spontaneous magnetization. T_c is usually higher than T_N, and in the range $T_N < T < T_c$, super-paramagnetism (i.e., unstable magnetization direction) occurs.

Table 7.10 Characteristics of ferrimagnetic materials: T_c the Curie temperature; M_s the saturation magnetization

Matter	T_c/K	M_s/(emu/cm^3)	μ/μ_B	Matter	T_c/K	M_s/(emu/cm^3)
Fe_3O_4	848	484	4.2	γ-Fe_2O_3	873	~ 300
$MnFe_2O_4$	573	410	5.0	$CoFe_2O_4$	792	~ 400
$NiFe_2O_4$	858	270	2.4	$Ba_{0.6}Fe_2O_3$	723	382
$CuFe_2O_4$	728	290	1.3	$Y_3Fe_5O_{12}$	560	~ 100
$MgFe_2O_4$	858	143	1.1	Fe_3S_4	606	~ 100

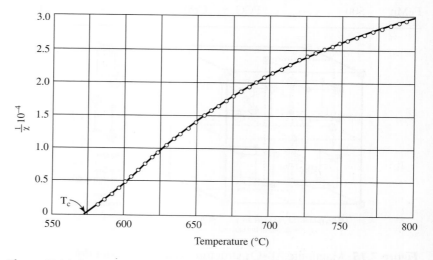

Figure 7.16 Inverse susceptibility of Fe_3O_4 (based on Kittel, 1986)

Above the Curie temperature, magnetite is typically paramagnetic. The Weiss-like molecular fields in magnetite can be written as:

$$H_A = -\lambda_{AA}M_A - \lambda_{AB}M_B \quad H_B = -\lambda_{BA}M_A + \lambda_{BB}M_B \quad (7.61)$$

where $\lambda_{AB} = \lambda_{BA}$, and all the λ parameters are greater than zero. The susceptibility can then be found by following the Brillouin theory:

$$\frac{M_A}{H_{tot}} = \frac{M_A}{H_{ext} - \lambda_{AA}M_A - \lambda_{AB}M_B} = \frac{n_A\mu_A^2}{3k_BT}$$

$$\frac{M_B}{H_{tot}} = \frac{M_B}{H_{ext} - \lambda_{BA}M_A + \lambda_{BB}M_B} = \frac{n_B\mu_B^2}{3k_BT}$$

$$\chi = \frac{M_A + M_B}{H_{ext}} = \frac{(n_A\mu_A^2 + n_B\mu_B^2)(T - \theta_3)}{3k_B(T^2 + \theta_1 T - \theta_2^2)} \quad (7.62)$$

Eq. (7.62) has the form $\chi^{-1} \propto (T - T_c)$ when $T \to T_c$, and the Curie temperature is

$$T_c = \frac{1}{2}\left[-\theta_1 + \sqrt{\theta_1^2 + 4\theta_2^2}\right]; \quad T_c > \theta_3 \tag{7.63}$$

$$\theta_1 = \frac{n_A \mu_A^2}{3k_B}\lambda_{AA} - \frac{n_B \mu_B^2}{3k_B}\lambda_{BB}$$

$$\theta_2^2 = \frac{n_A \mu_A^2}{3k_B}\frac{n_B \mu_B^2}{3k_B}(\lambda_{AA}\lambda_{BB} + \lambda_{AB}^2)$$

$$\theta_3 = \frac{n_A \mu_A^2}{3k_B}\frac{n_B \mu_B^2}{3k_B}(2\lambda_{AB} - \lambda_{AA} + \lambda_{BB})/\left(\frac{n_A \mu_A^2}{3k_B} + \frac{n_B \mu_B^2}{3k_B}\right)$$

where the ratio of the exchange coupling is $\lambda_{AB} : \lambda_{AA} : \lambda_{BB} = 22 : 11 : 3$; and the Curie temperature $T_c = 575°C$ in the magnetite, as shown in Figure 7.16.

7.3 Spins Interacting with Fundamental Particles

In this section, the effect of the scattering process of fundamental particles on spins in solids is discussed. Magnetic neutron diffraction, neutron inelastic scattering of magnons, nuclear magnetic resonance, and electron spin resonance are important related topics, in which the atomic magnetic moments and nuclear magnetic moments can be directly studied by observing the electromagnetic interactions and spin-dependent strong interactions.

7.3.1 Magnetic Neutron Diffraction and Magnetic Structure

The atomic spins in solids can be directly measured by magnetic neutron diffraction, which was developed by C. G. Shull in the late 1940s. As discussed previously, the neutron has a magnetic moment $\mu_z^n = g_n \mu_n H \sigma_z$, where $g_n \simeq -3.83$. The scattering length f_m of magnetic neutron diffraction can be written in the form given by O. Halpern and M. H. Johnson in 1939, where all neutron moments are in the same direction $\hat{\sigma}_n$ in the polarized neutron beam:

$$I = |f_{\text{non-m}} + \eta f_m|^2; \quad f_{\text{non-m}} = f_0 \sum_i \exp(-i\vec{s} \cdot \vec{r}_i) \tag{7.64}$$

$$f_m = \sum_i (\hat{\sigma}_n \cdot \vec{\mu}_i)\exp(-i\vec{s} \cdot \vec{r}_i); \quad \hat{\sigma}_n \cdot \vec{\mu}_i = \pm|\vec{p}_i| = \pm\mu_i \sin\theta_i$$

$$\tag{7.65}$$

$f_{\text{non-m}}$ is contributed by the neutron-nucleon strong interaction; f_{m} is caused by the neutron-atomic moment interaction, where $\vec{p}_i = \vec{\mu}_i - \hat{s}(\hat{s} \cdot \vec{\mu}_i)$ is the atomic moment perpendicular to the outgoing-incoming wavevector difference $\vec{s} = \vec{k} - \vec{k}_0$ ($\hat{s} = \vec{s}/|\vec{s}|$), the magnitude of the atomic moment is $\mu = g_J \mu_B \sqrt{J(J+1)}$ or simply $\mu = 2\mu_B \sqrt{S(S+1)}$, and and θ_i is the angle between $\vec{s}$ and $\vec{\mu}_i$ of the i'th atom or ion.

In the magnetic neutron diffraction of a ferromagnetic crystal, all atoms have a constant moment μ_0, and the total atomic scattering length is

$$f = f_{\text{non-m}} + \eta f_{\text{m}} = \sum_i (f_0 + \eta \hat{\sigma}_n \cdot \vec{p}_0) \exp(-i\vec{s} \cdot \vec{r}_i)$$
$$= (f_0 + \eta \hat{\sigma}_n \cdot \vec{p}_0) FS \qquad (7.66)$$

where $\vec{p}_0 = \vec{\mu}_0 - \hat{s}(\hat{s} \cdot \vec{\mu}_0)$; and S and F of elastic scattering are as discussed in Chapter Three. All the diffraction peaks are the same as those in normal diffraction.

In the paramagnetic phase, the angles $\{\theta_i\}$ of magnetic moments are randomly distributed, thus the total diffraction intensity is

$$I = |f_{\text{non-m}} + \eta f_{\text{m}}|^2$$
$$= f_0^2 S^2 F^2 + N\eta^2 \vec{\mu}^2 \langle \sin^2 \theta_i \rangle$$
$$= f_0^2 S^2 F^2 + \frac{2}{3} N \eta^2 \vec{\mu}^2 \qquad (7.67)$$

where the magnetic interaction part $|f_{\text{m}}|^2$ contributes a weak uniform background, and the non-magnetic neutron diffraction intensity $|f_{\text{non-m}}|^2$ still leads to normal diffraction peaks in line with Bragg's law.

In 1951, C. G. Shull and his colleagues performed magnetic neutron diffraction on manganese (II) oxide (MnO) with a sodium chloride (NaCl) structure. They found additional diffraction peaks for the MnO crystal in the antiferromagnetic phase at 80 K compared to those in the paramagnetic phase at room temperature, as shown in Figure 7.17. The most important difference is that there is an extra diffraction peak appearing at about half of the Bragg angle of (111) crystal planes, which is caused by the magnetic structure of antiparallel magnetic moments in the alternate (111) planes:

$$I = |f_{\text{non-m}} + \eta f_{\text{m}}|^2$$
$$= f_0^2 S^2 F^2 + \eta^2 \vec{\mu}^2 \sin^2 \theta \left| \sum_i \exp\left(-i(\vec{s} + \frac{1}{2}\vec{G}'_{111}) \cdot \vec{r}_i\right) \right|^2$$
$$= f_0^2 F^2 S^2 + \eta^2 \vec{\mu}^2 \sin^2 \theta F^2 S_{\text{m}}^2; \qquad (7.68)$$

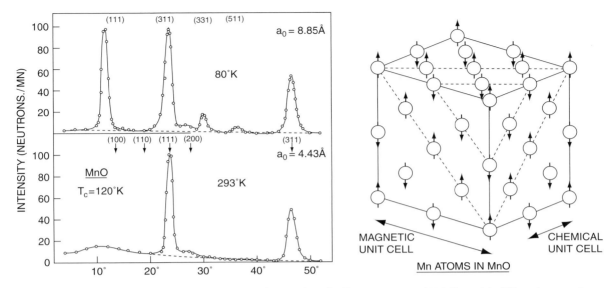

Figure 7.17 Magnetic neutron diffraction of MnO by Shull, Strauser and Wollan: (a) diffraction peaks above and below the Néel temperature; (b) magnetic order in the antiferromagnetic phase of MnO (based on Shull, Strauser and Wollan, 1951)

$$S \neq 0 \text{ (when } \vec{s} = \vec{G}'_{hkl}); \quad S_m \neq 0 \left(\text{when } \vec{s} = \vec{G}'_{hkl} - \frac{1}{2}\vec{G}'_{111}\right),$$

(7.69)

Therefore the first peak of magnetic neutron diffraction appears at $2k_0 \sin\theta_{\text{Bragg}} = \frac{1}{2} G_{111}$, and there is still a "normal" peak at $2k_0 \sin\theta_{\text{Bragg}} = G_{111}$. Using this elastic neutron diffraction method, the magnetic structures of paramagnetism, ferromagnetism, antiferromagnetism, and ferrimagnetism can all be determined.

7.3.2 Spin Wave and Neutron Inelastic Scattering

In 1931, Felix Bloch proposed the concept of spin waves to explain energy bands in ferromagnetic materials. Bloch's conclusion about the exchange coupling J_e with respect to different types of magnetism is similar to the Heisenberg model. The difference between the two models is that, in the Bloch formula, the $M_s(T)$ at low temperatures is directly related to the J_e. This led to the the work on inelastic neutron-scattering of magnons (quasi-particles of spin waves) by B. N. Brockhouse in 1951.

The schematics of spin wave are given in Figure 7.18. At low temperatures, the deviations from the zero-temperature spin $|S(\vec{R}_n) - S|$ are quite small (S: spin quantum number of atom). Bloch expressed the

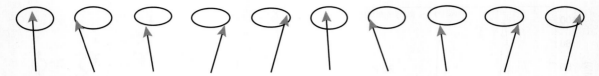

Figure 7.18 Spin wave: fluctuation of magnetic moment orientation in space

spin wave and its eigen-energy as:

$$S_{\vec{k}} = \frac{1}{\sqrt{N\bar{S}_{\vec{k}}}} \sum_n S_n \, e^{i\vec{k}\cdot\vec{R}_n}; \quad \vec{R}_n = \sum_\alpha n_\alpha \vec{a}_\alpha \qquad (7.70)$$

$$\mathcal{H} = \varepsilon_0 + J_e \sum_{\langle i,j \rangle} (\vec{S}_i - \vec{S}_j)^2, \qquad (7.71)$$

$$\varepsilon(\vec{k}) = \langle S_{\vec{k}} | \mathcal{H} | S_{\vec{k}} \rangle = \varepsilon_0 + 2 J_e S \sum_{\vec{d}} (1 - e^{i\vec{k}\cdot\vec{d}}), \qquad (7.72)$$

where $\vec{d}$ is the position vector between neighboring atomic sites, $\mathcal{H}$ is the Hamiltonian of the Heisenberg model, and $\langle i, j \rangle$ represents the sum over the pair of neighboring sites. The proof of the spin-wave energy $\varepsilon(\vec{k})$ is similar to the tight-binding band in Eq. (5.75).

The total free energy and total spin (z-component) of a ferromagnetic crystal with spin waves or magnons are:

$$E_{\text{tot}} = \sum_{\vec{k}} (\varepsilon(\vec{k}) - \varepsilon_0)$$

$$\simeq J_e S \sum_{\vec{k}} \sum_{\vec{d}} (\vec{k} \cdot \vec{d})^2 \simeq J_e S \sum_{\vec{k}} A(ka)^2 \qquad (7.73)$$

$$S_{\text{tot}} = \sum_{\vec{k}} N \bar{S}_{\vec{k}} = S(N - 2f), \qquad (7.74)$$

where all the magnons with momentums $\hbar \vec{k}$ can be treated as independent quasi-particles, the sum over $\vec{k}$ is finite, and there are $1, 2, \ldots, f$ excited magnons. In a ferromagnetic material, the exchange coupling $J_e > 0$, and the total energy E_{tot} minimizes when there are no magnons ($f = 0$). This condition is the same as that for the parallel-spin ground state in the Heisenberg model.

The average magnetization can be found using the Bose-Einstein statistics:

$$g(\varepsilon) = \rho(k)\frac{dk}{d\varepsilon(\vec{k})} \simeq \frac{1}{4\pi^2}\frac{1}{(J_e S A a^2)^{3/2}}\sqrt{\varepsilon} \quad (7.75)$$

$$\Delta \bar{M} = \int_0^\infty d\varepsilon\, g(\varepsilon)\frac{1}{\exp(\varepsilon/k_B T) - 1}$$

$$= \frac{1}{4\pi^2}\left(\frac{k_B T}{J_e S A a^2}\right)^{3/2}\int_0^\infty du\, \frac{u^{1/2}}{e^u - 1}$$

$$= \frac{1}{4\pi^2}\left(\frac{k_B T}{J_e S A a^2}\right)^{3/2}(0.2348\pi^2) \quad (7.76)$$

Therefore the low-temperature saturation magnetization can be expressed by the Bloch 3/2-power law:

$$M_s(T) = M_s(0)\left[1 - \frac{0.0587}{S(AS)^{3/2} a^3/\Omega_c}\left(\frac{k_B T}{J_e}\right)^{3/2}\right]$$

$$= M_s(0)\left[1 - \left(\frac{T}{\Theta}\right)^{3/2}\right] \quad (7.77)$$

where $M(0) = S/\Omega_c$ is the saturation magnetization in a Bravais lattice at 0K. If the constants $S = 1/2$ and $A = 2$ are chosen, the characteristic temperature Θ is:

$$\Theta = 10.5\frac{J_e}{k_B}\text{(FCC)}; \quad \Theta = 6.6\frac{J_e}{k_B}\text{(BCC)}. \quad (7.78)$$

In 1934, P. Weiss proved that the Bloch 3/2-power law is more accurate than Eq. (7.53) of the Weiss theory by measuring $M(T)$ of iron and nickel in the range 20 to 70 K. In Figure 7.19, the measured low-temperature magnetization of nickel is plotted as a function of $T^{3/2}$, and it can be seen that the Bloch 3/2-power law is valid for $T < 75$ K, except in the very low temperature range.

The spin wave spectrum can be measured using inelastic magnetic neutron diffraction, the theory of which is almost identical to that of the inelastic neutron scattering of phonons, discussed in Chapter Four. The diffraction intensity should include the time-term, and the extra structure factor of the spin wave $\vec{p}_i = \vec{p}_0 + \vec{A}\cos(\vec{q}\cdot\vec{R}_i - \omega t)$ is:

$$\Delta S = \left(\frac{1}{2}\hat{\sigma}_n \cdot \vec{A}\right)\sum_i\left[e^{-i(\vec{s}-\vec{q})\cdot\vec{R}_i - i(\Omega_0 + \omega)t} + e^{-i(\vec{s}+\vec{q})\cdot\vec{R}_i - i(\Omega_0 - \omega)t}\right]$$

$$(7.79)$$

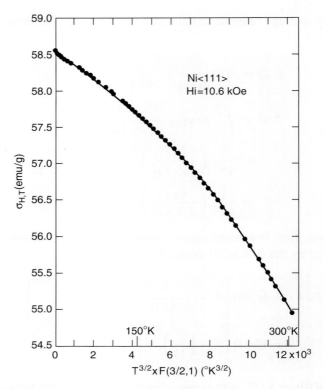

Figure 7.19 Bloch 3/2-power law in Nickel (based on Aldred, 1975)

The diffraction peaks of the magnetic inelastic scattering should appear at the new positions, satisfying the conservation law of energy and the conservation law of momentum in the neutron-magnon scattering process:

$$\vec{s} = \vec{k} - \vec{k}_0 = \vec{G}_{hkl} \pm \vec{q}; \quad \hbar\Omega - \hbar\Omega_0 = \pm\hbar\omega(\vec{q}) = \pm\varepsilon(\vec{q}). \quad (7.80)$$

During the measurement of a magnon spectrum, the intensity of neutrons at different energies should be counted near an elastic scattering diffraction point.

The magnon spectra of ferromagnetic matter and ferrimagnetic magnetite were given by Felix Bloch and H. Kaplan respectively in 1952:

$$\varepsilon(\vec{k}) = \varepsilon_0 + 2J_e S \sum_{\vec{d}}(1 - \cos\vec{k}\cdot\vec{d}); \quad (7.81)$$

$$\varepsilon(\vec{k}) = \{[5.5J_{AB}S_A S_B + J_{AA}S_A^2 + 2J_{BB}S_B^2]/[4(S_A + 2S_B)]\}(ka)^2. \quad (7.82)$$

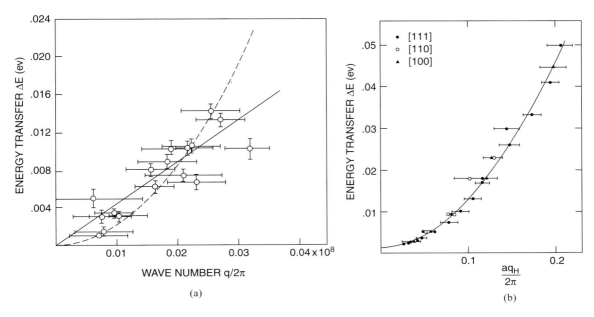

Figure 7.20 Measured spin wave spectra by Brockhouse: (a) magnetite; (b) Co (based on Brockhouse, 1957, and Sinclair and Brockhouse, 1960)

The above two equations have been proved by B. N. Brockhouse's inelastic neutron scattering experiments, as shown in Figure 7.20. The measured exchange coupling of cobalt is $J_e S = (1.47 \pm 0.15) \times 10^{-2}$ eV; if $S = 1/2$ is assumed, the measured $J_e/k_B T_c$ of cobalt is 0.245, which roughly agrees with the expectation $J_e/k_B T_c = 0.167$ for cobalt in the Heisenberg model given in Table 7.8.

Specific Heat of Spin Waves

In Chapters Four and Five, it was shown that the thermal properties of solids have three sources: (1) atomic vibrations, (2) conduction electrons in metals (only significant at very low temperatures), and (3) spin waves in ferromagnetic, antiferromagnetic, and ferrimagnetic materials (only significant at low temperatures).

The specific heat of magnons can also be found using the Bose-Einstein statistics:

$$\begin{aligned}
C_V^m &= \frac{d}{dT} \int_0^\infty d\varepsilon\, g(\varepsilon) \frac{\varepsilon}{\exp(\varepsilon/k_B T) - 1} \\
&= \frac{5 k_B}{8\pi^2} \left(\frac{k_B T}{J_e S A a^2} \right)^{3/2} \int_0^\infty du \frac{u^{3/2}}{e^u - 1} \\
&= 0.113 k_B \left(\frac{k_B T}{J_e S A a^2} \right)^{3/2}
\end{aligned} \qquad (7.83)$$

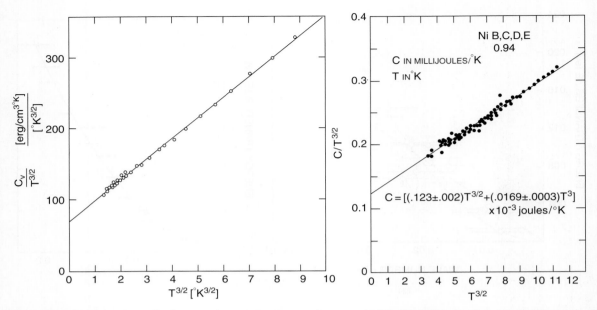

Figure 7.21 Low-T specific heat in ferrimagnetic insulators: (a) YIG; (b) NiFe$_2$O$_4$ (based on Edmonds and Petersen, 1959, and Pollack and Atkins, 1962)

This T-3/2 law of specific heat contributed by spin waves can be proved by the low-temperature specific heat in insulating antiferromagnetic or ferrimagnetic materials. The measured C_V at low temperatures consists of two terms:

$$C_V = 0.113 k_B \left(\frac{k_B T}{J_e S A a^2} \right)^{3/2} + \frac{12\pi^4}{5} k_B \left(\frac{T}{\Theta_D} \right)^3 \quad (7.84)$$

The equation is proved by the experiments shown in Figure 7.21. The exchange constant $J_e/a = 0.192 \times 10^{-6}$ erg/cm in YIG, and $J_e/a = 0.232 \times 10^{-6}$ erg/cm in γ-Fe$_2$O$_3$ (based on Edmonds and Petersen, 1959), which is about one-tenth of the exchange coupling in metallic iron, cobalt, and nickel.

7.3.3 Electron Spin Resonance and Neutron Magnetic Resonance

In 1944, the phenomenon of electron spin resonance (ESR) was discovered by E. K. Zavoisky and his colleagues at Kazan State University in the former Soviet Union. ESR, also known as electron paramagnetic resonance (EPR), provides a powerful tool for studying the unpaired electrons in condensed matter systems by the microwave resonance of Zeeman splitting of paramagnetic ions or molecules. ESR/EPR is widely

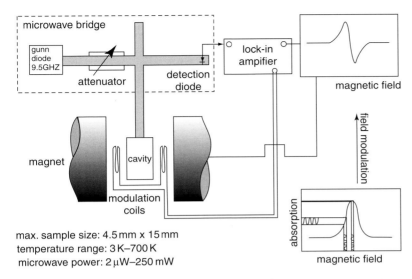

Figure 7.22 Electron spin resonance (ESR) or electron paramagnetic resonance (EPR)

applied in chemistry, physics, biology, and medicine: it can be used to investigate the static structures of solid and liquid systems, and it is also very useful for examining dynamic processes (i.e., chemical reactions).

The ESR system utilizes three electromagnets. The primary magnet is used to produce a large magnetic field (10^3–10^5 G), where a large current has to be provided, and this magnet must be water-cooled. The secondary magnet consists of Helmholtz coils between the two inner sides of the primary magnet, which produce small magnetic fields (10^1–10^2 G). The third magnet is the set of coils in the sample cavity, which are used for the production of tiny magnetic fields (10^{-2}–10^0 G).

The microwave radiation (between 10^2 MHz and 10^2 GHz) is sent to the sample by the waveguide. When an incident microwave has a photon energy equal to the Zeeman splitting of an atom or ion $\Delta E = g\mu_B H \Delta J_z$:

$$\hbar\omega = g\mu_B H_{tot} \Rightarrow \frac{f}{\text{MHz}} = g \times 1.40 \times \frac{H_z + H_{\text{eff}}}{\text{Oe}}, \qquad (7.85)$$

resonance absorption of the microwave occurs. The exact resonant frequency of the absorption is determined by both the external magnetic field H_z and interactions with magnetic moments in the systems. Here the interactions with magnetic moments can generate a local effective field H_{eff} (including the Weiss-type intrinsic magnetic field).

The measured ESR signals and susceptibilities of the antiferromagnetic MnF_2 are shown in Figure 7.23 at temperatures above and below the Néel temperature $T_N = 67.3$ K. Below T_N, the susceptibility

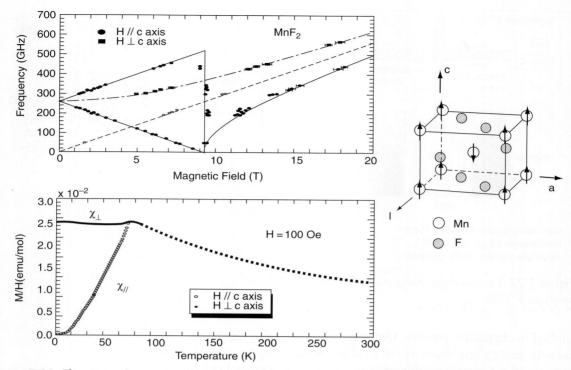

Figure 7.23 Electron spin resonance and susceptibility of antiferromagnetic MnF$_2$ (based on Katsumaka, 2000)

is anisotropic and the values are different for the parallel/perpendicular directions; the parallel/perpendicular ESR signals are non-zero even at zero external field H_z, due to the antiferromagnetic Weiss fields.

In 1945, Felix Bloch from Stanford University and Purcell, Torrey, and Pound from Harvard University discovered nuclear magnetic resonance (NMR), which is an electromagnetic resonance procedure used to study nuclear moments in solids, liquids, and gases. Bloch and Purcell won the Nobel Prize in Physics in 1952 for their investigations into nuclear magnetic resonance (although Bloch has made other contributions besides NMR.)

The system of nucleon moments is paramagnetic with the susceptibility

$$\chi = \frac{n(g\mu_n)^2 J(J+1)}{3k_B T} = \frac{291\,K}{T} \times 3.4 \times 10^{-10} \text{ (water)} \quad (7.86)$$

In water, at room temperature ($T = 291$ K), the constants $J = 1/2$, $n = 6.9 \times 10^{22}$ cm^{-3}, $\mu = g\mu_n J = 1.4 \times 10^{-23}$ erg/G. The paramagnetic susceptibility is so small that it cannot be observed easily and accurately.

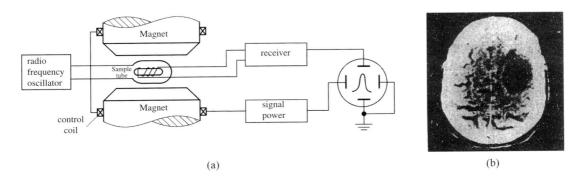

Figure 7.24 Nuclear magnetic resonance: (a) apparatus; (b) magnetic resonance image of a brain with cancer

In a method similar to ESR, a radio electromagnetic wave is sent to a sample using a waveguide. When an incident wave has a photon energy equal to the Zeeman splitting of a nuclear $\Delta E = g\mu_n H \Delta J_z$ (g-factor of proton $g_p = 5.5856912$):

$$\hbar\omega = g\mu_n H_{tot} \Rightarrow \frac{f}{\text{kHz}} = 4.26 \times \frac{g^*}{g_p} \times \frac{H_z}{\text{Oe}}, \qquad (7.87)$$

resonance absorption of the radio wave occurs. The resonant frequency is determined by the external magnetic field H_z, and the chemical shift $g \to g^*$.

The NMR frequency of protons is 42.6MHz ($H_z = 1$ T). However, in 1951, W. G. Proctor and F. C. Yu found that the nuclear resonance frequency of protons depends upon the chemical environment, in that the chemical shift is larger for higher electron densities around the hydrogen nucleon. Although the chemical shift $(g^* - g_p)/g_p$ of protons is only in the order of parts per million (ppm), NMR has significant applications in chemistry because it enables chemical clusters to be accurately determined.

In medical science, NMR is called magnetic resonance imaging (MRI). In practical MRI, a spatially linear varying magnetic field $H_z(x, y)$ is applied to the x-y surface, such that the resolution in the resonance frequency has one-to-one correspondence with the resolution in the space position. Software is used to reconstruct the NMR signals to the intensity distribution in the x-y plane. The NMR signal is proportional to the local proton density, therefore the medical condition of a human body (especially the brain) can be investigated using MRI, as shown in Figure 7.24(b). Paul C. Lauterbur and Sir Peter Mansfield were awarded the Nobel Prize in Physiology and Medicine 2003 for their discoveries in magnetic resonance imaging.

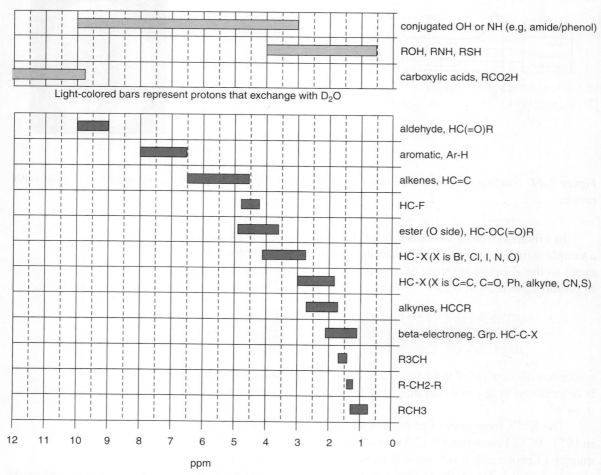

Figure 7.25 Chemical shift of protons (from http://www.csub.edu/chemistry/332/Spectroscopy/H1_shifts.gif)

Summary

This chapter has analyzed the origins of magnetism in solids based on the multi-electron Hamiltonian in electromagnetic field. The categories of magnetism have also been explained based on atomic moments; however the theory of ferromagnetism is not yet fully understood. Also explored are magnetic neutron diffraction, electron spin resonance, and neutron magnetic resonance, which are important techniques based on spin-particle interaction. The key areas covered in this chapter are summarized in more detail in the following:

1. Hamiltonian of multi-electron systems: the correct form of the quantum mechanical multi-electron Hamiltonian in the electromagnetic field was finally determined using relativistic quantum mechanics. In solid state physics, the low-speed limit of the relativistic Hamiltonian plays the dominant role.
2. Single atom limit: if the multi-electron Hamiltonian is applied to the electrons in an atom or an ion, the total atomic spin can be defined. If the total atomic spin is not zero, the free energy related to the magnetic field is simply the Zeeman energy, and the magnetic moment is proportional to the total atomic spin and the Landé g-factor.
3. Free electron gas limit: if the multi-electron Hamiltonian is applied in a 2D free electron gas, the eigen-energies of electrons can be quantized as Landau levels. The Landau level "energy step" is proportional to the magnetic field, based on which Shubnikov-de Haas oscillations can occur.
4. Diamagnetism: an atom or ion becomes diamagnetic when its total atomic spin becomes zero, i.e., the ion has a fully-filled electronic structure. The diamagnetic susceptibility is proportional to the number of electrons in an ion. Free electron gas has both diamagnetism and paramagnetism, and its diamagnetic susceptibility is due to fully-filled Landau levels.
5. Paramagnetism: paramagnetism occurs in a solid where some of the ions have non-zero atomic moments. Paramagnetic susceptibility is inversely proportional to temperature, as in Curie's law. According to experiments, the atomic magnetic moments are related either to the total atomic spin (rare earth elements) or the total electron spin (transition metal ions). The van Vleck paramagnetism of some rare earth ions comes from the second order perturbation term in the Hamiltonian. Free electron gas also has paramagnetism: the Zeeman splitting at the Fermi surface results in Pauli paramagnetism; while in the low-temperature high-magnetic field limit, the paramagnetic susceptibility exhibits de Haas-van Alphen oscillation, which is directly related to the Landau levels.
6. Ferromagnetism: ferromagnetism only occurs in a few elements, however it is the most important magnetic property of solids. The theory of ferromagnetism is not yet fully developed. The Weiss theory and the Heisenberg model describe the basic characteristics, but the saturation magnetization and exchange coupling constant cannot be precisely calculated.
7. Antiferromagnetism and ferrimagnetism: chromium is the only antiferromagnetic element. However, in magnetic oxides and ceramics,

some sublattices are antiferromagnetic but others might be ferromagnetic: this is known as ferrimagnetism. Both antiferro- and ferrimagnetism were discovered by Néel.
8. Magnetic neutron diffraction: elastic magnetic neutron diffraction can be used to determine the category of magnetism in a solid, and inelastic magnetic neutron diffraction can be used to determine the spin wave (low-temperature magnetization fluctuation) spectrum and magneton dispersion. The low-temperature specific heat of magnetic oxides is proportional to $T^{3/2}$, which is different from the Debye T3-law or the cT-type low-temperature susceptibility for free electron gas.
9. ESR and NMR: atomic moments can be measured directly using electron spin resonance due to the Zeeman splitting of atomic spins. Nuclear moments can be measured using nuclear magnetic resonance due to the Zeeman splitting of nuclear spins. Both ESR and NMR experiments are accurate, and have both engineering and medical applications.

References

Aldred, A. T. (1975). "Temperature Dependence of the Magnetization of Nickel," *Phys. Rev. B*, Vol. 11, 2597–2601.
Bitter, F. (1930). "On the Magnetic Properties of Metals," *Phys. Rev.*, Vol. 36, 978–983.
Bloch, F. (1931). "Zur Theorie des Ferromagnetismus," *Z. Physik*, Vol. 61, 206.
Bloch, F. (1946). "Nuclear Induction," *Phys. Rev.*, Vol. 70, 460–474.
Brockhouse, B. N. (1957). "Scattering of Neutrons by Spin Waves in Magnetite," *Phys. Rev.*, Vol. 106, 859–864.
Burns, G. (1985). *Solid State Physics*, Academic Press, Orlando.
Connolly, J. W. D. (1967). "Energy Bands in Ferromagnetic Nickel," *Phys. Rev.*, Vol. 159, 415–426.
de Haas, W. J. and van Alphen, P. M. (1931). *Leiden Comm. A*, Vol. 212, 215.
Dieny, B., Speriosu, V., Parkin, S., Gurney, B., Wilhoit, D. and Mauri, D. (1991). "Giant Magnetoresistance in Soft Ferromagnetic Multilayers," *Phys. Rev. B*, Vol. 43, 1297–1300.
Edmonds, D. T., and Petersen, R. G. (1959). "Effective Exchange Constant in Yttrium Iron Garnet," *Phys. Rev. Lett.*, Vol. 2, 499–500.
Föll, Helmut. "Ionic Crystals." Faculty of Engineering, University of Kiel, City of Kiel, Germany. Nov 20, 2007.
   <http://www.tf.uni-kiel.de/matwis/amat/def_en/kap_2/basics/b2_1_6.html>
Fowler, A. B., Fang, F. F., Howard, W. E., and Stiles, P. J. (1966). "Magneto-oscillatory Conductance in Silicon Surfaces," *Phys. Rev. Lett.*, Vol. 16, 901–903.
Halpern, O., and Johnson, M H. (1939) "On the Magnetic Scattering of Neutrons," *Phys. Rev.*, Vol. 55, 898–923.
Heisenberg, W. (1928). "Zur Theorie des Ferromagnetismus," *Z. Physik*, Vol. 49, 619.
Henry, W. E. (1953). "Some Magnetization Studies of Cr+++, Fe+++, Gd+++, and Cu++ at Low Temperatures and in Strong Magnetic Fields," *Rev. Mod. Phys.*, Vol. 25, 163–164.

Kaplan, H. (1952). "A Spin-wave Treatment of the Saturation Magnetization of Ferrites," *Phys. Rev.*, Vol. 86, 121.
Katsumata, K. (2000). "High-frequency electron spin resonance in magnetic systems," *J. Phys.: Condens. Matter*, Vol. 12, R589–R614.
Kaya, S. (1928). "On the Magnetization of Single Crystals of Co," *Sci. Repts. Tohoku Imp. Univ.*, Vol. 17, 1157.
Kittel, C. (1986). *Introduction to Solid State Physics*, Wiley, New York.
Knobel, R., Samarth, N., Harris, J. G. E., and Awschalom, D. D. (2002). "Measurements of Landau-level Crossings and Extended States in Magnetic Two-dimensional Electron Gases," *Phys. Rev. B*, Vol. 65, 235327.
Landau, L. D., and Liftshitz, E. M. (1987). *Quantum Mechanics*, Pergamon Press, New York.
von Laue, M. (1978), translated by Fan, D., and Dai, N. *History of Physics* (in Chinese), Commercial Press, Beijing.
Moyer, C. A., Arajs, S., and Hedman, L. (1976). "Magnetic Susceptibility of Antiferromagnetic Chromium," *Phys. Rev. B*, Vol. 14, 1233–1238.
Myers, W. R. (1952). "The Diamagnetism of Ions," *Rev. Mod. Phys.*, Vol. 24, 15–27.
Pollack, S. R., and Atkins, K. R. (1962). "Specific Heat of Ferrites at Liquid Helium Temperatures," *Phys. Rev.*, Vol. 125, 1248–1254.
Proctor, W. G., and Yu, F. C. (1951) "On the Nuclear Magnetic Moments of Several Stable Isotopes," *Phys. Rev.*, Vol. 81, 20–30.
Purcell, E. M., Torrey, H. C., and Pound, R. V. (1946). "Resonance Absorption by Nuclear Magnetic Moments in a Solid," *Phys. Rev.*, Vol. 69, 37–38.
Seitz, F. (1940). *The Modern Theory of Solids*, McGraw-Hill, New York.
Shull, C. G., Strauser, W. A., and Wollan, E. O. (1951). "Nucleon Isobars in Intermediate Coupling," *Phys. Rev.*, Vol. 83, 333–345.
Sinclair, R. N., and Brockhouse, B. N. (1960). "Dispersion Relation for Spin Waves in a FCC Cobalt Alloy," *Phys. Rev.*, Vol. 120, 1638–1640.
Stapleton, H. J., and Jeffries, C. D. (1961). "Paramagnetic Resonance of Trivalent Pm147 in Lanthanum Ethyl Sulfate," *Phys. Rev.*, Vol. 124, 1455–1457.
Squire, C. F. (1939). "Antiferromagnetism in Some Manganous Compounds," *Phys. Rev.*, Vol. 56, 922–925.
Tyler, F. (1931). "Magnetization-Temperature Curves of Iron, Cobalt, and Nickel," *Phil. Mag.*, Vol. 11, No. 7, 596–602.
Vuillemin, J. J. (1966). "De Haas-van Alphen Effect and Fermi Surface in Palladium," *Phys. Rev.*, Vol. 144, 396–405.
Weiner, D. (1962). "De Haas-van Alphen Effect in Bismuth-Tellurium Alloy," *Phys. Rev.*, Vol. 125, 1226–1238.
Xu, G. (1966). *An Introduction to the Matter Structure* (in Chinese), Higher Education Press, Beijing.
"Spectroscopic Structure Determination Strategy." Department of Chemistry, California State University, City of Bakersfield, USA. May 3, 2006. <http://www.csub.edu/Chemistry/332/Spectroscopy/H1_shifts.gif>

Exercises

7.1 In Figure 7.3(b), the anomalous Zeeman splitting of an isolated sodium atom is shown. When $H = 0$, the two $3p$-$3s$ lines have wavelengths of 589.6 nm and 589.0 nm respectively. If the 6-line

Zeeman splitting has a total splitting width $\Delta\lambda = 0.1\,nm$, what is the corresponding magnetic field?

7.2 In a 33kOe magnetic field along the [100] crystal direction of silicon, utilize the effective mass ($\perp \vec{H}$) given in Chapter Six to find the period of Fermi energy $\varepsilon_F^n - \varepsilon_F^{n-1}$ for the low-temperature Shubnikov-de Hass oscillation. What would the corresponding period of MOS gate voltage be if the approximate ratio of gate voltage increase versus the Fermi energy increase was about 100 V:23 meV?

7.3 Utilize the data of diamagnetic susceptibility in Table 7.3 to estimate the average orbital radius $\sqrt{\langle r^2 \rangle}$ of alkali metals, and compare the results with the ion-radius given in chemistry.

7.4 The $1s$ wavefunction of a hydrogen atom is $\psi(\vec{r}) = (\pi a_B^3)^{1/2} \exp(-r/a_B)$. Prove that $\langle r^2 \rangle = 3a_B^2$, and the corresponding diamagnetic susceptibility of a hydrogen molecule is $-2.36 \times 10^{-6}\,cm^3/mol$.

7.5 In the benzene structure, the edge of the hexagonal carbon ring is 1.4Å. In a carbon atom, three valence electrons stay in the sp^2 orbits and another valence electron is in the π bond extended over the whole ring. Roughly estimate the diamagnetic susceptibility of liquid benzene (density: 0.88 g/cm^3).

7.6 The paramagnetic susceptibility of CuSO$_4$ is mainly contributed by the Cu-ion with spin $J = 1/2$. If the average ion density is n, write down the expression for the magnetization $M(T)$ and its high-temperature expansion. In a magnetic field $H = 0.5\,T$, what is the lower boundary of the "high-temperature limit"?

7.7 Prove that, in a metal, the ratio of the diamagnetic susceptibility χ_d contributed by inner-core ions over the Pauli paramagnetic susceptibility χ_p of valence electrons is

$$\frac{\chi_d}{\chi_p} = -\frac{1}{3}\frac{z_i}{z_v}\langle (k_F r)^2 \rangle \qquad (7.88)$$

where z_i is the number of electrons in the inner-core ion and z_v is the number of valence electrons of an atom.

7.8 Prove that, in an ion with non-zero spin, the ratio of the paramagnetic susceptibility χ_p of valence electrons over the diamagnetic susceptibility χ_d contributed by inner-core ions is

$$\frac{\chi_p}{\chi_d} = -\frac{g_J^2 J(J+1)}{2Z k_B T}\frac{\hbar^2}{\langle mr^2 \rangle} \qquad (7.89)$$

where Z is the number of electrons in the inner-core ion and J is the atomic spin quantum number.

7.9 In a magnetic field, the Fermi energy and the effective mass of bismuth are different from that in zero field. In Figure 7.7, the de Haas-van Alphen oscillation period of bismuth is $\Delta H^{-1} = 7.8 \times 10^{-5}\,\text{G}^{-1}$. If the effective "Fermi energy" is $0.00839\,\text{eV}$, what is the ratio of the effective magneton over the Bohr magneton? What is the effective mass of electrons?

7.10 Use the saturation magnetization and the atomic moment (Ni: 0.68; Fe: 3) of Ni_3Fe in Table 7.6 to calculate the lattice constant a (in a cubic unit cell, Fe stays at corners, Ni stays at face-centers).

7.11 **a.** Prove that the Brillouin function $B_{1/2}(x) = \tanh(x)$ by Tyler expansions.

b. For ferromagnets with spin $J = 1/2$, the Weiss-Brillouin theory gives a self-consistent equation $M = (N\mu_B/V)B_{1/2}(\mu_B\lambda_E M/k_B T)$. Prove that the energy with respect to the Curie temperature $k_B T_c = N\lambda_E \mu_B^2/V$.

c. When the temperature $T \to T_c$, prove that $M = (3N\mu_B/V)(T_c/T - 1)^{1/2}$.

7.12 **a.** Utilize the neutron diffraction data of MnO in Figure 7.17 to calculate the lattice constant of MnO (neutron wavelength is $1.057\,\text{Å}$).

b. Based on the Curie-Weiss temperature of MnO in Table 7.9, and Eq. 7.60, calculate the mean field parameter λ_A.

7.13 Utilize the magnetization $M(T)$ in Figure 7.19 to calculate the exchange coupling J_e of nickel (assume $S = 1/2$, $A = 2$).

7.14 The Dysprosium metal has HCP structure. It is believed that the atomic moments on (001) surface are ferromagnetically aligned, but the magnetic moment has $40°$ rotation from one plane to the next. How can this magnetic structure be proved using magnetic neutron diffraction?

7.15 The low-temperature specific heat of $Y_3Fe_5O_{12}$ can be expressed in the form $C/T^{3/2} = a + bT^{3/2}$. What is the information given by parameters a and b?

7.16 Based on the chemical shift data of the CH_3 cluster in Figure 7.25 (last line), calculate the corresponding frequency shift of NMR.

Optical and Dielectric Properties of Solids

8

ABSTRACT

- Unification of optical, electrical, and magnetic properties
- Lorentz optical model and polarization process
- Laser: Einstein's stimulated radiation

Along with mechanics and atomism, the Ancient Greeks had also acquired a basic understanding of optics, particularly geometric optics. In Plato's Academy, for example, pupils were taught that the incident angle equaled the reflection angle of light. In ca. 300 B.C., Euclid of Alexandria in Egypt published the book *Optics*, in which he noted that a fire could be started by focusing light using a concave mirror. In about A.D. 139, Ptolemy of Alexandria determined that the refraction angle of light was basically proportional to the incident angle.

The next great breakthrough in optics came with Isaac Newton's explanation of the light spectrum. In 1672, he proved that white light could be produced by mixing all the colors together, which indicated that colored lights were simpler in nature. In 1800, Friedrich Wilhelm Herschel discovered that infrared light was outside the long-wavelength range of visual light, and in 1801, Johann Wilhelm Ritter and Willian Hyde Wollaston found ultraviolet rays in optical-chemical reactions. In 1885, Heinrich Rudolf Hertz discovered the electromagnetic waves in the radio frequency range 10 Hz–100 GHz. Then, in 1895, R. W. C. Roentgen made his epochal discovery of X-rays. "Maxwell's Rainbow" was completed in 1949 when Jesse DuMond identified γ-rays using diffraction methods.

The whole light spectrum is called Maxwell's Rainbow because, in 1865, James Clerk Maxwell predicted the existence of electromagnetic waves with fixed wave speeds and arbitrary wavelengths, and visual colored light was cited as an example. Maxwell's theory was developed through the unification of the basic sciences of electricity, magnetics, and optics.

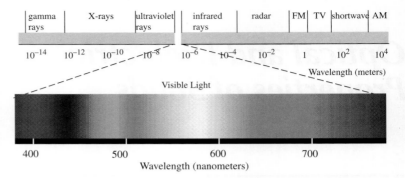

Figure 8.1 Maxwell's rainbow: the electromagnetic spectrum from γ-rays to radio waves (from Kaiser K., Peter. *The Joy of Visual Perception: A Web Book*, http://www.yorku.ca/eye/)

Maxwell's original theory did give a comprehensive description of the propagation of light in vacuum, even though the optical properties of matter were not well understood at that time. In 1870, Hermann von Helmholtz derived the correct laws of reflection and refraction from Maxwell's equations by using proper boundary conditions. Then, in 1880, the great Dutch physicist, Hendrik Anton Lorentz, analyzed the propagation of light in matter using Maxwell's equations. Lorentz found that the "co-vibration" or resonance of electrons in electromagnetic waves was the key to explaining the dispersion phenomena, i.e., the frequency-dependent refractive indices and velocity of light in matter. At the same time, Joseph Larmor at Cambridge produced similar results. In 1933, Drude and Zener described complementarity in metals. Lorentz won the Nobel Prize in Physics in 1902 for the Theory of Electrons and his work on light propagation.

Luminescence and laser emission are other important optical properties. In 1917, Einstein proposed the theory that the absorption or emission rate of photons from an exited atom was proportional to the radiation energy. In the 1950s, based on Einstein's theory, C. H. Townes, N. G. Basov, and A. M. Prokhorovmaser invented maser-laser technology, for which they were awarded a Nobel Prize in 1964.

Piezoelectricity and ferroelectricity are important dielectric properties of solids owing to the symmetry breaking, which is analogous to ferromagnetism. Piezoelectricity was discovered by Pierre Curie and his brother, Jacques, in 1880. They found that physical pressure applied to a crystal induced a small electric potential. Ferroelectricity with hysteric nonlinearity of polarization was found in Rochelle salt by a French physicist, Joseph Valasek, in 1920. These properties are not discussed further in this chapter, but interested readers may wish to explore related texts on ceramic materials.

8.1 Unification of Optical, Electrical, and Magnetic Properties

The optical, electrical, and magnetic properties of solids are unified by the Maxwell equations in continuous media. The Maxwell equations of microscopic electromagnetic fields in a vacuum were given in Eqs. (5.1) to (5.4); while the following Maxwell equations deal with the "macroscopic" electromagnetic fields, defined at scales above 1 nm (cgs units):

$$\vec{\nabla} \cdot \vec{D} = 4\pi \rho_0, \qquad (8.1)$$

$$\vec{\nabla} \cdot \vec{B} = 0, \qquad (8.2)$$

$$\vec{\nabla} \times \vec{E} = -\frac{1}{c}\frac{\partial \vec{B}}{\partial t}, \qquad (8.3)$$

$$\vec{\nabla} \times \vec{H} = \frac{4\pi}{c}\vec{j}_0 + \frac{1}{c}\frac{\partial \vec{D}}{\partial t}. \qquad (8.4)$$

where ρ_0 is the free charge density and $\vec{j}_0$ is the related current density, and the last term in Eq. (8.4) is the displacement current introduced by J. C. Maxwell. The above Maxwell equations have to be solved together with the electromagnetic characteristics of the continuous media:

$$\vec{D} = \vec{E} + 4\pi \vec{P} = \tilde{\epsilon} \cdot \vec{E} \qquad (8.5)$$

$$\vec{B} = \vec{H} + 4\pi \vec{M} = \tilde{\mu} \cdot \vec{H}. \qquad (8.6)$$

where $\tilde{\epsilon}$ is the dielectric constant matrix and $\tilde{\mu}$ is the permeability matrix. When an external $\vec{E}$-field is applied, the polarization $\vec{P}$ is induced in the dielectric; therefore the $\vec{D}$-field must be higher than the $\vec{E}$-field, and the difference is described by the $\tilde{\epsilon}$-matrix. Similarly, the difference between $\vec{B}$ and $\vec{H}$ is caused by the magnetization $\vec{M}$, and is described by the $\tilde{\mu}$-matrix, as illustrated in Figure 8.2.

In poly-crystalline materials or cubic crystals, both the dielectric constant matrix $\tilde{\epsilon}$ and the permeability matrix $\tilde{\mu}$ are diagonal, and can be replaced by the numbers ϵ and μ respectively. The electromagnetic wave equations can be derived from the Maxwell equations and Ohm's law $\vec{j}_0 = \sigma \vec{E}$:

$$\begin{cases} \vec{\nabla} \times \vec{E} = -\left(\frac{1}{c}\mu\frac{\partial}{\partial t}\right)\vec{H} \\ \vec{\nabla} \times \vec{H} = \left(\frac{4\pi}{c}\sigma + \frac{1}{c}\epsilon\frac{\partial}{\partial t}\right)\vec{E} \end{cases} \qquad (8.7)$$

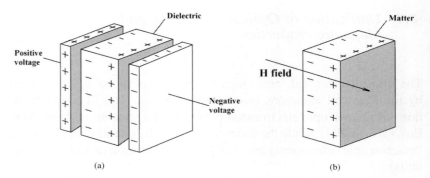

Figure 8.2 (a) Dielectric in an external $\vec{E}$-field; (b) matter in an external $\vec{H}$-field

The electromagnetic wave is a transverse wave, where both the $\vec{E}$-field and $\vec{H}$-field are perpendicular to the wavevector $\vec{k}$. The light velocity v and the refraction index n can be found for the material following Eq. (8.7):

$$\begin{cases} \vec{E} = \vec{E}_0 e^{i(\vec{k}\cdot\vec{r}-\omega t)} \\ \vec{H} = \vec{H}_0 e^{i(\vec{k}\cdot\vec{r}-\omega t)} \end{cases}$$

$$\Rightarrow \begin{cases} \vec{k} \times \vec{E}_0 = \mu \frac{\omega}{c} \vec{H}_0 \\ \vec{k} \times \vec{H}_0 = -\left(\epsilon + i\frac{4\pi\sigma}{\omega}\right) \frac{\omega}{c} \vec{E}_0 = -\epsilon^* \frac{\omega}{c} \vec{E}_0 \end{cases} \quad (8.8)$$

$$\vec{k} \times \vec{k} \times \vec{E}_0 = -\mu\epsilon^* \frac{\omega^2}{c^2} \vec{E}_0 = -k^2 \vec{E}_0$$

$$\Rightarrow n = \frac{kc}{\omega} = \frac{c}{v} = \sqrt{\epsilon^* \mu} \quad (8.9)$$

where $\epsilon^* = \epsilon + i\frac{4\pi\sigma}{\omega}$ is the general expression of the dielectric constant for insulators, semiconductors, or metals, and the effect of electrical conductivity is included.

The refraction index n has both the real part n' and the imaginary part n''. Because the conductivity σ might be non-zero, both the dielectric constant $\epsilon = \epsilon' + i\epsilon''$ and the permeability $\mu = \mu' + i\mu''$ could have the imaginary part:

$$(n' + in'')^2 = \left(\epsilon' + i\epsilon'' + i\frac{4\pi\sigma}{\omega}\right)(\mu' + i\mu'') = a + ib;$$

$$n' = \sqrt{\frac{1}{2}(\sqrt{a^2+b^2} + a)}; \quad a = \epsilon'\mu' - \epsilon''\mu'' - \frac{4\pi\sigma}{\omega}\mu''$$

$$n'' = \sqrt{\frac{1}{2}(\sqrt{a^2+b^2} - a)}; \quad b = \epsilon''\mu' + \epsilon'\mu'' + \frac{4\pi\sigma}{\omega}\mu' \quad (8.10)$$

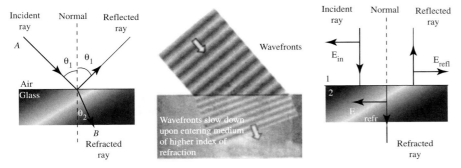

Figure 8.3 Reflection, refraction, and schematics of the Fresnel's formula

There are two other significant optical coefficients of materials: the absorption coefficient A and the reflection coefficient R, and both are related to $n = n' + in''$. The absorption coefficient can be found from the Poynting vector $\vec{P} = \vec{E} \times \vec{H}^*$:

$$P_{av} = \frac{1}{2}\Re(\vec{E} \times \vec{H}^*) = \frac{1}{2}E_0 H_0 \exp(-A\hat{k} \cdot \vec{r}) \Rightarrow A = 2k'' = 2n''\frac{\omega}{c}$$

(8.11)

The reflection coefficient at the interface of the non-ferromagnetic materials 1 and 2 can be found using Fresnel's formula at normal incidence, which can be derived by boundary conditions of electromagnetic fields together with Eq. (8.8) of electromagnetic waves, as shown in Figure 8.3(c):

$$\begin{cases} E_{in} + E_{refl} = E_{refr} \\ H_{in} + H_{refl} = H_{refr} \end{cases}$$

$$\Rightarrow \begin{cases} E_{in} + E_{refl} = E_{refr} \\ \frac{k_1}{\mu_1}E_{in} - \frac{k_1}{\mu_1}E_{refl} = \frac{k_2}{\mu_2}E_{refr} \end{cases} \quad (\mu_1 = \mu_2 = 1) \Rightarrow$$

$$E_{refl} = \frac{k_1 - k_2}{k_1 + k_2}E_{in}$$

$$\Rightarrow R = \frac{P_{av}^{refl}}{P_{av}^{in}} = \left|\frac{E_{refl}}{E_{in}}\right|^2 = \frac{(n_1' - n_2')^2 + (n_1'' - n_2'')^2}{(n_1' + n_2')^2 + (n_1'' + n_2'')^2}$$

(8.12)

The unification of the optical, electrical, and magnetic properties of matter is thus proved by the Maxwell equations, where the refraction index n, absorption coefficient α_0, and reflection coefficient R simply depend on the dielectric constant, conductivity, and permeability ϵ, σ, μ respectively.

8.2 Lorentz Optical Model and Polarization Process

The optical properties of non-ferromagnetic insulators are determined by the dielectric constant $\epsilon = \epsilon' - i\epsilon''$. The characteristics of the dielectric constant are influenced by three polarization processes, from high frequency to low frequency in Maxwell's Rainbow, as schematically shown in Figure 8.4, namely: (1) electron polarization near visual light frequency, as stated in the original Lorentz model, (2) ionic deviation polarization around infrared frequency, where the photons have resonance with the phonons in the optical branch, and (3) relaxation polarization in the microwave frequency, which is caused by the motion of defects.

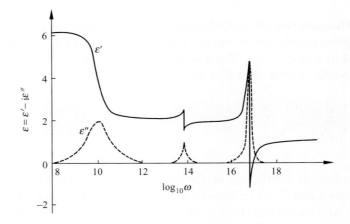

Figure 8.4 Real and imaginary part of a dielectric constant versus frequency

In 1880, Lorentz devised an optical model to explain the dispersion of insulators, which states that the resonance of the bound electron's oscillation in an atom with an electromagnetic wave is the key to polarization.

According to the Lorentz-Lorenz formula, devised by Danish scientist Ludwig Valentine Lorenz in 1869 and Dutch physicist Hendrik Antoon Lorentz in 1870, the macroscopic dielectric constant is related to the microscopic polarizability:

$$P = \sum_i n_i p_i = \sum_i n_i \alpha_i E^{\mathrm{loc}} = \sum_i n_i \alpha_i (E + N_{xx} P); \quad N_{xx} = 4\pi/3$$

$$\sum_i n_i \alpha_i = \frac{3}{4\pi} \frac{\epsilon - 1}{\epsilon + 2}; \quad \epsilon = 1 + 4\pi\chi = \frac{1 + (8\pi/3) \sum_i n_i \alpha_i}{1 - (4\pi/3) \sum_i n_i \alpha_i}$$

(8.13)

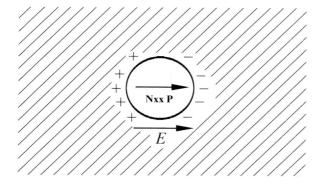

Figure 8.5 Local electric field: external field and depolarizing field

where n_i, p_i, α_i are the atomic density, dipolar moment, and polarizability of the i'th class of atoms or ions respectively; E is the external field, P is the polarization, and $\chi = P/E$ is the susceptibility of the solid; and E^{loc} is the local field felt by the ion or atom, including both the external field and the depolarizing field, where the depolarizing field is given by the mean-field expression $E^d = N_{xx}P$ ($\tilde{N}$ is called the depolarizing matrix, x is along the direction of the external field), as illustrated in Figure (8.5). The Lorentz-Lorenz formula is also called the Clausius-Mossotti relation, which was formulated independently by Rudolf Clausius and Mossotti in 1879.

Lorentz wrote the equation of motion of a bounded electron in an insulator as:

$$m\frac{d^2x}{dt^2} = -kx - \gamma m\frac{dx}{dt} - eE_0 e^{-i\omega t} \qquad (8.14)$$

where m is the mass of the electron, x is the coordinate of the electron along the external field, k is the elastic constant describing the bound state of the electron, γ is the damping coefficient, and $E = E_0 \exp(-i\omega t)$ is the electric field of the electromagnetic wave. The polarizability of the bound electrons can thus be found:

$$\alpha_e(\omega) = \frac{-ex(\omega)}{E} = \frac{e^2}{m}\frac{1}{(\omega_0^2 - \omega^2) - i\gamma\omega}; \quad \omega_0 = \sqrt{\frac{k}{m}} \qquad (8.15)$$

The electron polarizability has resonance with the electromagnetic wave at frequency $\omega = \omega_0$. In quantum physics, $\hbar\omega_0$ is simply the energy difference before and after the transition, therefore there might be several values of ω_0 resonant with the electromagnetic wave. This will be discussed further in the next section.

The dielectric constant, refraction index, and absorption coefficient due to the electron polarizability in an insulator with $\mu' = 1, \mu'' = 0$,

$\sigma = 0$ can be found from Eq. (8.15) and the Lorentz-Lorenz formula in Eq. (8.13):

$$\epsilon_e(\omega) = \frac{1 + (8\pi/3)n\alpha_e(\omega)}{1 - (4\pi/3)n\alpha_e(\omega)} = \epsilon'(\omega) + i\epsilon''(\omega);$$

$$\epsilon'(\omega) = \frac{\left(\omega_0^2 + \tfrac{2}{3}\omega_p^2 - \omega^2\right)\left(\omega_0^2 - \tfrac{1}{3}\omega_p^2 - \omega^2\right) + \gamma^2\omega^2}{\left(\omega_0^2 - \tfrac{1}{3}\omega_p^2 - \omega^2\right)^2 + \gamma^2\omega^2};$$

$$\epsilon''(\omega) = \frac{\omega_p^2 \gamma \omega}{\left(\omega_0^2 - \tfrac{1}{3}\omega_p^2 - \omega^2\right)^2 + \gamma^2\omega^2}; \quad \omega_p^2 = \frac{4\pi n e^2}{m} \quad (8.16)$$

$$n'(\omega) \simeq \sqrt{\epsilon'} = \sqrt{\frac{\left(\omega_0^2 + \tfrac{2}{3}\omega_p^2 - \omega^2\right)\left(\omega_0^2 - \tfrac{1}{3}\omega_p^2 - \omega^2\right) + \gamma^2\omega^2}{\left(\omega_0^2 - \tfrac{1}{3}\omega_p^2 - \omega^2\right)^2 + \gamma^2\omega^2}} \quad (8.17)$$

$$A(\omega) = 2n''\frac{\omega}{c} \simeq \frac{\omega}{c}\sqrt{\frac{\epsilon''^2}{\epsilon'}}$$

$$= \frac{\omega}{c}n'(\omega)\frac{\omega_p^2 \gamma \omega}{\left(\omega_0^2 + \tfrac{2}{3}\omega_p^2 - \omega^2\right)\left(\omega_0^2 - \tfrac{1}{3}\omega_p^2 - \omega^2\right) + \gamma^2\omega^2} \quad (8.18)$$

where ω_p is called the plasma frequency, which is in the order of $10^{16} s^{-1}$. In silicon, the refraction index $n = n'$ and the scaled absorption coefficient $\kappa = n'' = A/(2\omega/c)$ near the visual light frequency are shown in Figure 8.6(a); the calculated ϵ', ϵ'', n, κ versus the scaled ω are plotted in Figure 8.6(b). This explains the optical properties reasonably well (although in real solids, there are several sets of ω_0).

In covalent silicon crystals, the dielectric constant $\epsilon(\omega)$ simply equals the static dielectric constant $\epsilon_s = \epsilon(0)$ below the visual light frequency:

$$\epsilon(0) = \frac{\omega_0^2 + \tfrac{2}{3}\omega_p^2}{\omega_0^2 - \tfrac{1}{3}\omega_p^2} = 11.7; \quad \omega_0 = 0.653\omega_p \quad (8.19)$$

This is the reason why Figure 8.6(b) with coefficients $\omega_0/\omega_p = 7.2/11 = 0.654$ gives a reasonable explanation of the refraction index of silicon. In the quantum mechanical band theory, the resonant frequency of silicon (with respect to a transition keeping $\Delta k \simeq 0$) should be equal to

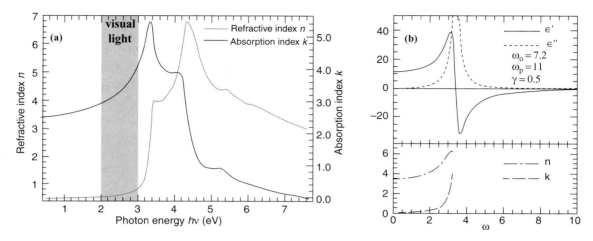

Figure 8.6 (a) real and imaginary part of the refraction index of silicon (from www.ioffe.rssi.ru); (b) calculated ϵ', ϵ'', n, κ versus the scaled frequency ω with respect to one ω_0

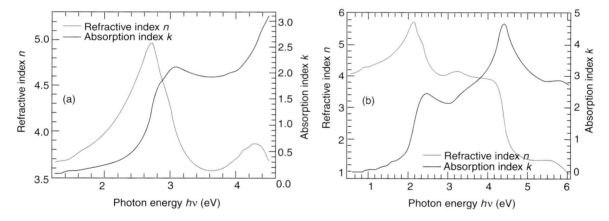

Figure 8.7 Normal and abnormal dispersion (a) GaAs; (b) Ge (from www.ioffe.rssi.ru)

$\omega_1 \simeq 3eV/\hbar$, $\omega_2 \simeq 4eV/\hbar$ and so on, as shown in the bands of silicon in Figure 6.8(a). The quantum mechanical expression of the dielectric constant will be discussed further later in this chapter, but it can be concluded that the classical frequency ω_0 and ω_p in the Lorentz model are not equivalent to the quantum frequency $\omega = \Delta E^{\hbar}$, where ΔE is the transition energy in electron bands.

The color of light corresponds to the frequency of the electromagnetic waves. Therefore, if the refraction index is dependent on the light frequency, different colors in visual light are due to different refraction indices. This reflects the light dispersion described by Isaac Newton, namely that violet light with a higher frequency is refracted further away than the red light with lower frequency. This is called "normal dispersion," and is due to the refraction index of silicon lying in the visual light range (2–3 eV), as shown in Figure 8.6. However, some solids have a resonant frequencies that are located in a range that their refraction indices decrease with a higher frequencies of incident light. This is called "abnormal dispersion," and can be seen in the refraction indices for gallium arsenide and germanium, as shown in Figure 8.7.

The colors of solids are related to their absorption coefficients and the reflection indices in the visual light region. Pure diamond or Al_2O_3 crystals are colorless, i.e., their absorption coefficients of visual light are very small. If there are impurities in crystals, an impurity level is introduced into the energy band, and the color of the crystal changes. For example, if chromium is doped in the transparent aluminum oxide (Al_2O_3) crystal, a new absorption peak appears at the green light frequency: this is the reason why the ruby has a red color. Similarly, if boron impurities are doped in diamond, a new absorption peak appears at the red light frequency, and the crystal has a deep blue color.

The Lorentz optical model can also be used to explain the ionic deviation polarization in ionic crystals or covalent crystals with ionic bonds:

$$M\frac{d^2x}{dt^2} = -Kx - \Gamma M \frac{dx}{dt} + qE_0 e^{-i\omega t} \quad (8.20)$$

$$\alpha_i(\omega) = \frac{-ex(\omega)}{E} = \frac{q^2}{M} \frac{1}{(\omega_{0i}^2 - \omega^2) - i\Gamma\omega}; \quad \omega_{0i} = \sqrt{\frac{K}{M}} \quad (8.21)$$

The resonant frequency ω_{0i} of ionic deviation polarization is the $\omega_j(0)$ of the optical branches discussed in Chapter Four. In Figure 8.8(b), the reflectance R of sodium chloride peaks at $185\,\text{cm}^{-1} = 5.55\,\text{THz}$ and $265\,\text{cm}^{-1} = 7.95\,\text{THz}$, which basically agrees with $\omega_j(0) = 4.8\,\text{THz}$ and $7.5\,\text{THz}$ of the optical branches shown in Figure 4.13(b).

In sodium chloride, the dielectric constant $\epsilon'(\omega)$ is about 2.8 to 3.0 in the microwave frequency range 0.4 to 4 GHz (based on Komarov, Mironov, and Romanov, 1999), but the static dielectric constant is 5.895 (based on Andeen, Fontanella, and Schuele, 1970). The reduction of $\epsilon'(\omega)$ at the radio frequency must be due to the relaxation polarization of defects. The relaxation polarization is also called the dipolar polarization, with respect to the rotation of molecular dipoles in liquids or gas. The dispersion of the dielectric constant can be described by the Drude formula and the motion equation for the polarization P:

$$\frac{dP}{dt} = -\frac{P}{\tau} + LE \Rightarrow \chi(\omega) = \frac{P(\omega)}{E(\omega)} = \frac{L\tau}{1 + i\omega\tau}$$

$$\epsilon'(\omega) = \epsilon_{\text{m.w.}} + \frac{\epsilon_s - \epsilon_{\text{m.w.}}}{1 + \omega^2\tau^2}; \quad \epsilon''(\omega) = (\epsilon_s - \epsilon_{\text{m.w.}})\frac{\omega\tau}{1 + \omega^2\tau^2}$$

$$(8.22)$$

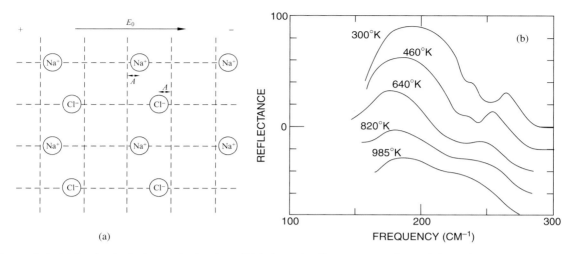

Figure 8.8 (a) ionic deviation polarization; (b) infrared reflectance R of NaCl (based on Hass, 1960)

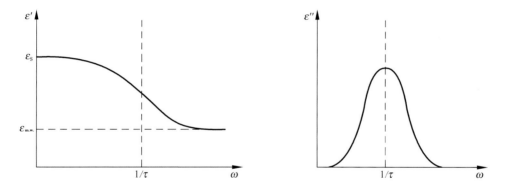

Figure 8.9 Relaxation polarization: real and imaginary dielectric constant

The dielectric dispersion near frequency $\omega \sim \tau^{-1} \sim 10^{10}\, \text{s}^{-1}$ is shown in Figure 8.9.

The covalent crystals, such as diamond or silicon, have no relaxation polarization with respect to the motion of defects, because there are few defects in these crystals. As a result, the refraction index of a covalent crystal is almost a constant below the infrared frequency. On the other hand, the crystals with ionic bonds have charged ions, thus both the ionic polarization and the relaxation polarization exist in the dispersion phenomena; therefore, their refraction index n_{opt} at the visual light frequency must be lower than the static refraction index n_s, as listed in Table 8.1.

In summary, following the Lorentz model, in semiconductors or insulators, the dielectric constant or refraction index has important changes at certain frequencies of electromagnetic waves, where the

Table 8.1 Refraction index n_s at zero-frequency and n_{opt} at visual light frequency (based on Andeen, Fontanella, and Schuele, 1970, and www.ioffe.rssi.ru)

	n_s	n_{opt}		n_s	n_{opt}		n_s	n_{opt}
diamond	2.38	2.41–2.46	LiF	3.006	1.39–1.40	CaF$_2$	2.60	1.43–1.44
SiC	3.11	2.65–2.75	NaF	2.252	1.45–1.46	BaF$_2$	2.71	1.47–1.48
Si	3.42	3.80–5.20	KF	2.345	1.49–1.51	glass	3–10	1.4–3
GaAs	3.59	4.0–4.95	NaCl	2.428	1.50–1.7	SiO$_2$	2.07	1.45–1.47
InSb	4.24	4.1–3.3	KCl	2.194	1.45–1.7	TiO$_2$	9.70	2.5–3.3

photon has resonance with quasi particles in solids. At the visual light frequency or a photon energy of 1–10 eV, the photon might be resonant with the Bloch electrons in energy bands. At the far infrared frequency or a photon energy of 10–100 meV, the photon would be resonant with the phonons in the optical branch. In the radio wave frequency or a photon energy of 0.1–10 μeV, the photon would be resonant with the motion of defects in solids or the rotation of molecular dipoles in liquids and gases.

Drude Optical Model of Metals

The Lorentz optical model was extended to explain the optical properties of metals by Drude and Zener in 1933 and Kronig in 1934. The intrinsic oscillation term in the Lorentz model is discarded, so the motion equation and the polarizability of free electrons in a metal become:

$$m\frac{d^2 x}{dt^2} = -\gamma m \frac{dx}{dt} - eE_0 e^{-i\omega t} \Rightarrow \alpha_e = -\frac{e^2}{m}\frac{1}{\omega^2 + i\gamma\omega} \quad (8.23)$$

The dielectric constant, refraction index, and absorption coefficient in a non-ferromagnetic metal with $\mu' = 1$ and $\mu'' = 0$ can be found from Eq. (8.23):

$$\epsilon_e(\omega) = 1 + 4\pi n \alpha_e(\omega) = \epsilon'(\omega) + i\frac{4\pi\sigma}{\omega}; \quad \omega_p^2 = \frac{4\pi n e^2}{m}$$

$$\epsilon'(\omega) = 1 - \frac{\omega_p^2}{\omega^2 + \gamma^2}; \quad \sigma(\omega) = \frac{\omega_p^2 \gamma/4\pi}{\omega^2 + \gamma^2} \to \frac{ne^2}{m\gamma}(\omega \to 0);$$

(8.24)

$$n'(\omega) \simeq \sqrt{\epsilon'} = \sqrt{\frac{\omega^2 + \gamma^2 - \omega_p^2}{\omega^2 + \gamma^2}} \quad (8.25)$$

$$A(\omega) \simeq \frac{\omega}{c}\sqrt{\frac{(4\pi\sigma/\omega)^2}{\epsilon'}} = \frac{\omega}{c}n'(\omega)\frac{\omega_p^2 \gamma}{\omega(\omega^2 + \gamma^2 - \omega_p^2)} \quad (8.26)$$

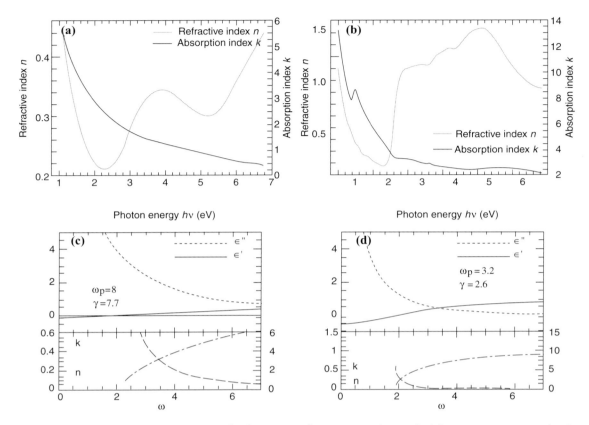

Figure 8.10 Real and imaginary parts of refraction indices versus the scaled frequencies: (a) Li; (b) Cu (from www.ioffe.rssi.ru), calculated n and κ by the Drude optical model; (c) Li; (d) Cu

where γ^{-1} is analogous to the relaxation time τ in the Drude model. The parameters in the Drude optical model and Drude model are compared in Table 8.2.

The Drude optical model can explain the real part $n = n'$ and the imaginary part $\kappa = n''$ of the refraction index reasonably well for alkali metals, as shown in Figure 8.10(a)(c); however, γ^{-1} is 1% of the relaxation time τ in the free electron Drude model, as listed in Table 8.2, which might be related to the short interaction time during the phonon-electron scattering in metals:

$$\gamma^{-1} \simeq \frac{\delta}{c} \simeq \frac{300\text{Å}}{3 \times 10^8 \text{ m/s}} = 10^{-16} \text{ s}; \quad \delta = \frac{\lambda}{4\pi\kappa} \simeq 300\text{Å} \quad (8.27)$$

where δ is the penetration depth of the electromagnetic wave into the metal. Due to the existence of the d-bands in crystals like cooper, the calculated n and κ differ significantly from the measurements for noble metals, as shown in Figure 8.10(b)(d).

Table 8.2 Drude model compared to the Drude optical model: $n/(10^{22}\,\text{cm}^{-3})$ is the free electron density, $\tau/(10^{-14}\,\text{s})$ and $\gamma^{-1}/(10^{-14}\,\text{s})$ are the relaxation times of electrons, $\omega_p^0/(10^{16}\,\text{s}^{-1})$ and $\omega_p/(10^{16}\,\text{s}^{-1})$ are the plasma frequencies

	n	τ	ω_p^0	$\hbar\gamma/(\text{eV})$	γ^{-1}	$\hbar\omega_p/(\text{eV})$	ω_p
Li	4.70	0.88	1.22	7.7	0.85×10^{-2}	8.0	1.21
Cu	8.37	2.66	1.63	2.6	2.53×10^{-2}	3.2	0.48

8.3 The Laser: Einstein's Stimulated Radiation Theory

In 1917, Albert Einstein studied the interactions between electromagnetic waves and quantum systems using thermodynamics. Einstein's conclusion regarding the two-dimensional quantum system ($\Delta E = E_b - E_a > 0$) suggested that the atoms and molecules could in fact amplify the electromagnetic energy (based on Townes, 1964):

$$\frac{dI}{dt} = AN_b - BN_aI + B'N_bI \Rightarrow I = \frac{A}{B}\frac{N_b}{N_b - N_a}[e^{B(N_b - N_a)t} - 1] \tag{8.28}$$

The terms A, B, B' in Eq. (8.28) stand for spontaneous emission, spontaneous absorption, and stimulated emission respectively, as seen in Figure 8.11; $B = B' = \eta_0|\langle b|\Delta\mathcal{H}|a\rangle|^2$ is usually true; N_a and N_b are the number of microscopic particles remaining in the non-degenerate lower energy level E_a and upper energy level E_b respectively.

In ordinary systems, Boltzmann's law requires $N_b = N_a e^{-\beta\Delta E}$ to be lower than N_a, thus the solution to Eq. (8.28) will finally stabilize at (based on Einstein, 1917):

$$I \to \frac{A}{B}\frac{N_b}{N_a - N_b} = \frac{A}{B}\frac{e^{-\beta\Delta E}}{1 - e^{-\beta\Delta E}} = \Delta\nu\frac{8\pi h\nu^3}{c^3}\frac{1}{e^{\beta\Delta E} - 1} \tag{8.29}$$

If the population inversion (or the negative temperature state) $N_b > N_a$ somehow occurs, the solution to Eq. (8.28) will display the characteristics of an electromagnetic energy amplifier:

$$I = I_0(e^{ut} - 1); \quad I_0 = \frac{A}{B}\frac{N_b}{N_b - N_a}, \quad u = B(N_b - N_a) \tag{8.30}$$

This is the basic mechanism of the laser, to which the population inversion on energy levels is the key.

8.3 The Laser: Einstein's Stimulated Radiation Theory

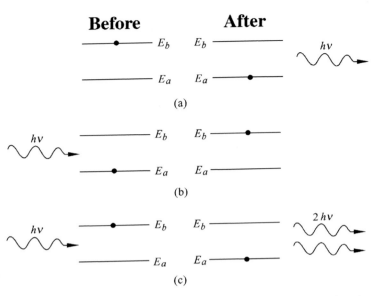

Figure 8.11 (a)(b) spontaneous emission and absorption; (c) stimulated emission

8.3.1 Quantum Mechanical Theory of Radiation

The parameter B of Einstein's radiation theory was explained after the establishment of quantum mechanics. Erwin Schrodinger, Oskar Klein, and Walter Gordon gave a semi-classical treatment of the quantum radiation theory in 1926.

The perturbation Hamiltonian related to radiation can be obtained by the Pauli Hamiltonian of an atom in an uniform magnetic field in Chapter Seven:

$$\mathcal{H} = \sum_i \left[\frac{\left(\vec{p}_i + \frac{e}{c}\vec{A}(\vec{r}_i)\right)^2}{2m} + v(\vec{r}_i) + \frac{1}{2}\sum_{j \neq i} \frac{e^2}{r_{ij}} + g_0 \mu_B \vec{s}_i \cdot \vec{H} \right]$$

$$= \mathcal{H}_0 + \mathcal{H}_1$$

$$\mathcal{H}_0 = \sum_i \left[\frac{\vec{p}_i^2}{2m} + v(\vec{r}_i) + \frac{1}{2}\sum_{j \neq i} \frac{e^2}{r_{ij}} + g_0 \mu_B \vec{s}_i \cdot \vec{H} \right]$$

$$\mathcal{H}_1 = \sum_i \left[\frac{e}{mc}\vec{p}_i \cdot \vec{A}(\vec{r}_i) + \frac{e^2}{2mc^2}\vec{A}^2(\vec{r}_i) \right] \simeq \sum_i \left[\frac{e}{mc}\vec{p}_i \cdot \vec{A}(\vec{r}_i) \right]$$

(8.31)

where the gauge $\vec{\nabla} \cdot \vec{A} = 0$ is used. In an electromagnetic plane wave radiation, the vector potential can be chosen as:

$$\vec{A}(\vec{r}) = \frac{c}{i\omega}\vec{E}(\vec{r}) = \frac{c}{i\omega}\vec{E}_0[e^{i(\vec{k}\cdot\vec{r}-\omega t)} - e^{-i(\vec{k}\cdot\vec{r}-\omega t)}]; \quad (8.32)$$

$$\vec{\nabla} \times \vec{A} = \mu\vec{H}_0[e^{i(\vec{k}\cdot\vec{r}-\omega t)} + e^{-i(\vec{k}\cdot\vec{r}-\omega t)}] = \vec{B}(\vec{r}) \quad (8.33)$$

The eigen-function and the eigen-values can be found by the perturbation theory:

$$|\psi\rangle = \sum_{n=0}^{\infty} a_n e^{-iE_n t/\hbar} |n\rangle; \quad (a_0 \simeq 1); \quad (\mathcal{H}_0 + \mathcal{H}_1)|\psi\rangle = i\hbar\frac{\partial}{\partial t}|\psi\rangle$$

$$(8.34)$$

$$a_n \simeq \langle n|\Delta H|0\rangle\frac{1 - e^{i(E_n-E_0-\hbar\omega)t/\hbar}}{E_n - E_0 - \hbar\omega} + \langle n|\Delta H^\dagger|0\rangle\frac{1 - e^{i(E_n-E_0+\hbar\omega)t/\hbar}}{E_n - E_0 + \hbar\omega}$$

$$(8.35)$$

Then the transition probability between the 0'th and n'th states should be

$$P_n(t) \simeq \int_0^\infty \frac{d\omega}{2\pi}|\langle n|\Delta H|0\rangle|^2 \frac{2\pi t}{\hbar}\delta(E_n - E_0 \pm \hbar\omega) \quad (8.36)$$

$$\langle n|\Delta H|0\rangle = \langle n|\frac{e}{mc}\frac{c}{i\omega}\sum_i \vec{p}_i \cdot \vec{E}_0 e^{i\vec{k}\cdot\vec{r}_i}|0\rangle$$

$$\simeq \frac{E_n - E_0}{\hbar\omega}\langle n|\sum_i (-e\vec{r}_i)\cdot\vec{E}_0|0\rangle \quad (8.37)$$

$$\left|\frac{1 - e^{i(E_n-E_0\pm\hbar\omega)t/\hbar}}{E_n - E_0 \pm \hbar\omega}\right|^2 \simeq \frac{2\pi t}{\hbar}\delta(E_n - E_0 \pm \hbar\omega); \quad (8.38)$$

This transition probability can be related to the Einstein coefficient B (based on Einstein, 1917; Schrodinger, 1926; Gordon, 1926; and Klein, 1926):

$$P(t) = B_{n0}\frac{\vec{E}_0^2}{2\pi}t = B_{n0}It; \quad B_{n0} = \frac{2\pi}{\hbar^2}\left|\langle n|\sum_i(-e\vec{r}_i)\cdot\hat{E}|0\rangle\right|^2$$

$$(8.39)$$

where $I = \vec{E}_0^2/2\pi$ is the radiation intensity and $\hat{E} = \vec{E}_0/E_0$ is the polarization direction. The coefficient B is related to the atomic dipolar transition matrix.

The average atomic polarizability in the electromagnetic wave can also be found:

$$\begin{aligned}
\alpha &= \frac{1}{E_0} \int \frac{dt}{T} \left\langle \psi \left| \sum_i (-e\vec{r}_i) \cdot \hat{E} \right| \psi \right\rangle \\
&\simeq \frac{1}{E_0} \int \frac{dt}{T} \sum_n \left[a_n \left\langle 0 \left| \sum_i (-e\vec{r}_i) \cdot \hat{E} \right| n \right\rangle \right. \\
&\quad \left. + a_n^* \left\langle n \left| \sum_i (-e\vec{r}_i) \cdot \hat{E} \right| 0 \right\rangle \right] \\
&\simeq \sum_n \frac{E_n - E_0}{\hbar \omega} \left| \left\langle n \left| \sum_i (-e\vec{r}_i) \cdot \hat{E} \right| 0 \right\rangle \right|^2 \\
&\quad \times \left[\frac{1}{E_n - E_0 - \hbar\omega} - \frac{1}{E_n - E_0 + \hbar\omega} \right] \\
&\simeq \sum_n \frac{B_{n0} \hbar \omega_0 / \pi}{\omega_0^2 - \omega^2}; \quad \hbar\omega_0 = E_n - E_0
\end{aligned} \qquad (8.40)$$

This quantum polarizability has the same form as the electron polarizability in the Lorentz optical model, except that there is a dissipation term in Eq. (8.21).

8.3.2 Masers and Lasers

The critical requirement for laser generation, recognized by C. H. Townes, N. G. Basov, and A. M. Prokhorov, is that a positive feedback has to be produced by some resonant circuit to ensure the gain from the stimulated radiation is greater than the circuit loss. The energy transfer rate (gain power) from the stimulated radiation of molecules can be found by Eq. (8.39):

$$W_s = (N_b - N_a) P(t) h\nu = (N_b - N_a) \frac{p^2 E_0^2}{3\hbar^2} \frac{h\nu}{\Delta\nu};$$

$$p^2 = \left| \left\langle b \left| \sum_i (-e\vec{r}_i) \right| a \right\rangle \right|^2, \qquad (8.41)$$

where p is the dipolar transition matrix element of the molecule, which ensures the selection rule of the transition; and $\Delta\nu = 1/t \sim 10^8$ Hz is the transition width.

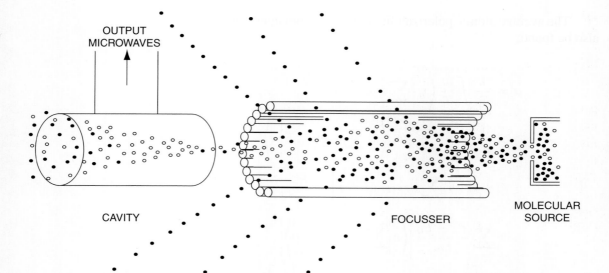

Figure 8.12 Maser system by Townes: directional emission of microwaves (based on Townes, 1964)

If the net power gain is positive in the maser/laser system,

$$W_s > W_c \Rightarrow (N_b - N_a)\frac{p^2 E_0^2}{3\hbar^2}\frac{h\nu}{\Delta\nu} \geq \frac{E_0^2 V\nu}{4Q} \qquad (8.42)$$

where Q is the Q-factor of the circuit, and V is the volume of the molecule system. The population reversion must satisfy these criteria to create a maser or laser:

$$N_b - N_a \geq \frac{3Vh\Delta\nu}{16\pi^2 p^2 Q} \qquad (8.43)$$

In the first maser, the transition in ammonia at $\nu = 23870$ MHz produces microwave emission, as shown in Figure 8.12. At room temperature, the thermal energy $k_B T \simeq 260\, h\nu$ is much larger than the microwave photon energy $h\nu$, therefore an extremely low temperature $T \simeq 300/260 = 1.15$ K has to be used to thermally excite the molecules and produce a low-noise maser of stimulated radiation.

Townes invented the maser in 1954 when he was at Columbia University doing research into microwave physics. Basov invented the semiconductor laser in 1961 while working at the P. N. Lebedev Physical Institute, Moscow in the field of quantum radiophysics. There, three different methods were proposed for obtaining a negative temperature state (or the population reversion) in semiconductors, in the presence of both direct and indirect transitions.

8.3 The Laser: Einstein's Stimulated Radiation Theory

Table 8.3 Semiconductor lasers: materials, structures, and mechanisms (based on Basov, 1964)

Matter	Wavelength $\lambda/$(nm)	Structure and population reversion mechanism
PbSe	8500	p-n junction, electric excitation
PbTe	6500	p-n junction, electric excitation
InSb	5300	p-n junction and high speed electron beam
InAs	3200	p-n junction and high speed electron beam
GaSb	1600	p-n junction and high speed electron beam
InP	900	p-n junction, electric excitation
GaAs	850	p-n junction, electric and optical excitation
CdTe	800	high speed electron beam
CdS	500	high speed electron beam
GaAs-GaP	650–900	p-n junction, electric excitation
GaAs-InAs	850–3200	p-n junction, electric excitation
InAs-InP	900–3200	p-n junction, electric excitation
$Hg_{1-x}Cd_xTe$	828–10^6	p-n junction, electric excitation

Most semiconductor lasers are built upon the structure of p-n junctions, as listed in Table 8.3. The physics of p-n junctions based on semiconductors was discussed in Chapter Six, however, the p-n junctions utilized in lasers have special characteristics for creating population inversion of electrons or holes.

In a p-n junction, in order to obtain a population reversion of electrons or holes, high-density doped impurities are necessary to create a large concentration of carriers. If a positive voltage is applied to such a p-n junction, the Fermi-Dirac statistics of electrons will be disturbed when the current flows through the semiconductor. Electrons flow into the region with a large concentration of holes. Before the annihilation, these high-density electrons injected from the n-region will create a local area with an inverse population (several microns thick at the p-n junction), i.e., more electrons at the conduction band edge than holes at valence band edge, as shown in Figure 8.13. In this local area, the electromagnetic wave can be amplified.

In the semiconductor laser, mirrors are added to introduce feedback coupling into the system. In the resonant cavity between the two mirrors, the number of photons will increase until it reaches equilibrium, in which the quantity of stimulated electrons equals the number of photons emitted from the system.

In 1962, practical gallium arsenide (GaAs) laser diodes were demonstrated by four groups from General Electric (which included Robert H. Hall and Nick Holonyak), the IBM Research Laboratory, and

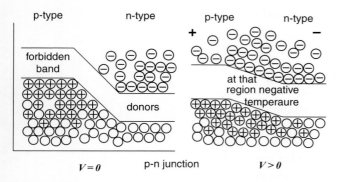

Figure 8.13 Population inversion in a p-n junction by Basov (based on Basov, 1964)

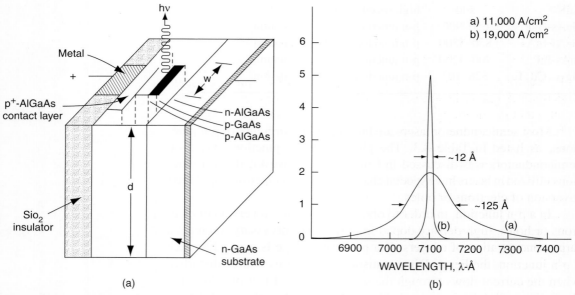

Figure 8.14 (a) III-V compound laser diode with the Fabry-Perot mirrors (from http://ocw.mit.edu); (b) spectral distribution of Holonyak's $Ga(As_{1-x}P_x)$ diode at 77 K (based on Holonyak and Bevacqua, 1962)

the MIT Lincoln Laboratory. Hall's mirror approach was to lap and polish the GaAs diode facets (based on Hall, Fenner, Kinsley, Carlson, and Carlson, 1962), while Holonyak tried to cleave the Fabry-Perot mirrors for his GaAs-GaP optical cavity. The Fabry-Perot mirrors, which enabled high resolution observation of spectral features, were developed by French physicists Marie Fabry and Jean Perot in 1899. The mirrors used in laser diodes are made of SiO_2 and n-type GaAs, as illustrated in Figure 8.14.

Generally, a laser system includes the working substance and a pumping system to introduce the population inversion. The working materials could be solids, liquids, or gases. The most common pumping systems are based on electrical excitation, optical excitation, thermal excitation, chemical excitation, and nuclear energy excitation. All these lasers are based on the same physical mechanism first introduced by Einstein in 1917 and used in the maser and laser devices of the 1950s and 1960s.

Summary

This chapter has focused on the unified optical, dielectric, and magnetic properties of solids. The Lorentz-Drude optical model quantitatively explained the reflectivity and the transmission of light in solids. The maser and laser evolved from the combination of quantum mechanics and statistics and modern technology. The following summarizes in more detail the main areas covered:

1. Refraction index: the refraction index is the most important optical parameter, and is determined from the dielectric constant, conductivity, and permeability of a solid. The reflection coefficient can be found based on the two refraction indices at the two sides of an interface.
2. Lorentz optical model: the bound electrons in a solid might oscillate in the presence of an external electromagnetic field. The Lorentz optical model used a damped oscillation differential equation, together with the effective field treatment, to quantitatively explain the frequency dependence of the dielectric constant of semiconductors and insulators. In the microwave frequency region, the dielectric constant spectrum can be explained by relaxation, as in the Drude model.
3. Drude optical model for metals: the basic equation of this model is similar to the Lorentz optical model, except that the $-kx$ force is removed. The calculated dielectric constant of metals can only explain the experiments quantitatively, and other details relating to quantum transitions remain unexplained.
4. Laser and maser (basics): in 1917 Einstein formulated an equation for a quantum system in an electromagnetic wave, in which the spontaneous emission, spontaneous absorption, and stimulated emission were considered. Following Einstein's equation, if the number of particles at the higher-energy-level is greater than that at the lower-energy-level, the intensity of the electromagnetic wave can be exponentially enlarged. In more recent quantum mechanics, it has been proved that Einstein's stimulated emission coefficient is proportional to the square of the transition matrix element of the atomic dipole moment.

5. Laser and maser (devices): the microwave-frequency coherent radiation (maser) was achieved by Townes in a system of ammonia molecules. The visual light laser system was first realized by Basov as a specially designed p-n junction, where the doping density had to be very high to achieve a "negative temperature state" for laser emission. The commercial laser diodes were made using GaAsX materials in the 1960s by Holonyak and others.

References

Andeen, C., Fontanella, J., and Schuele, D. (1970). "Low-frequency Dielectric Constant of LiF, NaF, NaCl, NaBr, KCl, and KBr by the Method of Substitution," *Phys. Rev. B*, Vol. 2, 5068–5073.
Basov, N. G. 1964. "Semiconductor Lasers," Nobel Lecture.
Einstein, A. (1917). "The Quantum Theory of Radiation," *Z. Physik*, Vol. 18, 121.
Gordon, W. (1926). "Der Comptoneffekt nach der Schrodingerschen Theorie," *Z. Physik*, Vol. 40, 117.
Hall, R. N., Fenner, G. E., Kingsley, J. D., Carlson, T. J., and Carlson, R. O. (1962). "Coherent Light Emission from GaAs Junctions," *Phys. Rev. Lett.*, Vol. 9, 366–368.
Hass, M. (1960). "Temperature Dependence of the Infrared Reflection Spectrum of Sodium Chloride," *Phys. Rev.*, Vol. 117, 1497–1499.
Holonyak, N. and Bevacqua, S. F. (1962). "Coherent (Visible) Light Emission from $Ga(As_{1-x}P_x)$ Junctions," *Appl. Phys. Lett.*, Vol. 1, 82–83.
Klein, O. (1926). "Quantentheorie und Fundimensionale Relativitaststheeorie," *Z. Physik*, Vol. 37, 895.
Komarov, S. A., Mironov, V. L., and Romanov, A. N. (1999). "Frequency Dispersion in Microwave for Complex Permittivity of Bound Water Stored in Soils and Wet Salts," *Geoscience and Remote Sensing Symposium Proceedings, IEEE 1999 International*. Vol. 5, 2643–2645.
Kronig, R. de L. (1934). *Nature*, Vol. 133, 211.
Landau, L. D., and Liftshitz, F. M. (1984). "Electrodynamics of Continuous Media," *Course of Theoretical Physics, Vol. 8*, Pergamon Press, New York.
von Laue, M., translated by Fan, D. and Dai, N.(1978) *History of Physics* (in Chinese), Commercial Press, Beijing.
Schrödinger, E. (1926). "Quantisierung als Eigenwertproblem (Vierte Mitteilung)," *Ann. Physik*, Vol. 81, 109.
Seitz, F. (1940). *The Modern Theory of Solids*, McGraw-Hill, New York.
Townes, C. H. (1964). "Production of Coherent Radiation by Atoms and Molecules," Nobel Lecture.
Zener, C. (1933). "Remarkable Optical Properties of the Alkali Metals," *Nature*, Vol. 132, 968.
"Photonic Materials and Devices." Kimerling, L. Department of Materials Science and Engineering, MIT, City of Boston, USA. 15 May 2006. <http://ocw.mit.edu/OcwWeb/Materials-Science-and-Engineering/3-46Spring-2006/Downloadthis-Course/index.htm>
"Semiconductors on NSM." Physico-Technical Institute of the Russian Academy of Sciences, City of St Petersburg, Russia. 3 May 2006. <http://www.ioffe.rssi.ru/SVA/NSM/nk/>

Exercises

8.1 The molar volume of a KCl crystal is 3.71×10^{-5} m^3. At visual light frequency, the electron polarizability of the K$^+$ and Cl$^-$ ions are 1.48 and 3.97 respectively (unit: 10^{-40} F·m^2).

   a. Use the Clausius-Mossotti relation to calculate the dielectric constant ϵ_{opt} at visual light frequency.
   b. In an external field of 10^5 V/m, what would the displacement of positive and negative ions be?

8.2 In Figure 8.6(b), the calculated dielectric constant and refraction index contributed by the first resonance frequency have been given for silicon crystals. If the second resonance peak appears at a frequency 4/3 times that of the first peak, calculate and draw the dielectric constant and refraction index using the Lorentz optical model, and compare with the experiment in Figure 8.6(a).

8.3 In Figure 8.8(b), the measured reflection coefficient of an NaCl crystal has been plotted. Using the ion deviation polarizability in Eq. (8.21), and the long-wavelength optical branch frequencies $\omega_j(0) = 4.8$ THz and 7.5 THz, calculate the reflection coefficient by the Lorentz optical model, and compare with experiments.

8.4 a. Prove that the polarizability of a metallic ball with a radius a is just a^3.
   b. If a is the atomic radius, estimate the order of the atomic polarizability.
   c. If these metallic balls with densities of n are embedded in an insulator with a dielectric constant ϵ_0, and the densities of the balls are small ($na^3 \ll 1$), prove that the dielectric constant of the composite material is $\epsilon_0 + 4\pi na^3$.

8.5 An insulator and a metal plate are put in a capacitor, where for the insulator plate, the dielectric constant is ϵ, the thickness is d; and for the metal plate, the dielectric constant is assumed to be zero, the conductivity is σ, and the thickness is ad. Prove that the two plates are equivalent to a uniform dielectric with a dielectric constant:

$$\epsilon^* = \frac{\epsilon(1+a)}{1 + ia\epsilon\omega/(4\pi\sigma)} \quad (8.44)$$

8.6 Using the molecular dipole moment of ammonia given in Table (2.5) of Chapter Two, and Eq. (8.39) in this chapter, estimate the Einstein coefficient B in the maser in which ammonia is the working substance.

8.7 Based on the two spectral widths of Holonyak's GaAs-GaP laser diode in Figure 8.14, estimate the two respective transition times t.

Index

Abrikosov lattice, 236
Abrikosov, Alexei, 236, 245
absorption coefficient, 311
active device, 193
alkali metals, 178
alkali-earth metals, 180
amorphous structure, 89
Ampere, Andre-Marie, 257, 260
analytical mechanics, 113, 123
Anaximander of Miletus, 2
Anaximenes of Miletus, 2
Aristotle, 4
arsenic structure, 58
Ashcroft, Neil W., 10
atomic solids, 33
atomism, 1, 4, 114
Avogadro, Amedeo, 4

Balmar, Johann Jakob, 17
band edge, 203
band electron, 185
band gap, 162, 174, 183, 196, 203, 204
    semiconductor, 201
band theory, 140, 148, 154, 162
Bardeen, John, 31, 193, 200, 223, 237, 247
Barlow, William, 48, 70
Basov, Nicolay Gennadiyevich, 308, 323, 324
BCC structure, 54
BCS theory, 237, 247
Bednorz, Georg, 194, 237, 238
Bernoulli, Daniel, 124
Bethe, Hans, 147
binding energy, 15
Bloch 3/2-power law, 293
Bloch electron, 164, 173
Bloch wave function, 167, 177, 184
Bloch's theorem, 164, 165
Bloch, Felix, 7, 140, 162, 164, 167, 184, 259, 291, 298
Bohr magneton, 260
Bohr model, 18, 141, 260
Bohr, Niels Henrik David, 6, 17, 147
Boltzmann equations, 155
Boltzmann factor, 112, 203

Boltzmann's law, 320
Boltzmann, Ludwig, 4, 112, 116
Born, Max, 8, 18, 23, 68, 113, 122, 123, 164
Born-Karman condition, 68, 118, 125, 165
Bragg planes, 66
Bragg's law, 79, 84
Bragg, William Henry, 7, 70
Bragg, William Lawrence, 7, 59, 70
Brattain, Walter, 193, 200
Braun, Karl Ferdinand, 199
Bravais lattice, 63
    2D clinic, 49
    2D rectangle, 50
    2D square, 49
    2D triangular, 49
    3D classification, 51
Bravais, Auguste, 7, 42, 48, 50
Brillouin function, 273
Brillouin zone, 63, 66, 162
Brillouin, Léon, 66, 147, 162, 272
Brockhouse, Bertram N., 78, 113, 131, 259, 291
Brownian motion, 114
Brugmans, Anton, 257
Buerger, Martin, 65
bulk modulus, 25

Callaway, Joseph, 182
Carnot, Sadi, 112
carrier, 160, 162
    electron, 204
    hole, 204
carrier density
    extrinsic
        freeze-out range, 217
        intrinsic range, 217
        saturation range, 217
    intrinsic, 214
Chadwick, James, 76, 131
chemical bond, 16, 176
chemical potential, 149
Clairault, Claude, 33
classical physics, 4
Clausius, Rudolf, 4, 111, 313

Clausius-Mossotti relation, 313
closed-packed structure, 54
cohesive energy, 16
    covalent, 28
    Ionic, 22
    metal, 31
color, 316
compass, 257
conductance quantum, 269
conduction band, 203
conductivity
    Drude, 144
    Hall, 161
    matrix, 196
    semiconductor, 209, 219
    Sommerfeld, 156
conductor, 162, 196, 203
conservation law of energy, 111
Coolidge, William, 71
Cooper, Leon, 237, 247
coordination polyhedron, 59
covalent bonds, 25, 58
covalent crystals, 25
crystal lattice
    see lattice, 42
crystal system, 51
crystallography, 6, 63
CsCl structure, 21
Cubic lattice, 51
Curie's law, 272
Curie, Jacques, 308
Curie, Pierre, 8, 258, 308
Curie-Weiss law, 280
Czochralski Process, 201
Czochralski, Jan, 201

Dalton, John, 4
Davisson, Clinton Joseph, 7, 74
de Broglie, Louis Victor, 6, 18, 74
de Broglie relationship, 74
de Broglie wave, 140
de Gennes, Pierre-Gilles, 96
de Haas-van Alphen effect, 179, 181, 277
Debye function, 153
Debye phonon model, 117, 199
    Debye frequency, 118, 119, 127
    frequency distribution, 118
Debye temperature, 120, 153

Debye, Peter, 8, 18, 35, 69, 85, 86, 112, 117, 147
Democritus of Abdera, 3
density functional theory, 20, 162
    see DFT, 176
density of states, 118
DFT, 163
    exchange-correlation energy, 177, 182
    Kohn-Sham equations, 177
    LDA, 177
    magnetism, 182
DIA structure, 54
diamond structure, 25
dielectric constant, 309, 313, 316
    semiconductor, 201
diffraction, 7, 69, 79
    amorphous structure factor, 90
    atomic scattering factor, 82
    crystal scattering factor, 82
    diffraction point, 83
    geometrical scattering factor, 82
    lattice structure factor, 82
    scattering length, 81
diffraction method
    Laue, 85
    powder, 85
    rotating crystal, 85
diffraction theory, 80
    inelastic, 132, 178, 293
diffusion coefficient, 114
    semiconductor, 229
Dirac, Paul, 140, 149, 259, 261
disorder in solids
    spin, 87
    substitutional, 87
    topological, 87
    vibrational, 88
dispersion, 308
    abnormal, 315
    light, 315
    normal, 315
drift velocity, 143
    semiconductor, 228, 229
Drude formula, 316
Drude model, 143, 319
Drude optical model, 319
Drude, Paul, 140, 141, 308, 318
Dulong-Petit law, 112, 117
DuMond, Jesse, 307

Edison, Thomas, 157
effective mass, 148, 154
Einstein coefficient, 322
Einstein phonon model, 116
Einstein relation, 226
Einstein, Albert, 6, 72, 74, 112, 114, 308, 320
electric component, 193
electron, 205
electron diffraction, 79
electron paramagnetic resonance, 296
electron spin resonance, 259, 296
electron-phonon scattering, 197
Elliott, Stephen Richard, 87
Elser, Veit, 94
Empedocles of Akragas, 2
energy band, 162, 177
    alkali-earth metals, 180
    APW, 177
    computational methods, 177
    conduction, 196
    DFT, 177
    insulators, 183
    LDA, 177
    noble metals, 179
    OPW, 177
    PSP, 177
    transition metals, 181
    trivalent metals, 181
    valence, 196
energy density of states, 118, 142, 148
energy spectrum, 66
Euclid of Alexandria, 307
Ewald structure, 84
Ewald, Paul Peter, 63, 84
exchange interaction, 19, 22

Faraday, Michael, 4, 8, 193, 257
FBZ, 66, 125, 162, 167, 170
FCC structure, 54
Feodorov, E. C., 7, 48
Fermi energy, 148, 149
Fermi sphere, 149
Fermi surface
    alkali metals, 179
    alkali-earth metals, 180
    insulators, 183
    noble metals, 179
    trivalent metals, 181

Fermi, Enrico, 77, 140, 149
fermions, 148
ferrite, 287
ferromagnet
    easy axis, 278
Feynman, Richard P., 4
Fock, Vladimir, 20, 177
Fourier space, 63
Fresnel's formula, 311
fundamental particle, 126

Galilei, Galileo, 4, 111
gallium structure, 57
Gauss, Carl Friedrich, 257
Gerlach, Walther, 259
Gibbs, Joseph Williard, 116, 148
Gilbert, William, 257
Ginzburg, Vitaly, 236, 245
Ginzburg-Landau theory, 236, 245
Gordon, Walter, 321
Gorter-Casimir model, 236, 238, 241
Goudsmit, Samuel, 259
graphite structure, 58
gyromagnetic factor, 263

Hall coefficient, 159
Hall, Edwin, 159
Hall, Robert H., 325
Hamilton, William Rowand, 124
hard sphere model, 59
hard sphere repulsion, 22
Hartree, Douglas Rayner, 20, 177
Hartree-Fock equations, 20
Hartree-Fock Hamiltonian, 19
HCP structure, 54
Heilmeier, George H., 96
Heisenberg model, 283
Heisenberg uncertainty principle, 31
Heisenberg, Werner, 147, 259, 279, 283
Heitler, Walter, 27
Heitler-London theory, 27, 283
Helmholtz, Hermann von, 116, 148, 307
Heraclitus of Ephesus, 2
Herschel, Friedrich Wilhelm, 307
Hertz, Heinrich, 257
Hertz, Heinrich Rudolf, 307
Hessel, Johann Friedrich Christian, 7, 47, 48, 50

Hexagonal lattice, 51
Hilbert space, 166
Hilbert, David, 67, 115
Hoerni, Jean, 201
Hohenberg, Pierre, 176
Hohenberg-Kohn theorem, 176
hole, 205
    bands, 206
    heavy, 206
    light, 206
    split-off, 206
Holonyak, Nick, 325
Hooke's Law, 123
Houston, William, 184
Huang, Kun, 10, 113
Hund's rule, 263
Hund, Friedrich, 263
hybrid orbits, 26, 171
hydrogen bond, 33, 35

IC chip, 225
information industry, 193
infrared reflection, 213
initial permeability, 281
insulator, 162, 183, 196, 203
integrated circuit, 193
International Tables for Crystallography, 63
ionic bonds, 20
ionic crystal, 20
    Pauling's rules, 60

Joliot, Frederic, 77
Joliot-Curie, Irene, 77
Josephson junction, 237
Josephson, Brian, 237
Joule, James Prescott, 111

Karman, Theodore von, 8, 68,
    113, 123
Kepler, Johannes, 6
Kilby, Jack, 193, 201
kinetic theory of gases, 112
Kittel, Charles, 10, 96
Klein, Oskar, 321
Klitzing, Klaus von, 269
Knoll, Max, 75
Kohn, Walter, 7, 20, 140, 162, 176, 223
Kohn-Sham equations, 177

Lagrange, Joseph Louis, 124
Landé g-factor, 263
Landé, Alfred, 263
Landau levels, 259, 266
    filling factor, 267
Landau vector potential, 262
Landau, Lev, 96, 236, 245, 259, 266
Landauer formula, 269
Landauer, Rolf, 269
Landshoff, R., 22
Langevin function, 272
Langevin, Paul, 74, 258, 260, 263, 272
Langmuir, Irvinga, 9
Laplace, Pierre de, 124
Larmor frequency, 270
Larmor theorem, 270
Larmor, Joseph, 270, 308
laser, 115
    diode, 325
    gain, 323
    loss, 323
    negative temperature state, 320
    p-n junction, 325
    population inversion, 320, 324
    pump system, 326
    resonant cavity, 325
    semiconductor, 324
    working substance, 326
lattice
    basis, 42
    Bravais lattice, 42
        position vector, 42
        primitive unit cell, 43
        primitive vectors, 42
    conventional unit cell, 44
    crystal direction, 44
    crystal planes, 44
    lattice constant, 44
    lattice plane separation, 45
    lattice site, 42
    Miller indices, 45
    non-Bravais lattice, 42
lattice dynamics, 113
    dynamic matrix, 125
    elastic coefficient, 123
    equation of motion, 124
    normal modes, 125
    potential, 123

lattice harmonic theory, 123
Laue's formula, 79
Laue, Max von, 7, 69, 85, 147, 157
Lauterbur, Paul, 299
Lavoisier, Antoine Laurent, 15
Le Bel, Joseph-Achille, 25
Lehmann, Otto, 95
Lenard, Philipp, 114
Lennard-Jones, John Edward, 36
Leucippus of Miletus, 3
Lewis, Gilbert Newton, 16
light spectrum, 307
liquid crystal, 95
    lyotropic, 97
    thermotropic, 97
liquid crystal display, 96
local field, 313
London equation, 236, 242
London, Fritz, 27, 36, 236, 242
London, Heinz, 236, 242
Lorentz force, 140, 187
Lorentz number, 157
Lorentz optical model, 312
Lorentz, Hendrik Antoon, 4, 114, 140, 219, 308, 312
Lorentz-Lorenz formula, 312
Lorenz, Ludwig Valentine, 312
Loschmidt, Joseph, 4

Madelung constant, 21
Madelung energy, 20
magnetic head
    GMR, 286
magnetic materials, 260
magnetic moment
    atom, 263
magnetism, 269
    antiferro-, 258, 285
    dia-, 258, 270, 272
    ferri-, 258, 286
    ferro-, 258, 278, 281
    para-, 258, 272, 274, 278
        Pauli, 275
magnetite, 257, 287
magnetization, 309
magnon, 78, 292
    spectrum, 294
Mansfield, Peter, 299

many-body effect, 18, 176
matter, 1
matter wave, 74
Matthiessen's rule, 197, 220
Maxwell Equations
    in media, 309
Maxwell rainbow, 307
Maxwell speed distribution, 112
Maxwell, James Clerk, 4, 114, 140, 257, 307
Mayer, Julius Robert, 111
mean free path, 140, 143, 144, 185
    phonon, 203
    semiconductor, 219, 220
Meissner effect, 235
Meissner, Walther, 235
Mendeleev, Dmitri Ivanovich, 15
Mermin, N. David, 10
metal bonds, 30
metal-semiconductor junction
    Ohmic contact, 230, 232
    Schottky barrier, 230
metals, 30
Meyer, Julius Lothar, 15
micromagnetic theory, 260
Miller indices, 84
molecular dynamics, 27
molecular solids, 33
Monoclinic lattice, 51
Moseley, Henry, 73, 148
MOSFET
    see semiconductor, 232
Mott insulators, 250
Mott, Nevill Francis, 250
Muller, Alexander, 194, 237, 238

Néel temperature, 285
Néel, Louis, 258, 285, 286
NaCl structure, 20
Nernst, Walther, 113, 119
neutron, 76
neutron diffraction, 79, 131
    magnetic, 289
Newton's second law, 185
Newton, Isaac, 4, 111, 113, 307, 315
noble metals, 179
Noyce, Robert, 193, 201
Nuclear magnetic resonance, 259, 298

Oersted, Hans Christian, 257
Ohm, George Simon, 7, 140
Onnes, Heike Kamerlingh, 193, 235
Onsager, Lars, 244
optical property
　　infrared reflection, 130
　　infrared transmission, 130
optics
　　geometric, 307
orbital angular momentum
　　quenching, 275
ordered solid, 54
Orthorhombic lattice, 51
oxide structure, 60

p-n junction
　　see semiconductor, 200
pair distribution function, 89
passive device, 193
Pauli Hamiltonian, 262
Pauli's exclusion principle, 18, 22, 152
Pauli, Wolfgang, 140, 147, 149, 259, 262, 275
Pauling's rules, 54, 58
　　radius ratio, 60
Pauling, Linus Carl, 58, 147
Peierls, Rudolf, 162
penetration depth, 319
Penrose, Roger, 93
periodic table, 15
permeability, 309
perturbation theory, 167, 172
　　degenerate, 174
phonon, 78, 113, 115, 116
　　acoustic branch, 127, 129
　　dispersion relation, 125, 126
　　optical branch, 127, 129
phonon spectrum, 128, 133
photoemission phenomena, 72
photolithography, 223
photon, 72, 114, 118
photon-phonon interaction, 130
Pippard, Alfred Brian, 244
Planck, Max, 6, 17, 69, 112, 116, 118, 230
Plato's Academy, 4, 307
point group, 47
　　international symbol, 48, 50
　　Shoenflies notation, 48, 50

polarization, 309
　　electron, 312, 313
　　ionic deviation, 312, 316
　　relaxation, 312, 316
potential
　　Bardeen, 31
　　Born-Mayer, 23
　　Lennard-Jones, 36
　　Stillinger-Weber, 29
　　three-body, 29
　　two-body, 29
Poynting vector, 311
Proctor, W. G., 299
Prokhorov, Aleksandr Mikhailovich, 308, 323
Prout, William, 15
Ptolemy of Alexandria, 307
Purcell, Edward Mills, 259, 298
Pythagoras, 3

quantum Hall effect, 269
quasi-particle, 126, 185
quasicrystal, 92
　　high dimensional projection, 94
　　tiling, 94

radiation
　　spontaneous absorption, 320
　　spontaneous emission, 320
　　stimulated emission, 320
Rayleigh-Jeans catastrophe, 118
reciprocal lattice, 63, 65
reciprocal primitive unit cell, 66
reciprocal space, 63
reflection coefficient, 311
refraction index, 311
Reinitzer, Friedrich, 95
relaxation time, 143, 187, 197
　　semiconductor, 220
resistivity, 140, 197
　　semiconductor, 201
Rhombohedral lattice, 52
Ritter, Johann Wilhelm, 307
Roentgen, Rector Wilhelm Conrad, 71, 307
Ruska, Ernst, 75
Rutherford, Ernest, 6, 17, 76
Rydberg unit, 18
Rydberg, Johannes Robert, 17

scattering, 79
  carrier-defect, 220
  carrier-phonon, 220
  electron-defect, 197
  electron-phonon, 197
Scherrer, Paul, 85, 86
Schoenflies, Arthur Moritz, 7, 48
Schottky, Walter, 200, 230
Schrieffer, John Robert, 237, 247
Schrodinger's equation, 164
Schroedinger, Erwin, 18, 74, 140, 162, 321
Seeber, Ludwig August, 7, 42
Seitz, Frederick, 10, 31, 54, 66, 162
selection rule, 173, 264
semi-insulator, 203
semiclassical model, 184
semiconductor, 162, 193, 196, 203
  acceptor, 215
  acceptor level, 216
  carrier density, 211
  conductivity, 209, 219
  cyclotron resonance, 207
  devices, 225
  diffusion coefficient, 229
  diode, 223
  direct, 203
  donor, 215
  donor level, 216
  drift velocity, 228, 229
  effective mass, 206, 212, 269
    conductivity, 212
    DOS, 207, 211
    heavy hole, 207
    light hole, 207
    longitudinal, 207
    split-off hole, 207
    transverse, 207
  energy band, 203
  extrinsic, 209, 212, 215
  Hall coefficient, 219
  hydrogenic state, 215
  impurity density, 217
  indirect, 203
  intrinsic, 209, 212, 214
  intrinsic density, 213
  ionized energy, 215
  law of mass action, 212
  life time, 229
  mean free path, 219, 220
  metal-semiconductor junction, 230
  mobility, 209
  MOS transistor
    drain, 232
    gate, 232
    gate voltage, 232
    n-channel, 233
    p-channel, 233
    source, 232
  n-type, 215
  non-degenerate, 210
  optical, 203
  p-n junction, 225, 226
    abrupt, 225
    band structure, 225
    bias, 227
    breakdown, 228
    build-in voltage, 225, 227
    depletion layer, 225
    depletion capacitance, 227
    depletion width, 227
    diffusion, 225, 228
    graded, 225
    I-V curve, 228
    laser, 325
    majority carrier, 225
    minority carrier, 226
    minority current, 228
    recombination, 225, 228
  p-type, 215
  physical properties, 201, 207
  processing, 225
  relaxation time, 220, 229
  Schottky diode, 232
  stimulus, 209
  thermal conductivity, 203
  transistor, 232
  triode, 223
Sham, Lu J., 177
Shechtman, Dan, 92
Shockley, William, 193, 200
Shubnikov-de Haas Oscillation, 207, 269
Shull, Clifford G., 77, 131, 259, 289, 290
silicon wafer, 201
Slater determinant, 18, 263

Slater, John Clarke, 19, 31, 177, 182, 247
solid electronic theory, 7
solid state physics, 6
solid-state device, 193
Sommerfeld expansion, 150
Sommerfeld model, 147, 184
Sommerfeld, Arnold, 7, 58, 66, 70, 117, 140, 147, 259, 260
sound velocity, 130
space group, 47, 63
specific heat
    coefficient, 154, 186
    electrons, 154
spectrum
    of gases, 17
spin, 116, 147, 149
spin wave, 291
spin-orbit coupling, 275
spin-valve, 286
static eigen-equation, 165
statistics
    Bose-Einstein, 116, 198, 292
    Fermi-Dirac, 140, 148, 184, 195, 198, 209
    Maxwell-Boltzmann, 131, 140, 142, 210
    non-equilibrium, 155, 156, 160, 184, 194
Stenson, Niels, 7
Stern, Otto, 259
structural chemistry, 59
structure
    semiconductor, 201
super-paramagnetism, 287
superconductivity, 235
    Cooper pair, 247
    critical field, 235, 238
    critical temperature, 235, 238, 249
    energy gap, 240, 249
    flux quantum, 243, 247
    flux tube, 247
    Ginzburg-Landau equation, 246
    Josephson junction, 241
    London length, 242
    London theory, 241
    Meissner effect, 242
    order parameter, 245
    penetration depth, 236, 242, 245
    Pippard length, 244
    S-I-N junction, 241
    superconducting density, 242
    theory, 236
    two-fluid model, 238, 242
    ultrasonic wave, 241
superconductor, 193, 235
    High-Tc, 237, 238
    physical properties, 238
    traditional, 236
    Type-I, 237
    Type-II, 237
symmetry of lattice, 46
    mirror image, 47
    rotational, 47
    rotational axis, 48
    transformation, 46
    translational, 47, 164

Tetragonal lattice, 51
Thales of Miletus, 1–3
thermal expansion, 25
thermal conductivity
    insulator, 201
    metal, 201
    metals, 144, 157
    semiconductor, 201
thermionic emission, 157
thermodynamics law
    first, 111, 156
    second, 112
Thomson, George Paget, 7, 75
Thomson, Joseph John, 5, 6, 17, 70, 75, 140
Thomson, William (Lord Kelvin), 111
tight-binding model, 163, 167, 178
tiling
    matching rule, 93
    Penrose, 93
    Wang, 92
Townes, Charles Hard, 308, 323, 324
transition metals, 181
Triclinic lattice, 51
Trigonal lattice, 51
trivalent metals, 181

Uhlenbeck, George Eugene, 259

Valasek, Joseph, 308
valence band, 203

van der Waals force, 22, 35, 58
    Debye force, 35
    Keesom force, 35
    London force, 35
van der Waals, Johannes Diderik, 35
van Helmont, Johann Baptista, 9
Van Vleck, John H., 223, 278
van't Hoff, Jacobus Henricus, 25

Wang, Hao, 92
wave
    electromagnetic, 63
    matter, 63
    mechanical, 63
wave-particle duality, 6, 18, 72, 74, 125, 164
weak potential approximation, 163, 172, 178
Weber, Heinrich Friedrich, 112
Weiss field, 280
Weiss, Pierre-Ernest, 8, 259, 279, 293
Weyl, Herman Klaus Hugo, 67
Wiedemann-Franz law, 144, 157, 197, 201
Wien's displacement law, 112
Wien, Wilhelm, 5

Wigner, Eugene Paul, 10, 31, 54, 66, 77, 167, 223
Wigner-Seitz cell, 44, 67, 167
Wollaston, Willian Hyde, 307
work function, 157
work-energy principle, 185
Wurtzite structure, 201

X-ray
    continuous spectra, 72
    line spectra, 72
X-ray diffraction, 79
    synchrotron, 89

Yu, Fu Chun, 299

Zavoisky, E K, 259, 296
Zeeman splitting, 297
    anomalous, 260
    normal, 260
Zeeman, Pieter, 259
Zinc Blende structure, 21, 201
Zworykin, Valdimir Kosma, 76

System of units: SI and cgs (E-electric, M-magnetic)

Quantity	SI	cgs	transformation
length l	m (meter)	cm (centimeter)	1m=10^2cm, 1in=2.54cm
time t	s (second)	s (second)	1Hertz=s^{-1}
mass m	kg (kilogram)	g (gram)	1kg=10^3g, 1pd=453.6g
temperature T	K (kelvin)	K (kelvin)	1°F=(5/9)K
current I	A (ampere)	esa [$g^{1/2}cm^{3/2}/s^2$]	1A=3×10^9esa
energy U, W, Q	Joule [kg m²/s²]	erg [g cm²/s²]	1J=10^7erg
force $\vec{F}$	Newton [kg m/s²]	dyne [g cm/s²]	1N=10^5dyne
charge q	Coulomb [A s]	esu [$g^{1/2}cm^{3/2}/s$]	1C=3×10^9esu
current density $\vec{j}$	A/m² [A/m²]	esa/cm² [$g^{1/2}/cm^{1/2}/s^2$]	1A/m²=3×10^5esa/cm²
E-potential ψ	Volt [kg m²/s³/A]	esv [$g^{1/2}cm^{1/2}/s$]	1V=$\frac{1}{3}\times 10^{-2}$esv
E-field $\vec{E}$	V/m [kg m/s³/A]	esv/cm [$g/cm^{1/2}/s$]	1V/m=$\frac{1}{3}\times 10^{-4}$ esv/cm
E-Displacement $\vec{D}$	C/m² [A s/m²]	esu/cm² [$g/cm^{1/2}/s$]	1C/m²=$12\pi\times 10^5$ esu/cm²
E-Polarization $\vec{P}$	C/m² [A s/m²]	esu/cm² [$g/cm^{1/2}/s$]	1C/m²=3×10^5esu/cm²
M-Flux Φ	Weber [kg m²/s²/A]	Maxwell [$g^{1/2}cm^{3/2}/s$]	1Wb=10^8 Maxwell
M-Induction $\vec{B}$	Tesla [kg/s²/A]	Gauss [$g^{1/2}/cm^{1/2}/s$]	1T=10^4G
M-field $\vec{H}$	A/m [A/m]	Oersted [$g^{1/2}/cm^{1/2}/s$]	1A/m=$4\pi\times 10^{-3}$Oe
M-magnetization $\vec{M}$	A/m [A/m]	emu/cm³ [$g^{1/2}/cm^{1/2}/s$]	1A/m=10^{-3}emu/cm³

Fundamental physical constants (formula are in cgs units)

Quantity	SI	cgs & other
Planck constant h	$6.62606896\times 10^{-34}$ J s	$6.62606896\times 10^{-27}$ erg s
Avogadro constant N_A	6.02214179×10^{23} /mol	6.02214179×10^{23} /mol
speed of light c	299792458 m/s	2.99792458×10^{10} cm/s
photon mass m_p	$1.672621637\times 10^{-27}$ kg	$1.672621637\times 10^{-24}$ g
atomic mass m_u	$1.660538782\times 10^{-27}$ kg	$1.660538782\times 10^{-24}$ g
Boltzmann constant k_B	1.3806504×10^{-23} J/K	1.3806504×10^{-16} erg/K
electron mass m_e	$9.10938215\times 10^{-31}$ kg	$9.10938215\times 10^{-28}$ g
elementary charge e	$1.602176487\times 10^{-19}$ C	$4.80320427\times 10^{-10}$ esu
Bohr radius $a_B = \hbar^2/m_e e^2$	$0.52917720859\times 10^{-10}$ m	$0.52917720859\times 10^{-8}$ cm
Bohr magneton $\mu_B = e\hbar/2m_e c$	$9.27400915\times 10^{-24}$ J/T	$5.7883817555\times 10^{-5}$ eV/T
Rydberg unit Ry= $2\pi^2 me^4/h^2$	$2.17987197\times 10^{-18}$ J	13.60569193 eV
molar gas constant $R = N_A k_B$	8.314472 J/mol/K	8.314472×10^7 erg/mol/K
magnetic flux quantum $\Phi_0 = hc/2e$	$2.067833667\times 10^{-15}$ Wb	$2.067833667\times 10^{-7}$ M
nuclear magneton $\mu_n = e\hbar/2m_n c$	$5.05078324\times 10^{-27}$ J/T	3.15245123×10^{-8} eV/T